- Contents -

One of the best on the Internet . . . for quality, accuracy of content, presentation and usability. - Britannica Presents Internet Guide Awards

Style is tight, clean and refreshingly free from scientific, mystical, philosophical and theological jargon. . . . It's the kind of book where you'll need to stop and do 15 minutes of deep thinking and pondering every few pages . . . Outstanding book, engrossing, a classic – Wendy Christensen, *Inscriptions*, the weekly e-zine for professional writers, (InscriptionsMagazine.com)

Selected by The McKinley Group's professional editorial team as a "3-Star" site . . . a special mark of achievement in Magellan, McKinley's comprehensive Internet directory of over 1.5 million sites and 40,000 reviews.

Found it very intriguing. A lot of your ideas had me nodding my head in agreement. - Karl B.

I was astonished to read your book non-stop over 7 hours last night. It was the best synthesis of the collective understanding I have read in my 48 years. - S. R.

I'm impressed by the breadth of your understanding and the amount of work you've obviously put into it. – R.W.K.

Wow! This is about all I can say at the moment. - Ella B.

Science Without Bounds web site: http://www.AdamFord.com
Free electronic copies available for personal and educational use.

ix

- Preface -

In 1970 I was twenty-two and looking for something to believe in, something to make sense of the world and my place in it, a world view. Years of religious elementary and high school had left me with a dislike of religion, a distaste for its irrationality, superstition, and guilt. Science had been much more to my liking; I had attended a state university and just received a degree in electrical engineering. Yet, religion had addressed, however ineptly, however superstitiously, some questions science ignored. What was my place in the world? Where had I come from? and Where was I going? Certainly these questions were as important to me as the voltage and current in an electrical circuit.

In the following years I attempted to find answers to those questions. I turned to philosophical, religious and spiritual books of all kinds. I learned to meditate. I spent years trying to live a monastic life, both alone and in community. I worked for a while and then returned to school. I received an M.A. in mathematics and spent two more years pursuing a Ph.D. but didn't finish. I married and was divorced.

Eventually, a body of ideas and concepts, many derived from my readings, a few perhaps original, coalesced into a world view. This world view is both simple and profound. It shares the single-minded dedication of mathematics and science to truth, as well as their rationality, logical methods, and rejection of lies and fantasy. Yet it addresses questions such as Who am I? Where have I come from? How should I live my life? and What is life's greatest good?

I've often wanted to communicate this world view to my friends, but the ideas aren't easy to explain in short, informal discussions. Besides, more than an understanding of the individual ideas is needed. For only when they are placed in proper relation to each other does the total picture emerge. The world view results not from a mere summing of its individual elements, but from their interplay, their fusing into a single, coherent whole, a logical system.

A similar fusing of individual truths occurred in geometry, many centuries ago. The Greeks are commonly said to have invented geometry. For example, William Dampier in *A History of Science and Its Relations with Philosophy and Religion*, considered "one of the outstanding histories of science," ([T07],239) writes:

> . . . [T]he first to create science . . . were the Greek nature-philosophers of Ionia. The earliest and most successful of such attempts was the conversion of the empirical rules for land surveying, mostly derived from Egypt, into the deductive science of geometry . . . ([D01],xiii-xiv).

Though the Egyptians had already discovered many geometrical facts, the Greeks are credited with creating the science of geometry. Why? Because they fused the individual discoveries of the Egyptians into a coherent whole, a logical structure, a science. For the Egyptians, geometrical claims were to be accepted on faith; for the Greeks, belief naturally followed understanding.

Similarly, the world view presented in this book depends not so much on faith as on understanding. Though most of its elements are drawn from the world's philosophies and religions, the world view itself aspires to the clean logic and crystal clarity of science.

I've never seen a book that presents this world view as a harmonious whole, though I've seen books that discuss bits and pieces, and even large sections. For a long time I wished such a book existed, a book I could give to my friends and say "Here. This is what I believe." For years I thought of writing such a book. Yet, for years I did nothing. In 1989, as my forty-first birthday approached, my father, who had seen over eighty birthdays, found himself in the hospital. His illness brought home the finiteness of life, particularly my own. In August of 1989 I decided to begin writing the work I had so often imagined. (The following June my father passed away.)

I write this book for myself, for my family, and for my friends, although I hope it eventually sees a wider circulation. For myself, I hope to finally record my ideas and insights, logically and coherently. Like a builder, I hope to draw the elements into a structure, an integral whole, that possesses a beauty, a truth, and a power exceeding the sum of its parts. For my family and friends, I offer this simply as a record of ideas that are important to me. And for those I've never met, I offer this as a gift. I hope those who hold a traditional, orthodox faith find much food for thought; no doubt not all of it to taste, but all of it, I hope, wholesome and true, ultimately strengthening understanding and deepening faith. And I hope the "New Age" believer also finds much that is rewarding, although there's little or nothing about astrology, numerology, channeling, crystal or psychic healing, magic or the occult. And finally, I hope the person who holds a rational, skeptical, scientific world view - especially someone with little or no interest in ultimate questions, who considers metaphysics and theology meaningless, who accepts no philosophy which treats ultimate questions - comes to see those ancient questions in a new light.

- Introduction -

Chapter Summary: This chapter discusses why science should investigate questions traditionally addressed by religion. Then, it introduces two basic and important ideas: world view and ways of knowing. A brief overview of coming chapters is next. The chapter closes by discussing some miscellaneous points.

> For a long time we have been accustomed to the compartmentalization of religion and science as if they were two quite different and basically unrelated ways of seeing the world. I do not believe that this state of doublethink can last. It must eventually be replaced by a view of the world which is neither religious nor scientific but simply our view of the world. More exactly, it must become a view of the world in which the reports of science and religion are as concordant as those of the eyes and the ears. ([W03],xviii)

I remember as a little boy learning of God from my mother, a religious woman with a life-long devotion to Mary, the mother of Jesus. The idea of God thrilled me, but I soon grew to dislike some of my religion's ideas. For example, I learned in a second grade Roman Catholic religion class that only people who are baptized and believe in Jesus can get into heaven. I recall thinking "What about Chinese who lived five thousand years ago? They had no chance of being baptized or believing in Jesus. Is it fair to keep them out of heaven for no fault of their own?" I remember suspecting that the teacher, a nun, was wrong about who could or couldn't get into heaven, that she didn't know what she was talking about.

As I grew up, I encountered other things I didn't believe. I found some of the ideas very odd, and wondered how anyone in their right mind could believe them. For example, I was taught that anyone who dies with an unforgiven serious sin spends the rest of eternity in Hell. In those days, intentionally eating meat on Friday or missing Mass on Sunday was a serious sin. So, a child who knowingly ate a hot dog on Friday, or skipped Mass and went fishing on Sunday, might die and spend the rest of eternity in hell, horribly tortured, in the company of murderers and devils. Strange.

But even stranger was the behavior of people who, supposedly, believed those ideas. Their words said the ideas were true but their actions said otherwise. They acted as if they themselves suspected they didn't know what they were talking about.

For example, in third or fourth grade, a classmate died of appendicitis. Though some fellow classmates worried if he was in heaven, no adult seemed concerned in the least. Of course, the adults were sorry for the little boy and his family. But none showed any real worry about the fate of his eternal soul. They all assured us (glibly, I thought) that our deceased classmate was in heaven with God and the angels. Since then, I've never attended a funeral where anyone, clergy included, seem the slightest bit worried about the eternal fate of the deceased. They act as if no one goes to hell, as if hell really doesn't exist.

Science is different; scientists act as if they believe what they say. If science says plutonium is deadly, you won't find a scientist with plutonium in his pockets. And scientists seem to know what they are talking about. When astronomers say an eclipse will happen, it does. But when some religious group predicts the world will end by September fifteenth or April tenth, it doesn't. It seems science is truer than religion, more to be taken seriously, more real.

But why compare science and religion? Why not leave science to scientists and religion to religious people? What's to be gained?

Well, a person might reasonably have a more than passing interest in what really happens after death. They might wonder Where do I come from? How should I live my life? and What really happens when I die? Religion discusses those questions but, for many people, its answers are not believable. Science, on the other head, ignores such questions. It has nothing to say about them.

To use an analogy, it's as if science has food of all kinds, wholesome, true, healthy food, but no water. And as if religion has water, brackish water, polluted with confusion, fantasy, contradiction and lies. Seeing the quality of religion's water, some people decide to eat only the healthy, clean food of science - until their thirst drives them back to religion. Religion fulfills a deep need so they eventually participate, sometimes in spite of themselves, sometimes with the excuse "Well, children need something to believe. It's better for them to grow up with religion than without it."

If only science had water of its own, pure, clean water. Or, dropping the analogy, if only science had answers to questions like Where do I come from? How should I live my life? and What happens when I die? If it had such answers, science would have a religion of its own, a religion as true, as powerful, and as accurate as the rest of science. The reports of science and religion would be as concordant as sight and sound. Science would finally have a *comprehensive* world view.

Science Without Bounds

What is a world view? It's our explanation of ourselves and the world around us. It's what we believe to be true. It's our estimation of "what is what."

Most people have some explanation of themselves and the world around them. They have some idea of who they are and how they fit into the world. But ask them "ultimate" questions such as *Did you exist before you were born? Will you still exist after you die? Is there an overall purpose for your life and, if so, what it is?* and they usually give a standard religious answer, or say "I don't know." That is, their world view is either religious and non-scientific, or it's incomplete.

Does anyone have a scientific world view that's comprehensive, that answers ultimate questions? Probably not, because science itself doesn't have a comprehensive world view. Science's world view is incomplete. Science is very good at explaining part of ourselves - our liver and heart function, for example - and part of the world around us - the behavior of electricity. But science has little to say about really important questions, about ultimate questions. What the great physicist Erwin Schrodinger wrote in 1948 is just as true today.

> . . . [T]he scientific picture of the real world around me is very deficient. It gives a lot of factual information, puts all our experience in a magnificently consistent order, but it is . . . silent about all . . . that is really near to our heart, that really matters to us. . . . [T]he scientific world-view contains of itself no ethical values, no aesthetical values, not a word about our own ultimate scope or destination, and no God . . . Science is reticent too when it is a question of the

great Unity . . . of which we all somehow form part, to
which we belong. . . . Whence come I and whither go
I? That is the great unfathomable question, the same
for every one of us. Science has no answer to it.
([S05],93,95,96)

Schrodinger says science *hasn't* investigated ultimate questions. Other writers believe science *can't* investigate them. For example, M.I.T. philosophy professor Huston Smith believes:

Strictly speaking, a scientific world view is impossible;
it is a contradiction in terms. The reason is that
science does not treat of the world; it treats of a part of
it only. ([S14],7).

He continues:

Values, life meanings, purposes, and qualities slip
through science like sea slips through the nets of
fishermen. Yet man swims in this sea, so he cannot
exclude it from his purview. This is what was meant . .
. that a scientific *world* view is in principle impossible.
([S14],16).

But is it? Will science's world view always be limited, always less than comprehensive? Or will science someday develop a comprehensive world view, a world view that explains our place in the universe, our origin and destiny? Can science investigate questions it has ignored for centuries? Or has it ignored those questions for good reason? Certainly, some early scientists had good reason to ignore ultimate questions - their own survival. The most famous is, perhaps, Galileo, who had to answer to the Inquisition for teaching the earth revolved around the sun. Galileo escaped with his life. Other early scientists were not so fortunate.

In its struggle to be born in the 16th and 17th century, science wisely decided not to investigate certain religious, philosophical, or metaphysical questions. Rather, it limited itself to the natural world, within bounds set by organized religion. Today, science still lies within those boundaries, certainly no longer out of necessity, perhaps only out of habit. Einstein describes such science as

. . . the century-old endeavor to bring together by
means of systematic thought the perceptible
phenomena of this world into as thorough-going an
association as possible. ([E03],44).

"This world" seems to limit science's domain. It seems to bar science from investigating the possibility of existence before birth or after death. It sets up the "perceptible phenomena of *this* world" as a boundary which science shouldn't cross.

But did Einstein think science should forever remain within that boundary? Perhaps not, for in the very next sentence he offers a broader description of science's scope.

> To put it boldly, it is the attempt at the posterior reconstruction of existence by the process of conceptualization. ([E03],44).

"Existence" is a much broader term than "this world." It includes any and all worlds - whatever exists. It suggests that even if our origin, meaning and ultimate destiny is in any way "supernatural," it's nonetheless a part of existence and a valid object of scientific inquiry.

Is Einstein's wider definition appropriate for science? Is it reasonable? I believe it is. Moreover, it better agrees with science's original goal as established by the ancient Greeks: the making of

> . . . a mental model of the whole working of the universe. ([T02],21).

And it agrees with contemporary physicist Stephen Hawking's description of science's purpose.

> [O]ur goal is a complete *understanding* of the events around us, and of our own existence. ([H02],169).

Certainly, our understanding is incomplete if we don't know who we are, where we came from, and where we are going.

Today, science is no longer struggling to be born. Rather, it's a mature, growing culture, the only world-wide culture, and the greatest intellectual achievement of the last four centuries. If it wished, it certainly could investigate ultimate questions. Moreover, Schrodinger believed it should.

> . . . I consider science an integrating part of our endeavour to answer the one great philosophical question which embraces all others . . . *who are we*? And more than that: I consider this not only one of the tasks, but *the* task, of science, the only one that really counts. ([S06],51).

Certainly, many people - especially those unconvinced by religion's answers - would welcome any light science could offer on such questions.

Carl Sagan has observed ([K01],37) there are in the United States 15,000 astrologers but only 1,500 astronomers. Many newspapers that don't have any sort of daily science column carry a daily horoscope. Irrational, superstitious beliefs are easy to criticize, but what does science offer in their place? What is our place in the universe? Why are we here? How should we live our lives? Must science forever ignore these questions? Or can it break its centuries-old bounds and investigate questions of ultimate importance?

Ways of Knowing

But *how* can science investigate ultimate issues? How can it find answers as true, accurate, and reliable as science itself? How can it create a scientific religion?

Suppose all the world's scientists decided that Jesus (or Mohammed or Krishna or Buddha) is right. Suppose they unanimously voted to adopt Christianity (or Islam, Hinduism, or Buddhism) as science's official religion. That would certainly provide science with a religion, but would the religion be scientific? Would its answers be as true, accurate, and reliable, as open to question, criticism, revision, and improvement, as the rest of science? No.

Why not? What makes something a science? Something is a science when it uses science's *way of knowing*. Chemistry uses science's way of knowing so chemistry is a science. Palmistry doesn't, and therefore is not. No existing religion uses science's way of knowing, so no existing religion is a science. So, even if all the world's scientists accepted a particular religion, that religion would still not be scientific.

But what is a way of knowing? It's a way of deciding if something is true or not. An illustration may be helpful.

Astrology teaches that a Cancer is sensitive and reserved, and a Gemini is communicative and witty. How can we decide if astrology is true? What way of knowing shall we use? Personal experience is one way - my Aunt Alice is a Cancer and she's sensitive, my friend Tom is a Gemini and he's witty, so astrology must be true. Another way is authority or faith - someone I respect believes in astrology and says it's true, so I believe in it, too. Yet another way is tradition - it's thousands of years old and millions of people have believed in it, so it must be true.

Judging from personal experience, judging from authority or faith, judging from tradition - each is a way of knowing, a way of deciding what's true and what's not. But none of those ways is scientific. How can we decide scientifically if astrology is true? How can we judge astrology with science's way of knowing?

Imagine giving a hundred people three horoscopes, their own and two others. Ask them which horoscope describes them best. Ask their family, their friends, their co-workers. Do they pick their "true" horoscope more than one-third of the time? Do their family and friends? If they do, then that's scientific evidence that astrology is true. But if the "true" horoscope is picked about thirty-three times in a hundred, if no one can tell the "true" horoscope from the "false" ones, then we have scientific evidence that astrology doesn't describe personality any better than flipping a coin. Heads you're sensitive, tails you're witty.

Of course, deciding scientifically is work. Deciding by faith, authority, tradition, or what other people believe, may be easier. But is it as good? Each way of knowing has its advantages and disadvantages. For a person eager to be accepted by some group of people, accepting what those people think may be the best way of knowing. Faith, authority, or tradition may be the best way for someone who wants to practice some religion. Science's way of knowing has proven very useful for understanding the natural world.

Yet, while each way of knowing has its own advantages and disadvantages, they all aren't equally good. Some are better than others. And picking the best way can be important. In fact, it can be a life-and-death decision. The history of medicine offers a poignant example.

Western medicine once used a way of knowing remarkably similar to science's.

> Looking nature full in the face, without being blinded by either the divine or the customary, Greek intellectuals sought rational explanations of all within man's ken. In the medical field perhaps this was exemplified best by the followers of Hippocrates (born ca. 460 B.C. on Cos). Their best writings and practices showed the fundamentals of the scientific method - observations and classification, rejection of unsupported theory and superstition, and a cautious

generalization and induction that remained open to
critical discussion and revision. . . . ([R02],8).

Medicine was on the road to understanding and curing disease. It took a detour in the sixth century when the bubonic plague hit the Roman Empire. The plague - which would ravage Europe again in the 14th century - struck about 540 C.E., during the reign of Justinian, and raged until about 590 C.E. At its height it claimed over 10,000 victims a day. Its total toll is estimated at one hundred million. Because contemporary physicians couldn't understand or stop the disease, many people turned to religion.

The effect of the plague of Justinian on the field of
medicine is unarguable, and was unfortunate. The
Christian Church rushed in to fill the medical void,
becoming doctor to the soul *and* the body. Progressive
Greek and Roman physicians had taught that disease
was caused by pathogenic agents; they were slowly,
but correctly, creating the discipline of medical
science. The church, however, in its new role as
healer, equated disease with vice and sin, the
punishment for leading an errant life . . . The brilliant
ideas of Galen and Hippocrates became heresies.
This repressive attitude lasted until the fourteenth
century and vastly altered what would have been a
very different course of medicine had it not fallen
under the domination of dogma and miracles.
([P02],225).

Medicine rejected a scientific way of knowing and understanding disease, and turned to a way based on faith and divine revelation. No longer need it laboriously search for the cause of disease; divine and unerring scripture had the answers. The "answers," however, aren't very good. Medicine based on scripture doesn't work, medicine based on science does. So, it's fortunate that medicine eventually returned to the "heresies" of Galen and Hippocrates. Medicine abandoned its faith-based way of knowing and understanding disease, and returned to a scientific way. As a result, someone you know is alive today who would otherwise be dead, perhaps one of your parents or children. Perhaps you.

Science's and Religion's Ways of Knowing

We'll see more about ways of knowing in the first two chapters; the first chapter explores religion's way of knowing while the second explores science's. We'll see that religion decides something is true because some authority or book says so. Therefore, the religious way of knowing demands blind acceptance; its beliefs are "set in stone," fixed, not open to criticism, revision and improvement. On the other hand, we'll see how science's way of knowing - often called the scientific method - discourages blind acceptance, and welcomes discussion, criticism and improvement. As we'll see, when a scientist theorizes that atoms behave one way, or electromagnetic fields behave another, other scientists don't blindly accept the theory. Rather, they test it with the scientific way of knowing. Once the theory is proven, engineers exploit it by making useful devices based on the theory. In doing so they further test and prove the theory. A radio, an automobile, and a computer are more than useful tools; they're living proof of the accuracy of the scientific theories and engineering principles they're based on. Scientific beliefs work, they prove themselves in daily life.

In medicine, in astronomy, in history, and in numerous other fields, science's way has yielded more and better knowledge than religion's way. We'll see why science's way of discovering and testing truth - its way of knowing - has generally been more accurate than religion's. Of course, some questions seem to demand religion's way of knowing. *Are Jews God's chosen people? Is Jesus the Son of God? Is Krishna God? Is Mohammed God's messenger?* Could science ever answer such questions? No. But science could investigate and answer other important, and perhaps more relevant, questions, such as *Who am I? Why am I here? How should I live my life?* and *What happens when I die?*

How could science investigate and answer such questions? That takes a few chapters to explain, but in a nutshell it's this: it could apply its way of knowing to them. Describing how science might do so is the task of chapters three, four, five and six. Chapter three discusses what science studies, with emphasis on an area that religion studies, too. Chapter four discusses what religion studies, with emphasis on the same area viewed from a religious perspective.

Chapter five discusses types of knowledge, and people who claim direct knowledge of God or "ultimate reality."

Chapter six builds on ideas of previous chapters to describe how science can apply its way of knowing to ultimate questions, how it can create a scientific explanation of our place in the universe, an explanation that's as verifiable, as open to question, disagreement and improvement, as true, and perhaps as useful, as science's explanation of physical, mechanical and electrical phenomena. Like any other knowledge produced and tested with the scientific method, a scientific religion would be an extension of science, an integral addition. Science's world view would finally include a religion of its own, a religion not merely compatible with science, but thoroughly scientific, a branch of science in its own right, a scientific religion and religious science. Science would finally have answers to questions such as *Why are we here? What is the purpose of our life? Where did we come from?* and *What happens when we die?* answers as thoroughly tested, true, and accurate as science itself.

A Scientific/Religious Comprehensive World View
In the first six chapters fundamental ideas are explored. Based on those ideas, Part II - chapter seven, eight, nine and ten - presents a world view that's deeply religious yet quite compatible with science, a world view that discusses the external world we live in, our individual internal world of thoughts and feelings, and the world of the "supernatural." And while our world view doesn't agree with all religions (it couldn't because religions themselves don't agree), and doesn't agree with any single religion, it does substantially agree with the so-called "perennial philosophy," the philosophy that Aldous Huxley considers ([S18],12) the "Highest Common Factor" of the world's religions, and that Huston Smith calls ([S14],x) "the primordial tradition."

Our world view is an example of applying science's way of knowing to certain religious, philosophical, and metaphysical questions. But is it fully scientific? No, because it's the world view of just one person. Yet because the world view is presented in the scientific spirit, as a hypothesis that others may criticize, correct, amend and extend, it's also potentially a starting point for deeper inquiry. Just as other sciences grow, develop, and evolve, the world view presented in these pages may, too. Rather than a dogmatic, fixed, answer-for-all-time, it's a seed that may one day become a comprehensive, fully scientific world view. And it's certainly not the only world view that could be derived by applying the scientific way of knowing to spiritual and mystical insights. Other world views are possible.

In Part I and II, we'll meet ideas of great richness and beauty, a few perhaps truly difficult, others not so much difficult as unfamiliar. In the third and final part - chapters eleven, twelve, thirteen, fourteen, and fifteen - we'll apply those ideas. We'll attempt to answer the questions "So what? How can all this affect me and my daily life?" by deriving practical consequences.

A short chapter concludes the book, followed by a bibliography.

Odds and Ends

We'll end the introduction by discussing some miscellaneous points.

First, a brief point. I've chosen to follow what seems common usage and label years *C.E.* Years *C.E.* correspond to years *A.D.* Therefore, 1234 *C.E.* and 1234 *A.D.* are the same year. Why not *A.D.*? Because *A.D.* is Anno Domini, Year of Our Lord, and is therefore appropriate only for Christians. *C.E.*, however, can mean either *Christian Era* or *Common Era* and so is appropriate for Christians and non-Christians alike.

Now, a more extended point. As you read this book you'll see many bits and pieces of other books; that is, you'll see many quotations. Quotes have full references, including page numbers, and they're exact: any italics shown are also in the original. I do, however, occasionally correct punctuation, change an uppercase letter to lower or vice versa. These changes are indicated in the usual way, with square brackets. I also use square brackets at the end of the quote to indicate its source. For example ([M12],9) indicates page nine of book [M12] in the bibliography.

Why all the quotations?

First, since this book is an exposition and synthesis of mostly non-original material, it seems appropriate to include the original sources, to present direct evidence. Rather than being told what someone thinks, you see for yourself what they wrote. Second, quotations sometimes make the point vividly and forcefully. Third, they introduce you to authors and books with whom you may want to become more acquainted. Fourth, translations differ. So it's wrong to say the Bible says this, or the *Tao Te Ching* says that. When a book wasn't originally written in English, it's more accurate to say this translation of the Bible says this, and that translation of the *Tao Te Ching* says that. In such cases I present a direct quotation.

Lastly, quotations allow me to steer a middle course between two unsatisfactory extremes. On one hand, if I present an idea but neglect to mention some religious analogue, I could be justly criticized for presenting an idea of Jesus or Buddha as my own. On the other hand, if I say that Jesus or Buddha taught a certain idea, I could again be criticized for overstepping myself and acting as an official religious spokesman. To avoid these two extremes, I present my ideas as simply and clearly as possible, along with quotations from the world's religious, philosophical, scientific, and mystical writings. I'll let the experts decide whether any of my ideas are actually identical to, similar to, or entirely different from someone else's.

Now let's discuss one purpose quotations aren't meant to serve. When I quote the religious or philosophical opinions of a world famous scientist, I don't mean to imply their achievement in science somehow guarantees they're right. So, suppose Einstein thought religions should give up belief in Gods who are Persons. Or suppose Schrodinger decided mystics of different cultures and times had essentially similar experiences. And suppose Einstein and Schrodinger were great scientists. They may still be wrong. In fact, there may be equally great scientists who disagreed with them.

So why present the quotations at all? Because the quotes do demonstrate that the beliefs aren't inherently absurd to the scientific mind. It would probably be hard to find a reputable, much less accomplished, scientist who believes in leprechauns, elves, or unicorns. But if some scientists believe the idea of God as a Person should be abandoned, or that mystics often experience the same thing, then perhaps these ideas are worth examining further.

In addition to quotations, you'll also meet a definition now and then. Words and concepts are often defined before they're used. The definitions are probably a result of my years in mathematics. In higher mathematics, ideas and concepts are almost always defined when they are introduced, before they're used. There are two other reasons for definitions. First, they're often essential for clarity and understanding. Many words have multiple meanings or an unfamiliar meaning. Or I may use a word in an unfamiliar way. In these cases, definitions aid understanding.

Clear definitions also help avoid futile arguments. Consider the following illustration.

You agree to participate in an experiment. You are asked to hold out your hand. A heavy weight is placed in it. You are also asked to keep your hand absolutely still for 10 minutes. By a great effort you manage to do so. Your back, shoulder, and arm ache. Finally, the weight is removed. Have you just done any work? The answer depends entirely on the definition of "work." One definition of work is "effort, labor, toil." Using this definition, you've just done work. The physicist, however, defines "work" in an entirely different way: work equals force multiplied by distance. Using this definition, you have done no work at all since the distance you moved the weight was zero - and zero distance multiplied by any force you exerted still gives zero work.

So there are two answers; you've done a fair amount of "everyday work" and exactly zero "physicist's work." We could waste a lot of time and energy arguing if any "real" work has been done, but I intend to waste none - there are two different ideas, everyday work and physicist's work.

Substitute an emotionally charged word such as "God" for "work" and even more time and energy - and even blood - could be wasted.

To help avoid such waste, I define what I mean. If you understand my meaning, then we have a basis for discussion. If, on the other hand, you just don't like my definition and insist, for instance, real work has been done no matter what the physicists say, or God is only as you conceive God to be and no one has the right to use the word "God" in any other way, then we have no basis for discussion.

Even with definitions, of course, you may not agree with what I say, but at least you'll understand it. If I've done my job, the following chapters will be lucid; if you as a reader have done yours, you'll have a clear idea of what I said. From that point, comments will be appreciated; I welcome any insights and criticisms.

Aside from the use of definitions and quotations with references, you'll find the discussions informal, not highly technical or scholarly.

Two factors force this informal level of presentation. First, necessity - I don't have the education needed to discuss science, philosophy, metaphysics, religion, and theology on the professional level. Second, readability - this work may already be too obscure and pedantic for some people. A book that demanded the learning of a professional scientist, philosopher, metaphysician, and theologian would be, for almost everyone, an incomprehensible book. Instead, I've tried to write the clearest book I could. Nonetheless, you may find some ideas unfamiliar and a few, difficult. When you do, feel free to skim ahead, at least on first reading. After you've seen the overall picture, things may fall into place.

Enjoy.

Part I: Fundamentals

1

- Religion's Way of Knowing -

Chapter Summary: This chapter examines how religion decides what is true. It begins by describing religion's way of knowing and then examines four claims religion usually makes for scripture: consistency and truthfulness, completeness and finality, necessity for salvation, and divine or inspired authorship. The chapters point out some flaws of religion's way of knowing.

Who are we? Where did we come from? How should we live our life? What happens when we die? How can such questions be answered? People have traditionally turned to religion for answers. And religion has usually answered in theological terms: Who are we? We are children of God. How should we live our life? As God wills. What happens when we die? We go to heaven or hell.

How good are religion's answers? How accurate? How true? Deciding can be difficult or impossible if the answers are stated theologically. It's hard to imagine how such answers can be investigated and tested, scientifically or any other way. But how does religion know? How does it find the answers? Usually by using the revelational way of knowing.

Though we can't directly test its answers, we can examine and evaluate religion's way of knowing. That is, we can investigate how good the revelational way of knowing is at knowing, at finding answers. We can ask how good of a way of knowing is it. We can ask how well, how accurately, it decides what's true. We can ask if the knowledge that the revelational way of knowing has produced is truthful, consistent and comprehensive.

This chapter examines the revelational way of knowing, the way of knowing used by religion. It identifies some of its shortcomings and shows why it's an inferior way of deciding what's true. The next chapter explores the scientific way of knowing, it identifies some its shortcomings, and shows why it's nonetheless a superior way of deciding what's true. In this and the next chapter, we'll find that the revelational way of knowing is faulty and that science's way of knowing is superior. Subsequent chapters will attempt to apply science's better way of knowing to ultimate questions.

The Revelational Way of Knowing

What is the revelational way of knowing, the way of knowing used by religion? Briefly, it's a way of knowing based on revelation, on scripture. It decides some writings are inspired, are ultimately written by God, and then follows them without question or criticism. Religions don't often describe their way of knowing so directly, however. Rather, they proclaim beliefs about scripture, beliefs from which their way of knowing naturally follows. Let's examine some of these beliefs.

The Roman Catholic Church teaches that

> . . . the books of the Bible are the inspired word of God, that is, written by men with such direct assistance of the Holy Ghost as to make God their true Author. ([N08],177).

Similarly, the Seventh-day Adventists believe that

> [t]he Bible's authority for faith and practice rises from its origin . . . The Bible writers claimed they did not originate their messages but received them from divine sources. ([S10],7).

From such beliefs it naturally follows that revelation should be accepted without question. Since God wrote it, revelation is not to be criticized, judged, or changed. Therefore, Seventh-day Adventists teach that

> [j]udging the Word of God by finite human standards is like trying to measure the stars with a yardstick. The Bible must not be subjected to human norms. ([S10],13).

Another consequence of divine authorship is that revelation is error-free. For example, the Catholic Church teaches that the books of the Bible

> . . . teach firmly, faithfully and without error all and only those truths which God wanted written down for man's salvation. ([D09],12),

and a Seventh-day Adventists publication has:

> How far did God safeguard the transmission of the text
> beyond assuring that its message is valid and true? . .
> . while the ancient manuscripts vary, the essential
> truths have been preserved. ([S10],11).

Two more beliefs are usually part of the revelational way of knowing. One is necessity for salvation, deliverance, or enlightenment. For example, the Catholic Church teaches:

> Revelation is that saving act by which God furnishes
> us with the truths which are necessary for our
> salvation. ([M07],213).

The other belief is finality.

> Christians . . . now await no new *public revelation* from
> God. ([D09],4).

God's general public revelation is finished and done, even if private revelations to an individual are still possible.

These two beliefs - necessity for salvation and finality - are usually part of the revelational way of knowing even though they don't necessarily follow from divine authorship. After all, God could write many books, each helpful for salvation but not necessary. And God could write another public revelation in the future. Yet most religions claim that their revelation is final, not to be revised, extended or superseded, and that it's necessary - required - for salvation, deliverance, or enlightenment.

Of course, religions disagree over which writings are inspired. For example, the fourteen books of the Apocrypha were in the Bible for over 1,000 years. They're still in the Roman Catholic bible but other Christian groups reject them. They aren't included in many modern Bibles. Do they belong in the Bible or not?

Not only does the Catholic Church include books in its Bible that Protestants do not, that church also labels some of the writings of Athanasius, Augustine, John Chrysostom and others ([N09],20) as "Divine Tradition" and believes that

> . . . Divine Tradition has the same force as the Bible . .
> . ([N09],20).

Other Christian groups disagree. In fact,

> [p]recisely at this point the greatest division in
> Christendom occurs: the Bible as the final source
> (standard or authority), or the Bible as *a* source.
> ([P07],18).

4

Of course, different religions accept entirely different revelations. Islam holds the Koran to be revealed. Hindus believe God spoke the Bhagavad-Gita and other writings. Buddhist accept the Tripitaka.

Though all of the religions we've mentioned may reject the inspired writings of other religions, they believe their own scripture is divinely revealed. In particular, religion often makes the following four claims for their own scriptures: that scriptures:

(1) are consistent and truthful ("without error"),

(2) are complete and final ("all and only those truths . . . no new *public revelation*"),

(3) are necessary for salvation, enlightenment, or liberation ("necessary for our salvation").

(4) have an inspired or divine author ("God who is their true Author"),

Are these claims true? Again, theological claims are difficult to test. *Is* God the author of any particular book? That's beyond the reach of logic to decide. Nonetheless, the four claims can be rationally investigated. And, as we examine and test the four claims we'll come to a better understanding of the revelational way of knowing. Let's begin with the first claim, consistency and truthfulness.

Claim 1: External Consistency

An "external" contradiction is when a scripture contradicts something outside itself, either some common belief or practice, or another scripture. Let's examine some external scriptural contradictions, beginning with three where the Bible contradicts common Christian belief or practice.

First, Jesus says "Just as Jonas was three days and three nights in the whale's belly; so shall the Son of Man be three days and three nights in the heart of the earth." ([H08], Mt 12:40). According to a footnote in another Bible ([N02], for Mt 12,38ff), this quote contains an "allusion to Jesus' resurrection". However, common Christian belief allows less than 48 hours between the Crucifixion and Resurrection (Good Friday to Easter Sunday), two nights, not three.

Second, in Mark 6:3 the people of Jesus' country say: "Is not this the carpenter, the son of Mary, the brother of James, and Joses, and of Juda, and Simon? And are not his sisters here with us?" If Jesus actually had a brother, then either the Roman Catholic belief in the perpetual virginity of Mary is incorrect, or the standard Christian belief that Jesus is the *only* begotten Son of God is wrong.

Lastly, Jesus forbids swearing (Mt 5:34-37), saying at one point "But let your communication be, Yea, yea; Nay, nay: for whatsoever is more than these cometh of evil" ([H08],Mt 5:37). Nonetheless, it is common practice in some Christian countries for a court witness to swear on the Bible that their testimony shall be true.

Now let's turn to another type of external contraction, where one scripture contradicts another. The world has many "revealed" writings. If they are all, in fact, revealed then they should all agree with each other because they all have the same ultimate author - God. How well do revealed writings agree with each other? Not very well. Let's examine some examples.

Of the three major revelations of Western religion, the earliest is the Jewish Torah, which is also part of the Christian Old Testament. Later, the Christian New Testament was written. Later still, the Koran (Quran) of Islam. Are these three revelations consistent with each other? No. For example, the Koran says Jews and Christians disagree:

> The Jews say the Christians are misguided, and the Christians say it is the Jews who are misguided. (Sura 2:13, [K07],344).

And the Koran disagrees with both:

> . . . [T]he Jews say: Ezra is the son of Allah, and the Christians say: The Messiah is the son of Allah . . . How perverse are they! (Sura 9:30, [M10],148).

So, advises the Koran,

> . . . admonish those who say that Allah has begotten a son. (Sura 18:4, [K07],91).

Islam teaches that Jewish and Christian scriptures are only partially true. For instance, it teaches that Jews were one of the first peoples who

> . . . recognized God's oneness, and also God's law. ([S16],12).

Quite an accomplishment, because after that recognition the

> . . . doctrine of monotheism, established by Abraham, never again quite lapsed. ([S16],12).

Unfortunately, the Jewish people (according to the Koran) failed to accurately preserve God's words.

> . . . [I]n course of time they allowed their copies of the text . . . to become corrupted. Their "scripture" became inaccurate. . . . In due course, to correct this desperate error, God sent another messenger, Jesus. ([S16],12-13).

But the followers of Jesus erred, too, since they worshiped

> . . . the messenger, instead of heeding the message. . . . focused their attention on Christ to the partial neglect . . . of God, whose transcendence they thus compromise . . . ([S16],13).

Even worse, in their worship of Jesus they attributed

> . . . to him and his mother wild, even blasphemous and obscene, relations to God Himself. ([S16],13).

So, according to the Koran, God had to send another messenger, Muhammad.

> This time there was to be no error, no distortion, no neglect. ([S16],14).

Since Muhammad perfectly captured God's revelation in the Koran, no other messenger will be needed or sent. Therefore, Muhammad is called the "seal" of the prophets.

For Muslims, the Koran is the perfect and complete revelation of God.

> For the Muslim, God's Message is wholly contained in the Koran . . . This Book does not annul but rather confirms the Divine Message as preserved, though in a corrupt and distorted tradition, in the Holy Scriptures of the Jews and the Christians. ([A08],12).

Can Jewish, Christian, and Islamic scriptures all be true? Obviously not. At least one scripture is wrong, either the Koran in its fault-finding or Jewish and Christian scriptures in their teachings. At least one of these scriptures is incorrect, untruthful. We'll see how Jewish and Christian scriptures disagree later when we discuss scripture's finality and completeness. Now, however, let's discuss scripture's truthfulness.

Claim 1: Truthfulness

Revealed writings often describe historical and miraculous events. Did those events actually happen? They describe extraordinary people. Did those people actually live? In general, are revealed writings true?

Once, it was thought all events described in the Bible were historically true. Christian medieval Europe based cosmology on Genesis, the first book of the Bible. It based biological evolution on Genesis, too. History was based on the Bible; stories such as Noah and the Great Flood were accepted as historically true. Astronomy was also based on the Bible. In fact, the source of Galileo's conflict with the Roman Catholic Church was the church's belief in biblical teachings about the earth and sun.

Today, some religious people still believe the Bible gives a truthful picture of the natural world. Fundamentalist Christians, for example, still accept biblical teaching about cosmology, biology, history and astrology. For them biblical revelation is

> . . . the supernatural (metaphysical) process by which God penetrated man's senses to give him an external, objective world view. ([P07],13).

How such religious believers have fought the advance of science in biology, geography, astronomy, medicine, hygiene, history, anthropology, and other fields is well described in *A History of the Warfare of Science with Theology in Christendom* ([W09]) by Andrew White.

Fundamentalists (of any religion) who think revelation has accurate teachings about the natural world disprove a common idea: that the essential difference between science and religion is that science deals with this world and religion deals with the next. Fundamentalists show this opinion isn't true - some religions deal very much with this world. And science - as we'll see - could investigate the "next" world.

How, then, do science and religion differ? They fundamentally differ in *how* they know, not necessarily in *what* they know. Both can know the natural world and, as we'll see, both can know the "supernatural" world. Therefore, the fundamental difference between science and religion is their different ways of knowing. Science finds truth with the scientific way of knowing. Religion finds truth with the revelational way, by following scripture.

But is scripture truthful? Fundamentalist Christians believe the Bible is entirely truthful. More than that, they believe

> . . . the complete Bible . . . is the final authority for all truth. ([P07],21)

and that

8

> [a] problem of terminology and interpretation may exist
> between science and the Bible but the only difficulty is
> man's inability to resolve the problem, *not* any conflict
> of truth. . . . The superior credence for Scripture over
> science is clear. ([P07],31).

Other Christians, however, admit the Bible isn't entirely true.
They don't base their entire world view on revelation. For them
cosmology, biology, history, and astronomy are no longer based on
scripture. Such Christians view Genesis as mythological and accept a
scientific explanation of biological evolution and the origin of the
universe. Biblical stories once thought historically accurate are now
considered by many greatly exaggerated, if not mythological.
Astronomers no longer look to the Bible for information about the
sun, stars, and planets. And the Catholic Church now teaches that

> . . . the Bible is free from error *in what pertains to
> religious truth revealed for our salvation.* It is not
> necessarily free from error in other matters (e.g.
> natural science). ([D09],12).

Biologists and astronomers have found science's way of knowing
superior to religion's. But if science's way of knowing yields superior
knowledge about the natural world, could it yield superior knowledge
about the "supernatural" world, as well? If revelation is wrong about
the natural world, could it be wrong about the "supernatural" world,
too? We'll return to these questions later.

Claim 1: Internal Consistency
Whenever revelation contradicts some accepted fact, fundamentalists
can always say revelation is right and the accepted "fact" is wrong. If
scientists say the universe is fifteen to twenty billion years old, and
the Bible says it's a few thousand years old then, say fundamentalists,
science is wrong and the Bible right. But what happens when the fact
is in another part of the revelation? For example, what happens when
the Bible contradicts itself? This brings us to the question of internal
consistence: does the bible agree with itself?

Throughout the ages, many leading religious figures have said it
does. For example, in *Inerrancy And The Church* ([I03]) we read that

> Clement of Rome claimed that the Scriptures were
> errorless. ([I03],23),

that

9

> Tertullian was swift to argue . . . that the Scriptures
> contained no contradictory material nor error.
> ([l03],24),

that Origen

> . . . perceived the Scriptures as perfect and
> noncontradictory . . . ([l03],25),

and, finally, that

> [f]or Augustine, it was an article of faith that there is no
> real discrepancy or contradiction in all of Scripture.
> ([l03],49).

Augustine's definition of error was strict.

> When Augustine declared the Bible to be free from
> error, he explicitly rejected the presence of inadvertent
> mistakes as well as conscious deception. ([l03],53).

Yet he knew Matthew 27:9 attributes a quote to Jeremiah which is
actually Zechariah 11:13. If not a conscious deception, wasn't this at
least a mistake? Could Augustine avoid seeing it as one or the other?

He could. Augustine's explanation ([I03],44) was as follows.
Under the inspiration of the Holy Spirit, the name "Jeremiah" first
came to Matthew's mind. Then Matthew realized the quote was
actually Zechariah's but decided the Holy Spirit had allowed
"Jeremiah" to come to mind to indicate "the essential unity of the
words of the prophets." So Matthew bowed "to the authority of the
Holy Spirit" and wrote "Jeremiah" instead of the correct reference,
Zechariah.

Augustine illustrates how religious believers defend scripture's
"inerrancy" and "harmonize" its inconsistencies. Augustine knows
Matthew 27:9 is wrong. Yet he can't make a simple correction or
acknowledge a simple mistake. Why? Why can't he improve scripture
and make it more truthful and consistent by correcting a simple error?
Because his way of knowing doesn't allow it. The principle that
scripture is written by God and already error-free prevents him from
acknowledging and correcting a simple mistake. Instead, he's forced
to find an "explanation" that upholds the inerrancy of scripture.

Augustine takes the safe, though not entirely truthful, path. Rather
than admit a simple mistake he "explains" it. What would have
happened if he had admitted and corrected the mistake? I don't know.
But here's what happened to some unfortunate monks who dared to
correct, not even scripture itself, but merely a manual of blessings.

10

By the seventeenth century, errors had crept into ([M02],66) medieval Russia's translations of scriptures and other holy writings. Three monks decided to correct a minor holy writing. But

> [t]o correct any text that had been good enough for the great saints of early Russian Christianity was bordering on heresy. ([M02],66).

So

> [i]n gratitude for their corrections made, the three had been tried in . . . 1618; their corrections were declared heretical. ([M02],67).

One monk was

> . . . excommunicated from the Church, imprisoned in Novospasskij monastery, beaten and tortured with physical cruelties and mental humiliations. ([M02],67).

Mistakes Perpetuated

Anyone who denies the smallest part of "revealed" scripture risks humiliation, ostracism, and perhaps torture and death. This was true at many times in the past. And in some countries it's still true.

It would be wrong, however, to think that only dishonesty or fear prevents Augustine from acknowledging mistakes in scripture. There's a deeper reason: he is blinded by his way of knowing. Believing that scripture is penned by God and error-free prevents him from correcting simple errors. His way of knowing, which is supposed to help him find truth, hinders him. This illustrates a failing of the revelational way of knowing itself, as opposed to a failing of any individual.

To elaborate, people who follow a certain ideology or belong to a certain group and who happen to be untruthful, sadistic or murderous don't necessarily discredit the ideology or group. (If members of a knitting club decide to poison their spouses, that doesn't necessarily show there is something wrong with knitting.) On the other hand, when the ideology or group itself turns truthful, sane people into untruthful, sadistic or murderous persons, then something is wrong with the ideology or group. (Racism, for example, can have this evil effect on those whom it influences.)

Although Augustine's way of knowing didn't make him sadistic or murderous (I don't know if the same can be said for the architects of the Inquisition.), it did blind him to an untruth and force him to accept the false as true. The principle that God is scripture's author blinded Augustine to a simple fact - that scripture sometimes contradicts itself.

Therefore, the revelational way of knowing can enshrine error and hinder the search for truth. The reference in Matthew could be easily changed from Jeremiah to Zechariah, but belief in divine authorship doesn't allow it. Yet the Bible has been amended - not with the effect of reducing an error but of increasing it. Here's the story of an intentional mistranslation that persists even today.

Consistency versus Truthfulness

Christianity teaches that Jesus was born of a virgin. About the Virgin Birth of Jesus, Matthew writes:

> Now all this was done, that it might be fulfilled which
> was spoken of the Lord by the prophet, saying,
> Behold, a virgin shall be with child, and shall bring
> forth a son, and they shall call his name Emmanuel,
> which being interpreted is, God with us. ([H08],Matt
> 1:22-23).

One bible has a curious footnote to this verse.

> [T]his is a prophetic reinterpretation of Is 7, 14 in the
> light of the facts Matthew has outlined . . .
> ([N02],NT,6),

the facts being Jesus's virgin birth, messianic mission, and special relation to God. The footnote continues:

> All these things about Jesus that were faintly traced in
> Is 7, 14 are now seen by Matthew to be fully brought
> to light as God's plan. ([N02],NT,6).

It's not quite clear what "prophetic reinterpretation" and "faintly traced" means. Perhaps a reference to Isaiah will help. Turning to Isaiah 7:14, we read

> Therefore the Lord himself shall give you a sign;
> Behold, a virgin shall conceive, and bear a son, and
> shall call his name Immanuel. ([H08],Is 7:14).

(This verse is an intentional mistranslation of the original, as we shall soon see.) This verse, too, has a curious footnote.

> The church has always followed St. Matthew in seeing
> the transcendent fulfillment of this verse in Christ and
> his Virgin Mother. The prophet need not have known
> the full force latent in his own words; and some

12

> Catholic writers have sought a preliminary and partial
> fulfillment in the conception and birth of the future King
> Hezekiah, whose mother, at the time Isaiah spoke,
> would have been a young, unmarried woman
> (Hebrew, almah). The Holy Spirit was preparing,
> however, for another Nativity which . . . was to fulfill . .
> . the words of this prophecy in the integral sense
> intended by the divine Wisdom. ([N02],OT,832).

Again, a few things aren't clear. What does "transcendent fulfillment" mean? Why would the church have to choose to follow either Matthew (who never identifies the prophet he quotes) or Isaiah? Why would some Catholic writers seek a "preliminary and partial fulfillment" in King Hezekiah? How could a prophet fail to know the "full force latent in his own words"? What does "integral sense intended by the divine Wisdom" mean? The authors of the footnote seem to be half-heartedly trying to tell us something. Like Augustine, does their way of knowing prevent them too from acknowledging a plain and simple fact, plainly and simply? We'll see that it does.

Arsenal For Skeptics ([A09]) has selections of biblical criticism whose authors don't accept the absolute truthfulness and sacredness of every biblical verse. Therefore, one writer can present a much clearer explanation of the verses from Matthew and Isaiah.

> Isaiah's original Hebrew . . . falsely translated by the
> false pen of the pious translators, runs thus in the
> English: "Behold, a *virgin shall* conceive and bear a
> son, and *shall* call his name Immanuel." (Isa. VII, 14.)
> The Hebrew words ha-almah mean simply *the young
> woman*; and *harah* is the Hebrew past or perfect
> tense, "*conceived*," which in Hebrew, as in English,
> represents *past and completed* action. Honestly
> translated, the verse reads: "Behold, the *young
> woman has conceived* - (is with child) - and bear*eth* a
> son and call*eth* his name Immanuel."
>
> *Almah* means simply a young woman, of marriageable
> age, whether married or not, or a virgin or not; in a broad
> general sense exactly like *girl* or *maid* in English, when we
> say shop-girl, parlor-maid, bar-maid, without reference to or
> vouching for her technical *virginity*, which, in Hebrew, is
> always expressed by the word bethulah. ([A09],68).

13

Thus, the words of Isaiah are falsely translated even today, and Matthew quotes no known prophet.

The authors of the footnotes tried to tell the truth of the situation, but could not. Why? Because the belief that God is scripture's Author prevented them. That belief prevented them from communicating the plain and simple truth. Their way of knowing, in this case, prevented them from reaching truth.

For those interested in a contemporary discussion of biblical inerrancy there is *136 Biblical Contradictions* ([O01]) and *136 Bible "Contradictions"...Answered* ([M08]). I've found contradictions in other scriptures but don't know of any similar references although they may well exist.

The Erosion of Truthfulness

Martin Luther once said:

> We know, on the authority of Moses, that longer ago
> than six thousand years the world did not exist
> ([C05],3).

Today some people still believe the world is only a few thousand years old and like the Seventh-day Adventists, who follow a scriptural view of creation, still reject biological evolution. From a Seventh-day Adventist publication:

> Evolution in whatever form or shape contradicts the
> basic foundations of Christianity . . . Christianity and
> evolution are diametrically opposed. ([S10],92).

Other religions, however, over the past few centuries have finally realized the Bible is less than perfectly true. The realization hasn't come cheaply. For centuries, anyone who dared disagree with the Bible risked exile, torture or death. Only the martyrdom of numerous men and women, in the Inquisition and other religiously-inspired pogroms, finally eroded belief in total biblical accuracy. Because of their sacrifice, today some Christian groups can admit that scriptures don't contain the absolute, complete and final truth. For example, Leonard Swidler writes:

> Until the nineteenth century truth in the West was
> thought of in a very static manner: if something was
> found to be true in one place and time, then it was
> thought to be true in all times and places . . . [I]f it was
> true for St. Paul to say that it was all right for slaves to

14

be subject to their masters (in fact, he demanded it!),
then it was always true.

But no Christian theologian today would admit the
truth of the Pauline statement. . . . [O]ur understanding
of truth statements in the West has become historical,
perspectival, limited, interpretive - in a single word:
relational. And that means deabsolutized. . . . Text can
be properly understood only within *context*; given a
significantly new context, a proportionately new text
would be needed to convey the same meaning.
([F02],xii).

The modern world is certainly a significantly new context. How
might a proportionately new text be written? By the continued
martyrdom of men and women? By taking some contemporary
writing, declaring it divine revelation, and blindly following it? Or by
employing science's way of knowing?

Claim 2: Attaining Completeness and Finality

Not only are scriptures said to be truthful and consistent, they're
thought to be complete and final, too. The second claim of the
revelational way of knowing is that scripture is complete - that it has
everything God wants to write - and that it's final - that no new
general revelation is in store. Of course, while it's being written
scripture isn't complete and final. Let's examine that period.

Scripture has been written over varying amounts of time. In the
West, it took about a thousand years to complete the Old Testament.
The New Testament, however, was accomplished in a few hundred
years. And the Koran was written within the lifetime of Muhammad.
While it's being written, scripture is often influenced by contemporary
beliefs, both foreign and local.

When Judaism was young, for example, its scriptures were
influenced by the older religion of Zoroastrianism, which especially
in its

. . . demonology, angelology, and eschatology,
influenced Judaism from the time of the exile onward.
([N04],v23,1013).

It seems to have influenced the Jewish conception of Satan, for
instance.

> Before the exile - for example, in the prologue to Job
> (1:6-12) and in the mouth of Zechariah (3:1-2) - Satan
> was no more than the servant of God, acting on his
> orders as prosecutor; after the exile he is portrayed as
> God's adversary. ([N04],v23,1013).

As another example, there is a story that's told twice, in

> . . . II Sam. 24:1 and I Chron. 21:1. In the first, the
> preexilic version, the Lord incites David to wickedness
> so that he may wreak vengeance on the Israelites; in
> the second it is Satan, not God, who is responsible for
> the calamity. ([N04],v23,1013).

(Yet another instance of scriptural inconsistency.)

How much did Zoroastrianism influence Judaism and
Christianity? *The Ethical Religion of Zoroaster* ([D05],xxi-xxiv) lists
similarities in Zoroastrian, Jewish, and Christian scripture, doctrine
and practice. The list is four pages long. Writers have pointed out
other pagan influences. Powell Davies, for instance, writes:

> Mithras was a Redeemer of mankind; so were
> Tammuz, Adonis and Osiris. . . . Jesus as a
> Redeemer was not a Judaic concept; nor was it held
> by the first Christians in Palestine . . . ([D03],90).

It was only, continues Davies, when Christianity spread to pagan
culture that

> . . . the idea of Jesus as a Savior God emerged. This
> idea was patterned on those already existing,
> especially upon Mithras. . . . [T]he birthday of Mithras,
> the 25th of December (the winter solstice), . . . was
> taken over by the Pagan Christians to be the birthday
> of Jesus. Even the Sabbath, the Jewish seventh day
> appointed by God in the Mosaic Law and hallowed by
> his own resting on this day after the work of Creation,
> had to be abandoned in favor of the Mithraic first day,
> the Day of the Conquering Sun. ([D03],90).

Davies continues:

> In the Mediterranean area during the time of Christian
> expansion, nowhere was there absent the image of
> the Virgin Mother and her Dying Son. . . . ([D03],90).

So it seems scripture when it's being written is liable to be
influenced by contemporary beliefs. Not only that, it may also be
influenced after it's been written. That is, scripture may undergo

editing and revision (the technical term is "redaction") by other than its original author. "Editors" and "compilers" may alter scripture to suit their beliefs. For example, Jewish and Christian scriptures are widely believed to have been redacted. Certainly there was much opportunity to alter Christian scriptures.

> The earliest manuscripts we have . . . are no earlier than the fourth Christian century, and by then - indeed, considerably before - there had been time for the church fathers to make many redactions in accordance with the outcome of theological controversy. ([D03],88).

So, scripture is sometimes changed by other than its original author. Or, to be precise, by other than its original human author, since it could be said God wrote it and later God changed or "redacted" it. Yet it certainly seems strange God wouldn't get it right on the first try and need to edit His own work!

However, inconsistencies aren't always redacted. Sometimes they're allowed to remain but explained away. They are said to be apparent, not real.

> While many differences at first existed among the rabbis as to the actual meaning of the various contradictory stories of . . . revelation, the overriding belief . . . that all of the five books of Moses were divinely inspired and thus incapable of self-contradiction finally gave rise to the consensus that every verse of those books had been revealed by God to Moses on Sinai . . .
> An attempt was made to explain the seeming contradictions among the various versions of the revelation - as also among individual laws - through the utilization of certain hermeneutic principles. ([N04],v22,87).

Hermeneutics is the science of interpreting a writing, usually a revealed writing. We've already seen an example of Augustine's use of hermeneutics.

Claim 2: Completeness

Eventually, scripture becomes fixed and final, beyond the reach of change; there's no possibility of adding anything. If the scripture is complete, if it already contains everything God wanted to say, there's no need of adding anything. Otherwise, it's incomplete, and some important truths are missing.

We'll examine finality in the next section. Now, let's examine completeness. Is scripture complete? Does scripture contain everything believers need to know? In the theological sense, scripture may be complete - that is, it may provide everything a believer needs for salvation, liberation or enlightenment. Or it may not. Again, theological statements are difficult to test.

What about the non-theological sense? As we've seen, some believers think scripture has truths about the natural world. We've already discussed if those "truths" are always true. Are they complete?

No. They aren't. Here's an example.

Spain once pondered building a Panama Canal. Should the canal be built or not? Since Spain then thought scripture to be error-free and complete, it must have seemed logical to see what the Bible had to say about such a project.

> After consulting with his religious advisers (who reminded him of the scriptural warning: "What God has joined together let no man put asunder"), King Philip declared that "to seek or make known any better route than the one from Porto Bello to Panama (is) forbidden under penalty of death." ([C05],220).

It's easy to find the episode amusing (unless you're the unfortunate individual about to be put to death for advocating a canal), but remember the best theological minds of the Spanish empire, with God's eternal revelation to guide them, came to the above conclusion.

Today, a Panama Canal is no longer an issue, but genetic engineering and nuclear power is. They raise questions that scriptures don't address. So, scriptures may or may not be complete in a theological sense, but they certainly don't have all the wisdom we need to make decisions in today's world. Yet many religions insist their scripture is complete and final. Not only does such insistence prevent scripture from directly addressing new issues, it also denies the validity of earlier scripture. It implicitly sets one "perfect and complete" revelation against another.

For example, if the Old Testament was complete there would be no need for the New Testament. Consequently, Christianity - almost of necessity - should teach that the Old Testament is incomplete. It does. For example, a Roman Catholic publication has:

> The knowledge of God, as being just and merciful in His dealings with men was . . . taught to them little by little, in keeping with their developing religious understanding. ([D09],15).

Therefore, God revealed

> . . . His truth slowly and piecemeal and patiently through the ages. As a result, the doctrine in some parts of the Old Testament is more developed than in other parts dating from an earlier period. At times, temporary and incomplete things are found which give way later to fulfillment and completion. ([D09],15).

Similarly, if the New Testament was complete, there would be no need for the Koran. However, Islamic scripture teaches Jewish and Christian revelation isn't complete but incomplete and inaccurate, as we've seen.

It's odd that a religion which thinks God has revealed truth "slowly and piecemeal and patiently through the ages" in keeping with "developing religious understanding," would deny revelation is still occurring today. There certainly seems to be a need for a continuing revelation. For was religious understanding so developed fifteen hundred or two thousand years ago that truth could be revealed once, totally, and for all time? Indeed, is it now at such a level? Rather, it seems likely there would be periodic revelations, more and more divine truth revealed slowly and patiently through the ages, until the entire human race had been raised to intimate union with God.

A few religions do acknowledge the dangers of thinking scripture complete and final. For example, a publication of the Society of Friends, also called Quakers, has:

> Among the dangers of formulated statements of belief are these: (1) They tend to crystallise thought on matters that will always be beyond any final embodiment in human language; (2) They fetter the search for truth and for its more adequate expression . . . ([F01],52).

And the Quaker writer Rufus Jones writes:

> If God ever spoke He is still speaking. . . . He is the *Great I Am*, not a Great He Was. ([F01],51).

But does any religion actually have a continuing, evolving revelation, open to correction and improvement, able to address new issues? If it does, it's not using the revelational way of knowing. Perhaps it's using something like science's way of knowing, which has provided an ever-widening "revelation" about the universe we live in.

Claim 2: Finality

Scripture is eventually considered not only complete, but final and completed. As such, it's closed, immutable, frozen. Certainly, no additions to the Torah, Bible, or Koran are possible. These scriptures are closed.

Being closed and final has its advantages and disadvantages, its "yang and yin." An advantage is that scripture may serve as a constant beacon, an unchanging yardstick for measuring passing fads and temporary lunacies. A disadvantage is scripture can't adapt. Sooner or later, in a hundred years or ten thousand, some scriptural wisdom is no longer wisdom but merely tradition, or even foolish or dangerous. Yet because it's fixed forever in scripture, believers are still obligated to observe it.

Obsolete scriptural "wisdom" seems to be of three kinds: the cryptic, the innocuous, and the injurious. Let's examine examples, beginning with two examples of cryptic teachings.

In *Judaism, The Way of Holiness* Solomon Nigosian writes:
> The biblical injunctions against eating certain birds, or flying insects, are difficult to apply since the species are not always identifiable from the biblical name or description. ([N13],178).

Believers should observe scriptural rules, but how can a rule be followed when it can't be understood? What could be the meaning of useless rules? Or of useless groups of letters?
> Here a word should also be said about the cryptic Arabic letters which head certain chapters of the Koran. Various theories have been put forward by Muslim and Western scholars to explain their meaning, but none of them is satisfactory. The fact is that no one knows what they stand for. ([K07],11).

No believer derives any meaning or benefit from these cryptic, obscure bits of scriptural "wisdom." Yet, the closed, immutable nature of scripture insures such phrases and prohibitions will remain forever, even if no one understands them.

Most scriptural rules can be observed, of course. And they often are, even if there is no longer any good reason to do so. In short, the rules are sacred cows, that is,

> . . . a person or thing so well established in and venerated by a society that it seems unreasonably immune from ordinary criticism even of the honest or justified kind. ([W05],1996).

The idea of a "sacred cow" comes from India, where killing a cow is a great sin, a greater sin than killing other animals. Why? Here's an explanation I once heard.

In ancient India, cows were used to plow the fields. During a famine, hungry people would naturally want to slaughter and eat their cows. But if they did, there would be no way to plow and plant after the famine. A temporary famine would become a permanent famine.

If this story is true, then there was once a very good reason to protect cows. The rule made sense. Eventually, however, the rule was included in divine and unchangeable scripture. The Vedas, one of India's ancient scriptures, refer to the cow as a goddess ([N07],v3,206), and identify it with the mother of the gods. This fixed the rule forever. What once served the welfare of society has become a religious prohibition independent of society's welfare. Today, an observant Hindu can't eat beef because of a religious rule that is, and always will be, fixed in scripture. Yet, the belief is innocuous. It doesn't hurt people and very much helps cows.

Another innocuous scriptural rule originated (I once heard) as follows. In the hot climate where Judaic scripture was written, meat and dairy products in the same dish were unhealthy because the combination easily spoiled. So, a taboo against eating meat and dairy together made good sense. Solomon Nigosian admits this possibility when he writes:

> The regulations about forbidden, *treyfah*, and permissible, *kosher*, foods may well have originated in association with taboos of antiquity. ([N13],178).

He continues:

> Whether or not health or hygiene determined the rules in the first place is little more than speculation, and is irrelevant to pious Jews who refuse to rationalize

> kosher laws. They accept them as part of a total
> system ordained by God. ([N13],178).

Today, an observant Jew can't eat a cheeseburger. Once the rule was written in scripture, it became forever binding. A rule that was originally wise will forever be binding, the invention of refrigeration notwithstanding. Due to the closed nature of scripture, it will forever be unlawful to eat certain healthy foods. A scripture written today might forever forbid fried foods, eggs, and red meat, and forever command a high fiber, low cholesterol diet, oat bran, and aerobic exercise.

The beliefs we've seen are innocuous and harmless. Even if Jewish people can't eat cheeseburgers, their health need not suffer. There are certainly many healthful diets that don't include meat and dairy combinations, or meat and dairy products of any kind. In fact, avoiding cheeseburgers may be healthy. Some people believe there are very good health, ethical, and moral reasons for avoiding cheeseburgers and even all meat. One obvious reason is reverence for life, the wish to avoid unnecessary killing. Another is avoiding cholesterol. And yet another is that meat production ([L03],9) is inefficient: it takes about 16 pounds of grain - grain that could be feeding starving people - to produce 1 pound of meat. Eating less meat could help alleviate world hunger. These people follow dietary rules for solid health and humanitarian reasons. Though the rules may be like religious rules, theirs is an important difference. Rules followed for rational reasons are open to change if the situation changes, or if new research suggests a better path. But religious rules are forever fixed. A faithful Jehovah's Witness believer can't accept a blood transfusion, even if their life depends on it.

Injurious beliefs are the last type of scriptural beliefs we'll discuss. We'll examine three examples.

The Koran's Sura 4:34 says women are inferior to men.

> Men have authority over women because Allah has
> made the one superior to the other . . . ([K07],370).

Will women ever achieve equality in Islamic countries? If they do, they'll have to overcome a divine affirmation of their inferiority and subjugation to men - no easy task. Yet, though they're subject to men, at least women aren't slaves. Is slavery permitted by God? It was for

many centuries in Christian countries. Why? Perhaps because the
Christian Bible has:

> Slaves, obey your human masters with fear and
> trembling . . . ([G02],188, Eph 6:5).

During the 19th century in the United States, people in favor of
slavery used such biblical verses to show that it wasn't against the will
of God. The King James Version, however, has "servants" rather than
"slaves," a crucial difference. Which word did God mean to write?
The revelational way of knowing can't answer since it has no way of
verifying if a writing is actually penned by God, no independent way
of testing if a writing is inspired or not. We'll return to this point later.

Like slavery, the caste system of India is rooted in scripture, has
existed for millennia, and injures the society that tolerates it. Caste
rules were at one time extremely brutal and oppressive: a lower caste
man ([N05],v16,858) who struck, or merely threatened to strike, a
higher caste man might lose a hand or foot. Recognizing the evils of
the caste system, social reformers have worked to abolish the near
slavery of the lower castes. They've been opposed by people who use
scripture to show that God himself supports the caste system. In the
Indian scripture *Bhagavad-Gita*, for example, God in the form of
Krishna declares:

> I established the four castes . . . ([S18],51).

Claim 3: Necessity for Salvation, Enlightenment

The third claim of the revelational way of knowing is that scripture is
necessary for salvation, enlightenment, or liberation. Are any of the
world's many scriptures necessary for salvation, enlightenment, or
liberation? Once again, the theological question can't be proven one
way or the other. We can, however, discuss its implications.

Logically, if scripture is *necessary*, not merely helpful, then
salvation can't be achieved without it - a sad situation for those who
either have never heard of scripture or don't believe it. But if
supernatural revelations are essential and necessary for achieving
life's greatest good, would not God have made them universally
available? Millions of people have lived and died with no opportunity
to read the Torah, the Bible, the Koran, the Vedas, or the Buddhist
scriptures. A person living four thousand years ago in South America,
for example, had no chance to read any of them. And even if

scriptures had been available, for many centuries the ability to read was rare. Millions of people have had no opportunity to read any scripture. Furthermore, if only one scripture is the full and complete Word of God, then many more millions (the past and present followers of other religions) must be numbered among those who had no access to the perfect, complete Word of God. In either case, most people who've lived have had no access to scripture. Yet, some religions teach that people who don't follow a particular scripture are infidels who won't reach salvation or enlightenment. But could God, the Father and Mother of all, have neglected to provide the vast majority of His and Her past and present children with a complete, perfect revelation if it was so vital, so important, so essential for them?

Claim 4: Reasons for Believing in Divine Authorship
The last principle of the revelational way of knowing is that God wrote scripture. Although we've reached it last, divine authorship comes first. The other claims derive from it. Because God wrote it, so it must be truthful, error-free and consistent. God wrote it, so it's complete and final. God wrote it, therefore it's necessary for salvation, deliverance or enlightenment. The principle of divine authorship is basic; it supports the other claims rather than vice versa. That is, scripture isn't so truthful, error-free, consistent, and complete that only God could have written it. Rather, because it's thought to be written by God, scripture is assumed to be truthful, error-free, consistent, and complete, as well as necessary for salvation. Yet, we've seen that scripture isn't entirely truthful, error-free, consistent, or complete. Therefore, it could be argued that God didn't write it. The argument would convince few believers, however, since their belief doesn't rest on logical evidence.

But if evidence doesn't support divine authorship, what does? Why would someone believe God wrote a book? We'll examine four possible reasons: authority, tradition, faith and pragmatism.

A person might believe in divine authorship because some authority who they respect says they should. For instance, they might believe God wrote a book because religious authorities require it as a condition of membership.

Tradition is another reason. People believe because their ancestors believed, because most of the people they know believe.

Yet another reason someone might believe is an inexplicable faith.

Pragmatism is one more reason someone might believe in divine authorship and in religion. After all, *Where did we come from? Why are we here? How should we conduct our lives? What happens after we die?* and similar questions demand answers. What answers are available? Religion's, mostly. Therefore, a deep need for answers to life's most important questions may drive a person to religion, to accepting its dogmas, even if the person isn't fully satisfied by religion's answers, or fully persuaded by its beliefs.

Authority, tradition, faith and pragmatism are ways of knowing, ways of deciding what's true and what isn't. How good are they? They have advantages and may often be better than using reason to decide what's true.

For example, suppose I had a disease and wanted to know how to treat it. How could I proceed? One way is authority; I follow the advice of some medical authority. Another is tradition; I take a traditional folk remedy. In either case, I need faith in the authority or tradition, faith that the remedy works. And in either case, I'm pragmatic; I look for something that works, that cures my disease. Reason is another way I can use. I can try to logically figure out why I'm sick and how to cure myself. Unless I'm an inspired medical genius, however, the purely logical approach wouldn't be a good one. That is, I'd probably do better accepting an established remedy than trusting one I'd invented.

The human race has gained a tremendous store of knowledge over the centuries through the individual and co-operative efforts of millions of men and women. Authority, tradition, and faith are ways of connecting with that knowledge. They are how knowledge is transmitted. Only a rash individual would reject all that authority and tradition have to offer and strike off on their own, intent on believing only what they had personally and independently rediscovered, tested and proved.

Society's Ways of Knowing

Authority, tradition, faith and pragmatism can be very good ways of deciding what's true. And a person who accepts religious beliefs on authority, tradition, faith or pragmatism may find the beliefs very satisfying and useful.

But there's a difference between "private" and "public" ways of knowing. A private way of knowing can be used by an individual to decide what's true. A public way is used by a society. Something that works for an individual may not work for society.

For instance, a person who is sick would be wise to have faith in the wisdom of medical authorities and follow their advice. But how can the world's medical authorities themselves decide what's true? Not by using authority because there are no other medical authorities to turn to. Could they use tradition? If medical authorities merely accepted traditional beliefs, medicine would stagnate. It would never advance if it just believed what had always been believed. Should they use faith? We've saw earlier the disastrous result when medicine turned to scripture, in Justinian's era. So how can society's authorities find truth?

Medicine uses science's way of knowing, a public way of knowing that avoids many of the revelational way's shortcoming. Therefore, medicine is free to correct its mistakes. It's free to acquire new knowledge, to adapt and evolve. And rather than thinking itself in possession of the final and complete truth, it sees itself in the endless pursuit of more and better knowledge.

Religion's authorities could also use science's way of knowing, as coming chapters describe. Like medicine, religion could correct its errors. It could grow, adapt and evolve, in an endless pursuit of knowledge and truth. But it usually doesn't. Rather, religious authorities find the truth by following tradition, by believing what their predecessors did.

But how did the original authorities know? God told them. Moreover, God wrote a book that is the consistent, complete, definitive expression of truth. It's a concrete expression of higher authority that keeps leaders, as well as followers, from going astray, much as the military code is a higher authority that even a ship's captain must obey.

God wrote the book, so follow it. Don't try to correct it, criticize it, or improve it. That's divine authorship in a nutshell.

Shortcomings of Divine Authorship

Because the principle of divine authorship is basic to the revelational way of knowing, many of the shortcomings we've already seen ultimately derive from it. First, God wrote scripture, so it must be necessary for salvation, enlightenment, deliverance. Therefore, people who don't have it are infidels, heathens, and will never be saved, enlightened, or delivered. Second, God wrote scripture, so it must be final and complete. Therefore, it's unable to grow, evolve and adapt. Third, God wrote scripture, so it must be consistent and truthful. Therefore, mistakes and inconsistencies must be explained away rather than acknowledged and corrected. Each shortcoming derives ultimately from the principle of divine authorship. But the principle of divine authorship has a few shortcomings of its own.

One shortcoming can be put theologically. Divine authorship is liable to lead to an idolatrous faith that worships mere written records, mere words, as if *they* are God. Jewish and Christian scriptures describe how Moses overthrew from its pedestal the golden calf ([H08],EX32) his people worshiped and gave them in its place the stone tablets containing the ten commandments. Had the people put those tablets upon the same pedestal and worshiped them in place of the golden calf, their faith would have been no less idolatrous and misplaced as that of many people today who worship Torah, Bible, Koran, Gita, or Tripitaka. Another shortcoming of divine authorship is that there's no way to independently prove that God is the author. Someone must accept the writings on faith.

Blindly accepting writings on faith is dangerous. To illustrate, some Christians regularly risk and sometimes lose their lives, motivated by words Jesus may or may not have spoken.

The gospel of Mark in the *New American Bible* ([N02]) has a "longer ending," a "shorter ending," and a "freer logion." Which ending belongs? The question is more than academic since in the longer ending, the ending accepted in the King James Version, Jesus says:

> And these signs shall follow them that believe; In my
> name shall they cast out devils; they shall speak with

> new tongues; They shall take up serpents; and if they
> drink any deadly thing, it shall not hurt them; they shall
> lay hands on the sick, and they shall recover.
> ([H08],MK 16:17,18).

Because of these words, members of some Christian groups
demonstrate their faith by handling snakes and drinking poisons.
Sometimes that faith costs them their lives.

Yet the longer ending isn't in some early manuscripts of Mark,
manuscripts that a footnote ([N02],NT,65) describes as "less
important." However the importance of an ancient biblical manuscript
is measured, the longer ending isn't in some very old biblical
manuscripts. Is God the author of the longer ending? Does it represent
the actual words of Jesus. Or is it rather a latter-day addition to Mark's
gospel? Should these verses be in the Bible at all? The revelational
way of knowing has no way of determining which ending belongs,
just as it has no way of determining which books belong in the bible.
(As we saw earlier, Roman Catholics have books in their bible that
non-Catholics do not.)

Another danger of blind acceptance is illustrated by the curious
story of Dionysius the Aeropagite.

"Dionysius"

Aside from troublesome passages and doubtful books in the bible,
there's another set of writings whose inspiration has puzzled
Christianity. They were written by someone called "Dionysius the
Aeropagite" and also "Pseudo-Dionysius." Although the writings of
Dionysius are unknown to many Christians, they deeply influenced
Christian mysticism and Christianity in general. We'll meet his ideas
often in the coming chapters, occasionally in his writings, more often
in the writings of those he influenced. The writings of Dionysius
again demonstrate a basic weakness of the principle of divine
authorship. They were accepted not because of their content but
because of their supposed authorship. In fact, Christianity had already
condemned very similar ideas. Yet, when the ideas re-appeared in
writings supposedly penned by a disciple of St. Paul, Christianity felt
obliged to accept them. The story follows.

The biblical book of Acts records that St. Paul spoke before the
"Areopagus," the council of Athens. Paul made believers of certain

Athenians, including a man whose name was "Dionysius". As a member of the Areopagus, Dionysius became known to history as "Dionysius, the Areopagite," just as John Smith of the Senate might be known as "John Smith, the Senator". The conversion of Dionysius may have given welcome publicity and prestige to the young Christian religion, just as the conversion of a senator or congressman might do today for some other emerging religion.

About four hundred years later, the Christian religion had all the prestige and exposure it could desire; it was the state religion. It used its power to destroy competing religious and philosophic systems. For example, in 527 C.E. the Christian church banned ([J03],78) Neoplatonism, a system derived, as its name suggests, from the ideas of Plato. About the same time, someone, probably a Syrian monk, wrote ([D08]) *The Divine Names, The Mystical Theology,* and other works under the pseudonym "Dionysius the Areopagite," that is, under the name of St. Paul's ancient Athenian convert. The works are filled with Neoplatonic ideas. Yet because they were believed to be the writings of St. Paul's convert they

> . . . had an immense influence on subsequent Christian thought. The medieval mystics are deeply indebted to him, and St. Thomas Aquinas used him as authority. ([D08],back cover).

A curious situation: ideas once banned by the Christian church are accepted. Neoplatonism is wrong, but Neoplatonic ideas with a Christian veneer are not. True, a few churchmen may have had their doubts, but

> [s]o long as his traditional identification with the disciple of St. Paul was maintained, and he was credited with being, by apostolic appointment, first Bishop of Athens, these distinctions made suspicion of his orthodoxy seem irreverent and incredible. But when the identification was questioned by the historical critics of the seventeenth century, and the tradition completely dispelled, then the term Pseudo-Dionysius began to be heard and to prevail, and criticism upon its orthodoxy arose . . . ([D08],212-3).

For over a millennium ideas are accepted, not because they pass any objective, verifiable test of truth but because they are believed to be the work of an authority. Yet when their authorship is questioned, the ideas themselves also come under question.

By the way, the Neoplatonic ideas of "Dionysius" are profound and valuable. In fact, it was probably their profundity and value that made them so attractive to the early Christian Church. The problem, however, was that the church's way of knowing prevented it from accepting ideas not ultimately derived from Jesus. Attributing the ideas to a prime disciple of Paul was a clever but not entirely honest solution. In contrast, science is free to find the truth anywhere and accept it from anyone.

Separating Truth and Lies, Wisdom and Nonsense

The revelational way judges a statement by judging the person who supposedly said it. If an authority - a saint or god - said something then it feels obliged to accept it. But if someone else makes a claim, it may feel unable to accept it. Because the revelational way has no method better than this, it's a flawed way of knowing in that it cannot separate truth and wisdom from inconsistency and nonsense. Moreover, this flaw is its fundamental weakness, a weakness that follows directly from the belief that God wrote all of scripture. Because it believes God wrote scripture it cannot acknowledge or correct inconsistencies and contradictions. Rather, it must deny their existence, insisting that all of scripture is true and consistent.

Yet, there are in the world many different "revelations," many different scriptures, supposedly written by God. They disagree with each other and even with themselves. If all were written by God, then how can we explain the inconsistencies and contradictions? If even one was written by God, then how can we explain its inconsistencies and contradictions?

If scriptures and other religious writings were entirely untrue and foolish, then the proper course would be easy - reject them entirely. But suppose they contain truth and error, wisdom and foolishness. Suppose they have much that is profound and enlightening, and much that is nonsense and wrong. Then, accepting them accepts nonsense with profundity, lies with truth. And rejecting them rejects truth and insight along with the lies and nonsense. The ideal is to take the truth and wisdom, and leave the lies and nonsense. But since it uses the revelational way of knowing, religion can't do that, because it has no independent way of deciding what belongs and what doesn't, what is true and what isn't. Therefore, it demands wholesale acceptance of scripture. Some people react against such demands by entirely rejecting scripture. Other people accept scripture entirely. Neither person follows the optimum course; neither takes only the useful, inspired material and leaves the rest.

Of course, some people (perhaps the majority) take what they wish from scripture and quietly ignore the rest. But how do they decide what is true and significant, and what is false or, at least, insignificant? By faith? By instinct? By whim? Could the process of extracting truth from falsehood be done more methodically and rationally? In particular, can any recognized way of knowing reliably separate truth and wisdom from nonsense and lies?

Yes, science's way. In fact, science routinely separates truth from falsehood, and has been doing so for centuries. Moreover, science's way of knowing allows scientists who disagree to co-operate in a common search for truth. Scientists throughout the world routinely work together, testing and extending human knowledge. And scientific knowledge always remain open to test, revision, and improvement. Science welcomes correction, improvement, and evolution.

In contrast, different religions have no way of jointly working towards a common truth, because each is bound to follow scripture it considers complete and final. Religion has no independent way of testing scripture, much less correcting it, much less extending it. Therefore, an eternal revelation that disagrees with another will disagree eternally, dividing forever it followers and the followers of the other eternal revelation.

Of course, not all religious people follow scripture narrowly and literally. Some accept science's view of the natural world and re-interpret scripture to fit by altering the meaning of scripture to accommodate the evidence. For example, if evolution contradicts the bible then they re-interpret the bible. Genesis isn't to be taken literally and scientifically, they say, but symbolically; evolution and Genesis are both true, each in their own way. These believers, who are often in liberal and progressive groups, accept a "de-absolutized" view of scripture. They don't use the revelational way of knowing exclusively to find religious truth. Rather, they indirectly use science's way, too, to decide religious dogma. Sometimes, such people are empirical and experimental, open to new truths and old verities, taking truth in whatever book, person or scripture they find it, rejecting untruth no matter who said it. In this, they approach the scientific way of knowing truth.

Conversely, some scientists hold scientific truth in a dogmatic, closed-minded way. In this, they fail to fully live up to science's way of knowing. We'll turn to the scientific way of knowing after we discuss a final point.

Reason Inadequate?

Science's way of knowing ultimately depends on reason, not faith. Is reason capable of answering life's most important questions? Many religions argue it is not. Let's examine a typical argument.

We're just fallible human beings. So how can we expect to attain higher truth, divine truth, salvation, or enlightenment without supernatural help? After all, we all make mistakes. Our senses are fooled by optical illusions. Our reasoning and understanding is limited. There are many things we don't fully or even partially understand. We err in common, everyday opinions and judgements. We'll never find religious, spiritual or metaphysical truth on our own.

Therefore, the Divine must actively reach out and reveal Itself to us if we are to be saved, delivered, or enlightened. But our imperfect minds may fail to understand or appreciate revelation. The revelation may seem imperfect, wrong, even foolish. If it does, the fault is ours, not revelation's. Therefore, even if divine revelation seems wrong or foolish - as it well might - we must nonetheless accept it and cooperate with God in our deliverance.

An early Christian who seems to have accepted such an argument is Justin Martyr. When Justin

> . . . recognized the great difference between the human mind and God, he abandoned Plato and became a Christian philosopher. ([P01],146).

Justin believed

> . . . that the human mind could not find God within itself and needed instead to be enlightened by divine revelation - by means of the Scriptures and the faith proclaimed in the church. ([P01],146).

32

Justin believed God had to answer ultimate questions in inspired scripture, because the human mind could never find the answers on its own. Only through an act of God could the human race come to know these essential truths, and be redeemed.

Justin decided - with one fateful, irrevocable decision - not to trust his own mind but scripture instead. But if the human mind is so faulty and liable to err, is it safe to make one irrevocable decision as Justin did? Is it safe to decide once and for all and then follow, regardless of any evidence that later comes to light? Wouldn't it be better to constantly test what is thought true? to correct errors when they become apparent? to constantly look for more accurate beliefs? In other words, wouldn't it be better to use science's way of knowing?

It is to science's way of knowing that we now turn.

2

- Science's Way of Knowing -

Chapter Summary: This chapter examines how science decides what's true. It begins by describing the four elements of science's way of knowing. It describes various kinds of sciences, and then discusses shortcomings of science's way. Finally, it compares science's ways of knowing with religion's.

Three principles of the revelational way of knowing make it a faulty way of knowing. First, the principle of divine authorship forces it to enshrine errors and mistakes. Second, the principle of consistency and truthfulness forces it to ignore or explain away inconsistencies and untruths. Lastly, the principle of completeness and finality makes it unable to evolve and adapt, to directly address new issues and problems. And while the fourth principle doesn't hinder the search for truth, it does deny heaven, salvation, or enlightenment to anyone who doesn't possess the one, perfect revelation, that is, to almost all of the people who've ever lived.

Is science's way better? Let's examine it and see. We'll begin with its history and then discuss its elements. We'll see there are various kinds of sciences. And after discussing some of its shortcomings, we'll compare the scientific way of knowing to the revelational way.

The Origin of the Scientific Way of Knowing

The ancient Greeks established science's goal: understanding the universe, what Einstein described as the ([E03],49) "rational unification of the manifold." The goal assumes, of course, that the universe is understandable. So science, as Schrodinger pointed out, rests upon the assumption, the belief - the faith, if you wish - that

> . . . *the display of Nature can be understood.* . . . It is the non-spiritistic, the non-superstitious, the non-magical outlook. ([S05],88).

Unlike religion, which demands belief before, independent of, or even contrary to, understanding, science seeks to understand first. With science, belief is based on understanding. With religion, it's based on faith.

Since science values understanding it must reject blind acceptance and insist on testing statements for itself, not accepting them on faith. It accepts as true only what it has tested and proven. Certainly, science admits that truths exist which aren't yet understood. And it may admit the existence of truths which are forever beyond the reach of human understanding. But, unlike Justin Martyr, science doesn't forever blindly accept "truths" which it can't test. Science is founded on understanding rather than blind faith.

Science's foundation was set many centuries ago when the ancient Greeks arranged geometric rules into a logical system of axioms, theorems and - most important - proof. Before the Greeks, Egyptians discovered many geometric facts but gave them no proof, no logical foundation. Something was true because it worked, that is, it gave a useful answer. Or it was true because some authority said it was. The Greeks replaced blind acceptance with reasoned proof. In their system, a few self-evident principles, *axioms*, were accepted as true. Other statements, *theorems*, were accepted only after they were deduced, i.e., logically derived, from the axioms. The logical derivation was the theorem's proof. An illustration may be helpful.

Suppose we accept as two axioms: that no person has more than one million hairs on their head, and that New York City has at least a million and two people. Then, like Sherlock Holmes, we can deduce - logically prove - the theorem that at least two people in New York City have exactly the same number of hairs on their head.

Why? Because there just aren't enough different numbers to go around. In the best case, the first person has zero hairs, the second person has one, all the way up to the millionth and first person who has a million hairs. Now the next person has either more than a million hairs or exactly as many hairs as a previous person. But no person has over a million hairs. Therefore, the last person's hair count must exactly match a pervious person's count. End of proof.

The Greeks established science's goal and, in their emphasis on the importance of proof, established part of its method. Science's explanations were to be logically deduced from its axioms, its laws, just as our theorem was logically deduced from our axioms. Geometric reasoning was, and still is, a model of scientific reasoning. But how can science find its axioms and laws in the first place? How can it find the laws which govern phenomena like motion, heat, and light? How can we be sure that no person has more than a million hairs?

It seems obvious that principles and axioms must be based on observation. So it seems logical to answer "How can we find the facts?" before we ask "How can we find the laws and axioms which describe the facts?"

How *are* we to discover what the universe does? A commonsense answer is: Look and see. In other words, observe, experiment.

Surprisingly, the ancient Greeks failed to see this. They disdained experimentation in practice:

> Simple experiments with tools and vessels and
> mechanical contrivances they felt to be slavish and
> degrading . . . ([T02],21),

and in theory:

> The Aristotelian tradition . . . held that one could work
> out all the laws that govern the universe by pure
> thought: it was not necessary to check by observation.
> ([H02],15).

In effect, before the Greeks decided "How are the facts to be found?" they asked "How are scientific laws and axioms to be found?" and came up with the wrong answer. Their answer was very much like the revelational way of knowing. Aristotle taught in effect that the laws and axioms that describe the physical universe were "revealed." The difference was that pure thought revealed the laws rather than some god. Had he understood the necessity of observing and experimenting, certain sciences might be much more advanced today.

> Aristotle greatly hampered physics and astronomy by
> building a system on two assumptions which he
> omitted to check by experiment. . . . the speed of fall
> of a body was (1) proportional to its weight, (2)
> inversely proportional to the resistance of the medium.
> . . . Consequently mechanics had to wait nearly two
> thousand years to make a start. ([T02],31-2).

Though flawed, the ancient Greek intellectual tradition greatly influenced the Roman, Byzantine, and Islamic civilizations which followed. Each of these civilizations made important contributions to the general store of knowledge. It was medieval Europe, however, that firmly united reasoning and proof with observation and experimentation. Welding reason and experiment created the scientific way of knowing.

> The complete scientific method, combining systematic experimentation with analysis and proof, has been used consistently only since the 16th Century. ([M04],77).

Or in Albert Einstein's words,

> . . . Development of Western Science is based on two great achievements, the invention of the formal logical system (in Euclidean geometry) by the Greek philosophers, and the discovery of the possibility to find out causal relationship by systematic experiment (Renaissance). ([M04],77).

Medieval Europe discovered a way of knowing which had largely eluded other civilizations, for example, those of Rome, Byzantium, Islam, China and India. It's interesting to speculate why. Three possible factors come to mind. They are scholasticism, alchemy and the invention of printing.

The first factor is Scholasticism, a philosophy based on the writings of Christian thinkers and Aristotle. During Europe's "Dark Ages," when much ancient learning was lost, Scholasticism preserved the belief that the universe was understandable. As Dampier writes:

> . . . Scholasticism upheld the supremacy of reason, teaching that God and the Universe can be apprehended, even partially understood, by the mind of man. In this it prepared the way for science, which has to assume that nature is intelligible. The men of the Renaissance, when they founded modern science, owed this assumption to the Scholastics. ([D01],xv).

Alchemy is the second factor. It gave Europe the practical techniques of experimentation which later proved so essential to scientific experimentation. Alchemy had long been practiced in Europe and other parts of the world. Alchemists had examined matter in depth, seeking the secret of turning lead into gold. Their experiments had taken almost every conceivable form. As a by-product ([M04],46-7), they developed some of the apparatus and experimental techniques later used in sciences such as chemistry.

The last factor is printing, which helped spread scientific ideas. For centuries, books had been copied by hand, a time-consuming process. They were so expensive that only a privileged few could afford them. The invention of the printing press in Germany, however, coupled with the Chinese invention ([F03],62) of inexpensive, quality paper, allowed books to be produced quickly and cheaply. Now, researchers could publish and share their experiments. Participation in science opened to practical people from the crafts and trades, who could now afford scholarly publications. The participation of craft and trade people was to have a radical and invigorating effect.

> The academicians, who had been arguing theories with brilliance and insight for hundreds of years, found their ranks infiltrated by a new type of practical personality, often of lowly birth, whose characteristic attitude was: "Let us cease arguing and find out. Let us experiment." ([M04],81).

Such people realized the importance of exact observations that establish the pure and simple facts. But they were, fortunately, too "unsophisticated" to tolerate torturous "prophetic reinterpretations" that deform facts to fit cherished theories. Rather, ideas and beliefs now became subject to what Dampier calls ([D01],xv) the "tribunal of brute fact." Thus, science

> . . . does not, like medieval Scholasticism, accept a philosophic system on authority and then argue from the system what the facts ought to be. ([D01],xv).

Rather,

> . . . observations or experiment is the starting-point of the investigation and the final arbiter. ([D01],xv).

The First Element: Observation and Experiment

Now that we've seen how science's way of knowing originated, let's examine how it works. We'll begin with its starting-point, observation and experiment. First of all, how do they differ? Observation is passive. When the sun rises, we observe it. On the other hand, experimentation is active. If we throw a ball and examine its behavior, we're experimenting.

Observation and experiment establish the facts, science's raw data. The facts science studies seem to be of two different kinds, objective and subjective, probably because we look out on two different worlds, an outer world of people and things and an inner world of emotions and thoughts, feelings and beliefs.

Inner world facts - subjective facts - vary from person to person. One person says an chocolate ice cream tastes good; someone else says it tastes terrible. Someone says a room is cold; someone else, warm. These are inner world facts even though they aren't stated as such. The ice cream doesn't taste good or bad. Rather, someone finds its taste good or bad. The room is a particular temperature, but is that temperature warm or cold? It depends on the person. When a person says good or bad, warm or cold, they're describing their feelings. They're making an inner world statement.

Unlike subjective facts, outer world facts - objective facts - are the same for everyone. After it's measured, everyone can agree a certain ice cream has this many calories and this much butterfat. With an accurate thermometer, everyone can agree a room is 20 degrees centigrade, even if they can't agree if that's cool or warm. Calories, butterfat, and temperature are objective quantities.

Sciences that treat objective facts have been more successful than those that treat subjective facts. For example, "hard" sciences like physics and chemistry have been very successful: they've achieved a deep understanding and can make precise predictions. In contrast, social sciences, which study subjective phenomena, haven't been as successful: they haven't the deep understanding and aren't as predictive as the hard sciences, probably because understanding subjective phenomena is a more difficult task. Strangely, mathematics, which treats numbers, ideas, and concepts - all far removed from the "real" world of hard objective fact - has enjoyed great success.

Because it's had more success with objective facts, science tends to turn subjective facts into objective facts whenever possible. For example, even though ice cream's taste is subjective, the fact that twenty people say it tastes good and five, bad is objective. So "In one taste test, 80% liked our ice cream." is an objective, outer world statement, while "This ice cream tastes good." is an inner world, subjective statement.

Another way of making a subjective fact objective is measuring it, attaching a number to it. For example, the feeling of oppression that often precedes a storm and the feeling of lightness that often follows one are subjective, even if people generally agree they exist. When atmospheric pressure is measured with a mercury barometer, however, they become objective facts. The number of inches a column of mercury rises in a tube is an objective indicator of air pressure, just as how high the mercury rises in a thermometer is an objective indicator of temperature. Moreover, facts like air pressure and temperature are more than objective, they're also "operationally defined" or "procedurally defined." Temperature is an operationally defined fact because it's defined and measured by following a procedure: by reading a thermometer. Similarly, air pressure is found by reading a mercury barometer. However, not all operationally defined facts are simple to measure: finding the charge/mass ratio of the electron was so difficult that Joseph Thomson ([M04],189) won the Nobel Prize for physics for doing so.

There's a gray area between objective and subjective facts, a kind of fact that seems partly subjective and partly objective. For example, imagine a image from a bubble or cloud chamber. Such images are used in sub-atomic research. To an educated physicist, the image may be objective proof of some sub-atomic particle interaction. To the uninitiated, however, the image is a curious graph of straight lines and spirals which proves nothing. Suppose a physicist says, "This is a picture of a sub-atomic interaction." Is that an outer world, objective statement of fact? Or is it an inner world, subjective statement of belief? If the physicist is wrong about the picture then the statement is an inner-world statement; it expresses a wrong opinion, an erroneous inner-world state of mind, rather than any objective, outer-world reality. But what if the physicist is right?

The Second Element: Hypothesis
As it's practiced today science's way of knowing has four parts: 1) observations and experiment, 2) hypothesis, 3) law, and 4) theory. After observation and experiment have established the starting-point, the facts, a scientist creates a hypothesis, a general rule that describes the facts. After a rule is tested and proven it becomes a scientific law.

Facts lead to hypotheses which lead to laws. Let's look at an illustration.

We put a thermometer in water and heat the water to boiling. The thermometer registers 100 degrees centigrade. We repeat the experiment, in many different rooms with many different pots on many different stoves, and observe that from day to day water boils at a temperature very close to 100 degrees. So, we hypothesize "Water boils at about 100 degrees with slight day-to-day variations due to cause or causes unknown."

Our hypothesis is a rule that describes many different facts, the result of many different experiments. And our hypothesis is tentative; it's a kind of educated guess that must be tested further. If it passes testing - that is, if it continues to describe the facts accurately - then it's promoted to scientific law. Before that can happen, however, other scientists must test it, too. Therefore, once we confirm our hypothesis we publish it so that other scientists may test and confirm it, too. Testing an hypothesis is called "replication" since the aim is to replicate, to reproduce, the facts that led to the hypothesis. Replication ensures the observations or experiments are real objective facts, external world facts that anyone can observe.

Of course, replication isn't always successful. Sometimes it disproves the hypothesis. Returning to our illustration, suppose a scientist in Denver, Colorado ([M04],33) finds water boils at about 95 degrees. Suppose one in Quito, Ecuador, sees it boil at about 90 degrees. Suppose one on Mount Everest sees it boil at about 71 degrees. And suppose each scientist sees the same day-to-day variation around the different temperature. Clearly, something is wrong with our hypothesis. So, we propose a better one: for any one place water boils when it's heated to near a certain unvarying temperature, but the temperature may vary from one place to another.

Our new hypothesis describes the facts better but also introduces some questions: Why does the boiling point of water vary slightly from day to day? What hidden factor or factors control it? Why does it vary from one place to another, for the same reason it varies day to day, or for entirely different reasons? The questions suggest more experiments.

Notice the interplay between hypotheses, and experiment or observation; not only do experiments and observations lead to hypotheses, but hypotheses lead to experiments and observations. In fact, all four elements of science's way of knowing interact. Science's way of knowing is more than four separate elements - it's the dynamic interplay of those four elements. Freshly uncovered facts suggest new hypotheses, or force changes in laws and theories. Hypotheses and laws suggest new theories, experiments or observations. Theories suggest new experiments as well as new or revised hypotheses. Rather than independent elements, the four "parts" are four sides of the single entity that is science's way of knowing.

After our hypothesis has been repeatedly tested and confirmed, it becomes a scientific law. But the process doesn't stop there, because we can still ask "Why? Why does the boiling point change from day to day and from place to place?" Science seeks connections, reasons. Let's continue our illustration. Suppose we know that air pressure also varies from day to day. We might wonder if air pressure is the hidden factor that causes the day to day variations, the place to place variations, or both. In other words, we'd hypothesize "Variations in air pressure cause variations in the boiling point of water" and then go about testing our hypothesis. First, we might observe if air pressure and boiling point always rise and fall together. If they didn't that would be strong proof our hypothesis is wrong. But suppose they did, suppose we perform experiments and find they do. Because air pressure at a particular place varies from day to day, we have an explanation of why water's boiling point varies. But air pressure also varies with altitude above sea level. As elevation rises, air pressure decreases. And water's boiling point decreases along with it. So our new hypothesis also explains the wide variations in boiling point from place to place, from Denver to Quito to Mount Everest. After experiment, hypothesis, replication, amended hypothesis, and further experimentation we have arrived at a much improved hypothesis: water's boiling point is governed by local air pressure. The hypothesis is much more general than the one we began with. It describes water's behavior at different places on earth, in space and on other planets, too.

Inductive Logic and Deductive Logic

In the illustration, our hypotheses were based on inductive logic, the type of logic that reasons from the particular to the general. Inductive logic's input is facts; its output is a general rule that describes the facts. This cat has four legs and a tail, that cat has four legs and a tail, any cat I've ever seen had four legs and a tail, therefore inductive logic says that all cats have four legs and a tail.

Hypotheses can also be based on deductive logic, the opposite of inductive logic. Deductive logic reasons from general principles to a particular fact. Suppose we know that all cats are mammals. And suppose we know that of all mammals only the duckbill and the spiny anteater are born as eggs. Then we can deduce with deductive logic that kittens are not born as eggs. Hypotheses are sometimes deductively derived from theories, which we discuss later.

With deductive logic we can be absolutely sure of the conclusion - *if* we are absolutely sure of our principles. On the other hand, with inductive logic we can never be absolutely sure. As we see more and more cats with four legs and a tail, we naturally feel surer that our hypothesis is correct. But we can never be absolutely sure that *all* cats have four legs and a tail. (In fact, some types of cats don't have tails.)

Because inductive logic can't yield absolute certainty, nature's "laws" - which are based on inductive logic - might better be called "habits." We know the "laws" describe what usually happens. But we can't be absolutely sure they describe what always happens, what must happen.

The inherent uncertainty of inductive logic is one reason for science's skepticism and humility. It's why science's beliefs remain open to challenge, criticism and improvement, and why science sees itself as moving closer and closer to, but not actually in possession of, the ultimate, unchanging, and absolute truth. It's also why science continually tests its hypotheses, and even its laws.

The Third Element: Law
Once an hypothesis has been repeatedly tested, refined, and confirmed, it becomes a law. At this point, inductive logic has fulfilled its mission. Now deductive logic is used to exploit the law and to re-test it, too. Let's look at an example.

James Maxwell discovered the laws that describe light, electricity, magnetism and other electromagnetic phenomena. Maxwell expressed those laws in the form of mathematical equations. By examining Maxwell's equations, Heinrick Hertz logically deduced that certain electromagnetic waves can travel not only through space, but also through solid material like wood. Using Hertz's work, Guglielmo Marconi devised a way to use electromagnetic waves for communication; his invention is called the radio. (Radio waves can travel through wood and brick - that's why a radio works inside a home.) Later, television was based on Maxwell's equations, too.

Our example shows not only how facts are logically deduced from laws, but the use of inductive and deductive logic, as well. Based on studies of light and magnetism, scientists used inductive logic to discover laws. Based on those laws, scientists used deductive logic to discover radio waves, a previously-unknown type of electromagnetic radiation. (By the way, some people fear the word *radiation* and believe all types of radiation are harmful. Not so. Light is a type of electromagnetic radiation.) Our example also shows an international group of scientists testing, replicating, building on and applying each other's work, in a joint search for truth.

Technological advances based on the systematic application of scientific laws are powerful proofs of the laws' validity and accuracy, are also one of science's most visible consequences. Scientific advances have touched all our lives and, literally, changed the face of the earth. The application of Newton's laws led to machines of all kinds. Maxwell's laws led to radio and television. Einstein's equations led to atomic energy as well as the atomic bomb. The laws of quantum mechanics led to the transistor, computer and laser. In each case, the applications are not only useful in themselves but further prove the law.

Nonetheless, laws always remain open to test, criticism, and challenge. If one ever fails then it's revised to describe the new facts, or it's abandoned entirely. Science accepts no sacred cows. Sometimes laws aren't so much abandoned as absorbed, as when an existing law is discovered to be special case of another, more general law. For example, Galileo found a law that describes the changing speed of a falling body. Later, his law was derived as a special case of Newton's law of universal gravitation. In a sense, Newton's more general law absorbed Galileo's law.

The laws of Galileo and Newton express measurable quantities, such as speed and time. Not all scientific laws describe such quantities, for example, evolution's "survival of the fittest" law does not. When a law does involve measurable quantities, however, it may be expressible as a mathematical equation. Newton's law is expressible as a mathematical equation, as is Einstein's famous $E=mc^2$ law that describes a relation between the measurable quantities of energy, matter, and the speed of light. Such equations not only describe the facts, but may also reveal unsuspected phenomena, just as Einstein's equation first predicted atomic energy. Einstein deduced the $E=mc^2$ law from theories he invented. Theories are the fourth and last element of science's way of knowing; we discuss them next.

The Fourth Element: Theory

Observation and experiment establish facts, and facts describe a particular *what*. It's a fact that the earth and moon have a mutual pull towards each other, an attraction. Hypotheses and laws describe whole classes of *whats*. Newton's law of gravitation precisely describes the attraction of all material bodies, including earth and moon. But it doesn't explain why the pull exists, or how it travels through empty space from one object to the other. Facts, hypotheses and laws give knowledge but not understanding. They describe the *what* but don't explain the *why*. Theories explain *why*.

> A theory makes sense of and explains a vast body of scientific knowledge, including both laws and the facts dependent on the laws. ([B12],16).

Einstein's theory of relativity, for example, explains gravitation as a kind of bending of space itself. (Nowhere is it written the explanation must be simple!)

While hypotheses and laws concern facts and express our knowledge, theories concern ideas and concepts, and express our understanding. Science's theories embody its mental model of the universe. They fulfill the original aim of science - partially, not completely because science doesn't completely understand ourselves and our world. Because science's theories are still incomplete, its understanding is still incomplete.

Yet, some theories are more complete and therefore more valuable than others. Science has various ways of judging the value of a theory. We'll discuss five: accuracy, simplicity, predictivity, invariancy, and scope.

Accuracy is the first and most important criterion. No matter how simple, predictive, or invariant a theory is, no matter how wide its scope, if it isn't accurate - if it isn't truthful to the known facts - then it isn't very valuable to science, which values truthfulness above all else. To illustrate, a few hundred years ago Nicolaus Copernicus devised a theory that explained the motion of the planets. The theory assumed the sun, instead of the earth, was at the center of the universe. It also assumed that the planets follow simple circular orbits around the sun. From these two assumptions, Copernicus explained (that is, logically deduced) the observed motion of the planets. Copernicus was attacked because his theory broke with the millennia-old theory of Ptolemy which placed the earth in the center of the universe. Many scientists fought long and hard for Copernicus' theory. Yet, when Johann Kepler advanced a more accurate theory, which replaced circular orbits with elliptical orbits, science accepted it.

But what if two theories are equally accurate? Then science values the simpler theory over the more complex one. This preference is often called "Occam's razor," and is ([C06],686) one of the "fundamental principles of the scientific method." Another way to express this principle is that science prefers the theory that has the fewest or simplest theoretical constructs; we'll discuss theoretical constructs soon. The story of Copernicus also illustrates science's preference for simplicity and its use of Occam's razor. Copernicus' theory - with the planets in simple circular orbits around the sun - is simpler than the theory of Ptolemy - which requires ever more complicated circles and circles upon circles, called epicycles, to explain the orbits. Though neither perfectly agrees with the facts, both theories come reasonably close. But Copernicus's theory is simpler and more elegant, so scientists prefer it over Ptolemy's.

Predictivity is yet another way of measuring the value of a scientific theory. Science values predictive theories above those that only describe or explain. Why? Because predictive theories are more testable and more powerful - testable because experiments can test their predictions and prove them right or wrong, powerful because such theories often predict unsuspected, useful facts, as did the theories of Maxwell and Einstein. Non-predictive theories are less valuable but they are still useful. For example, though the theory of evolution is non-predictive (because it doesn't predict how species will evolve in the future), it's useful because it explains why species have evolved in the past. It organizes the facts and gives them coherence. But a theory that explained *and* predicted would be more valuable still.

Our fourth criterion is invariancy. A theory that's true regardless of time, place and condition is more valuable than one that depends on time place, and condition. By the way, Einstein thought "Relativity Theory" a poor name for his ideas and preferred "Invariant Theory" because the theory concerns quantities (like the speed of light) that are the same for different observers, moving at different speeds. A simpler example is distance. Ships may take different routes from one island to another, but if they measure accurately they'll find the islands are the same distance apart. So length and time of travel may vary for different ships, but distance is invariant.

Even a theory that's entirely invariant may not have much scope, however: 2+2=4 is very invariant but not very broad. Scope, our last criterion, refers to how much a theory explains. The more facts a theory explains, the more valuable it is to science. For example, Newton's law of universal gravitation explains more than Galileo's law of falling bodies. Therefore, science values Newton's law more than Galileo's.

The quest for maximum scope motivates one of science's goals: the search for a "unified" theory, a single theory that absorbs the theories of gravitation, electromagnetism, and nuclear phenomena. Such a "super-theory" would explain almost all known physical phenomena. Its scope would be enormous and its value to science correspondingly high. A more distant goal is a single theory which explains *everything*. Even if this goal isn't reachable, it's still an ideal towards which science strives.

Theoretical Constructs

Though theories represent the pinnacle of science's understanding, they often contain "theoretical constructs," that is, ideas that require an unscientific, almost religious, kind of "faith." To understand this "faith" we must first understand theoretical constructs. So, what are theoretical constructs? Let's begin with an illustration.

Primitive people sometimes explain thunder as the sound of gods at war. Thunder is the fact. Theory is the explanation - the mental creation - that explains the fact. Here, theory contains an unproven idea: that gods are at war. Because no one has ever seen the gods at war, the idea is a theoretical construct, an unproven part of the theory.

Scientific theories often contain theoretical constructs, also called "inferred entities" for which there is no direct proof. In fact,

> [I]nferred entities . . . are usually a critical working part of . . . theory, despite their unverified status. The atomic theory of matter explains Dalton's law of fixed proportions, but at the time the theory was formulated and for long afterward, there was no direct evidence of the existence of atoms. Genes were first posited in theories about genetics long before their physical nature was discovered. ([B12],16-17).

No matter how much atoms and genes explained, they remained theoretical constructs until they were actually detected.

Today, science accepts other theoretical constructs, ideas for which there is no direct proof. Scientists accept them and work with them daily. The "faith" of scientists in theoretical constructs is in some ways similar to the faith of the religious believer. Yet there are important differences. The scientist's "faith" is tentative: if something seems to exist, if it explains the facts, then a scientist may assume it exists until proven otherwise. But the issue usually doesn't rest there; proof of existence or non-existence may become a major goal, as in the case of the neutrino. In the long run, science demands proof before belief. In the short run, it tentatively accepts theoretical constructs. Although such acceptance could be called a kind of faith, it radically differs from the kind of religious faith which is unbending, and, if need be, fact-ignoring.

It's interesting that theoretical constructs also occur in religion. For many believers, God is a theoretical construct, an idea that explains the world and the things they directly experience. Only the person who has had direct experience of God (for example, Moses before the burning bush), can know that God exists. For others, God is a theoretical construct.

It's interesting, too, that a few centuries ago a pair of Western philosophers, John Locke and George Berkeley, showed (refer [C06],206-210) that matter is also a theoretical construct! To paraphrase loosely: we never experience the external world directly but only experience our own senses. The senses are not matter.

Therefore, we do not directly experience matter! We directly experience the senses and create the idea of matter, a theoretical construct, to explain what we experience. To elaborate: when we see a seashell, we see light. Light is not matter. When we hold a seashell and feel its solidity, we feel a push, a force which isn't matter either. When we drop a seashell and hear it hit the ground, we hear sound, not matter. We have no direct and immediate experience of the seashell. We only have direct and immediate experience of light, touch and sound. But the sensations agree - we can touch what we see, and hear what we see hit the ground. So we theorize that something, a seashell, is the cause of the sensations. We invent a theoretical construct which neatly accounts for our sensations. Nevertheless, it remains a theoretical construct.

Kinds of Sciences

Now that we've seen how science's way of knowing originated and how it functions, let's discuss its scope, strengths and weaknesses. We've seen that the scope of a theory is how much the theory explains. What is the scope of science's way of knowing, that is, how much can it explain? In other words, how many fields can it be successfully applied to? So far, science's way of knowing has been very successfully applied to the physical world where it's uncovered a great deal of accurate, consistent knowledge. It's also been used to try to understand the human psyche, but there it hasn't been quite as successful. Why? Because the human psyche is too complex and science's way of knowing isn't equal to the task of exploring it? This question suggests another: Can science's way of knowing be used to explore any and all fields of knowledge?

Francis Bacon and Rene Descartes ([M04],82) were two early scientists who believe that it can. A contemporary author agrees, partially.

> In theory, almost any kind of knowledge might be made scientific, since by definition a branch of knowledge becomes a science when it is pursued in the spirit of the scientific method . . . ([M04],75).

The quote makes an important point so let's digress to consider it.

We've seen that the fundamental difference between science and religion is not so much *what* they know but *how* they know. That is,

49

the fundamental difference lies in their different ways of knowing, not in their fields of knowing. Any field of knowledge becomes a science if it uses science's way of knowing. Therefore, any religion could become a science - if it abandoned the revelational way of knowing and used science's way, instead. Or science itself could examine religious questions, questions of values and ethics, and ultimate questions - just as it has examined many other fields of knowledge, by using the scientific way of knowing. Beliefs that science's way of knowing proves are scientific beliefs, no matter what the field of knowledge. Therefore, in principle science's way of knowing can be applied to ultimate questions. Any difficulty lies not in principle but in practice. So, the question isn't "Can science's way of knowing be applied to ultimate questions?" but rather "*How* can science's way of knowing be applied to ultimate questions?" Subsequent chapters attempt to answer that question. Let's return now to the question of scope.

Bacon and Descartes thought science's way of knowing could be applied to any and all fields of knowledge. Experience has shown, however, that

> . . . not all the subjects practiced as sciences have
> proved susceptible to full treatment by the scientific
> method. ([M04],76).

Therefore, we must revise our definition of scope. Scope includes not only how many fields science's way of knowing applies to, but how well it applies to them, too. Depending on how well the scientific way of knowing applies to a particular subject matter, a particular field of study, a science may be classified as descriptive, experimental, explanatory, predictive, and/or exact.

All sciences are descriptive because they all make accurate, objective observations and classify the results. A paleontologist describes where a dinosaur bone was found, its appearance and condition, the surrounding geological environment, etc. An oceanographer describes ocean currents, temperature, depth, etc. In fact, there's a science, taxonomy, whose entire purpose is classifying living beings in terms of the kingdom, phylum, class, order, family, genus, and species hierarchy that many students encounter in high-school biology courses.

Some descriptive sciences aren't experimental. A geologist describes the creation and evolution of a mountain; an astronomer, the creation of a supernova. But the astronomer can't perform direct experiments with supernova, nor the geologist with a mountain (although computer simulation offers an indirect kind of experimentation). Other sciences allow direct experimentation; such sciences are experimental science, as well as descriptive sciences. In experimental sciences the investigator can devise experiments that answer questions and test hypotheses. In such sciences, experiments can test hypotheses about falling metal balls, chemical interaction, and (within the bounds of certain ethical and humanitarian limits) living organisms, including human beings.

Once an observational or experimental science begins to devise tentative explanations of the facts - theories - it becomes an explanatory science, as well. Its theories strive to account for the facts, to explain them. To illustrate, a paleontologist who often finds the bones of a certain species of dinosaur in pre-historic swamps may deduce the species often died in swamps. A few questions might naturally follow: Were most of the deaths natural? If so, why would such a dinosaur seek out swamps when it was old and ready to die? Because some physical characteristic, such as dry skin aggravated by age, made swamps attractive? Or did the species usually die in battle? Would the creature have been particularly vulnerable in swamps? Such questions lead to theories that explain the observations. Such theories make an observational science an explanatory science, too.

Sometimes theories merely explain what has already happened, without explaining or predicting what will happen. Evolution, for example, doesn't help scientists predict what new forms of life will arise. Other theories, however, give such insight and understanding that prediction becomes possible. Therefore, some sciences are predictive sciences, too. For example, meteorological theories are used to predict the weather - with varying success.

When a number can be attached to the results of observations and experiments, then hypotheses and laws may be expressible in the exact language of mathematical equations. In such cases, exact prediction is possible. Such a science is not only predictive but exact. Physics is such a science. Newton's law that force equals mass times acceleration expresses an exact, mathematical relationship. The great value of an equation is that it enables exact prediction. The force required for a given mass to achieve any acceleration is easily calculated. The ability to perform such calculations has directly led to the creation of new or improved machines of all kinds. Similarly, an exact understanding of heat, electromagnetism, and atomic phenomena has led to devices undreamt of just one or two centuries ago.

Shortcomings of the Scientific Way of Knowing
Exact sciences are fields of knowledge to which science's way of knowing fully applies. Not all sciences are exact, however. That is, science's way of knowing doesn't apply equally well to all fields of knowledge. This is one shortcoming of science's way of knowing. Let's examine a few more.

The scientific way of knowing is often called the scientific method but "method" implies a definite procedure or plan of operation, and is too strong a word. Rather than a sure and certain method, science's way of knowing is a philosophy, a value system, an attitude of approaching the unknown in a rational way, intent on uncovering its secrets, on discovering truth. It isn't a sure, cut and dried method of discovering truth. This is another shortcoming of science's way of knowing.

Another shortcoming is that science has no place for the emotional, poetic, dreamer type of personality that religion often attracts. True, it does allow participation to two quite different personalities: the practical, fact-oriented person and the abstracted, idea-oriented type. In fact, a scientist is often classified as either an experimentalist or a theorist. Experimentalists observe and experiment, verifying facts and testing hypotheses. Theorists spin the theories that explain the facts.

> It is a great triumph of the scientific method that it enables these two extremes of talent, the data-gatherers and theory-makers, to complement each other. ([M04],51).

Yet, science's way of knowing fails to make full use of some important human talents. Some types of persons feel excluded. Therefore, it's less than ideal.

Yet another shortcoming, or rather, entire class of shortcomings, involve not the scientific way of knowing itself, but science as it's practiced today. Once, scientific knowledge was pursued by the solitary scholar or small team, often inadequately equipped and funded. Starting about the time of World War II, however, science changed. Scientists charged with the development of an atomic bomb organized into large teams, generously supported with equipment and funds. Such an environment was then the exception; it eventually became the rule.

Today, scientific knowledge is pursued by full-time, career scientists, supported by grants from government, industrial, or academic institutions. Limited grant monies foster intense competition, a "publish or perish" environment where published papers establish the recognition so necessary to win financial support. *Betrayers of the Truth* ([B12]) describes the dishonest practices that intense desire for recognition and success sometimes foster. Such practices include concealment of raw data ([B12],76,78), slight "improvement" of raw data ([B12],30-31), unfair denial of credit to associates ([B12],ch.8), outright theft of other scientist's work ([B12],ch.3), and even wholesale fraud ([B12],ch.4,5,11) in the invention of data and experiments. Even the vital areas of food, drugs, and pesticides testing ([B12],81) aren't immune to such problems. Dishonesty isn't a recent problem, by the way. Ptolemy, Galileo, Newton, Dalton, Mendel, Millikan and others ([B12],22-3) falsified and misrepresented their research.

Scientific abuses are a cause for concern. During the flowering of monasticism in 11th and 12th century Europe, religion was the leading, most influential state-supported ideology. And religion's full-time practitioners, monks and nuns, were sworn before God to poverty, chastity, and obedience. Yet they often succumbed to the allurements of money and prestige. Eventually public support waned and religion ceased to be Europe's predominate ideology. Science today is the world's most influential state-supported ideology. And while medieval monks and nuns were sworn before their God to an above average morality, the personal morality of scientists is often no higher than average. Therefore, it's not surprising that science's full-time practitioners sometimes succumb to the allurement of money and prestige.

If not corrected, might not scientific abuses someday so erode public confidence and support that science ceases to be the leading and most influential state-supported ideology? Will the desire for money and prestige injure the scientific enterprise even as it injured Christianity and other religions in the past? Can anything be done to help scientists resist what monks and nuns could not? *Betrayers of the Truth* offers a few remedies.

The picture is much brighter when we turn from the failings of individuals to flaws inherent in the scientific way of knowing. The inconsistencies and lies of Ptolemy, Galileo, Newton, Dalton, Mendel and Millikan are acknowledged, not "reinterpreted." We've seen how the revelational way of knowing itself forces Augustine to ignore a clear contradiction. I know of no comparable instance where the scientific way of knowing itself - as opposed to the prestige and money that reward scientific accomplishment - forces, or even promotes, untruth, or blindness to contradiction and falsehood.

Scientists have their share of failings. But human failings don't invalidate science or its way of knowing, even as the failings of religious men and women over the ages don't invalidate religion or its way of knowing.

Science's way of knowing has some shortcomings: it's not a certain method, it denies full participation to the emotional, poetic, dreamer type of personality, it could do more to prevent fraud. How can it be improved? That's an interesting and important question that is, unfortunately, beyond the scope of this book.

Science and Divine Authorship

Both the scientific and the revelational ways of knowing have shortcomings; neither way is perfect. Which is the better way? How do the two ways compare? We'll compare the two ways of knowing by measuring science's way against the four claims made for the revelational way of knowing, beginning with the claim of divine authorship.

The revelational way of knowing claims that God wrote scripture, or caused it to be written. How does science regard it own expressions of truth? It claims no divine sanction or authorship for its beliefs.

A fundamental feature of science is its ideal of objectivity, an ideal that subjects all scientific

54

> statements to the test of independent and impartial
> criteria, recognizing no authority of persons . . .
> ([S03],1).

Scientific truth is discovered and tested through natural, human means, which are usually, but not necessarily, rational processes: flashes of intuition and insight are included, too. What science denies is not the existence of the prodigy or genius, but their supernatural origin.

For example, although Wolfgang Mozart was a musical genius, he was still a man fathered by another man - not an incarnation of Music. And although the six year old Carl Gauss devised a clever method of almost instantaneously calculating 1+2+3+...+99+100 (the answer is 5050) and turned out to be one of the greatest mathematicians who ever lived, when he died his body suffered the usual fate and did not ascend into higher Mathematical realms.

Science accepts no supernatural persons, authority or writings. In contrast to religion's claims of divine authority, science bases its claims on observation and reason, and demands no perpetual faith in "things unseen." Scientific truths can be demonstrated and checked by anyone with sufficient time, equipment, and education.

Moreover, science's way of knowing is the more mature way because it demands judgement and discernment. In contrast, the revelational way of knowing is the more juvenile way. A young child has only one way to decide if something is true or not - they ask someone they trust. Ask a child why they believe something, and they'll answer "Because Daddy said so" or "Because Mommy said so." Ask a believer why they believe something and they'll answer "Because God said so." Each person bases belief on authority, and has no way of finding or testing the truth for themselves.

To illustrate, Brad and Dan are seven years old. Brad believes his town has a very good mayor because his father says so; Dan believes the town's mayor is incompetent because that's his father's opinion. Brad and Dan have just discovered they hold different beliefs. Who's right? Can they decide rationally?

No, they cannot. They have no basis for deciding if the mayor is good or not. All they can say is "I'm right, you're wrong," or "My father's right, you're father's wrong." Not surprisingly, their discussions may lead to fighting. Not surprisingly, too, the same dynamics have often occurred on a much larger scale: religious disagreements have led to much bloodshed.

If they were more mature, Brad and Dan might decide to respect each others faith. "I believe this, you believe that. Each faith is worthy of respect. Let's not discuss the matter further." This is often the religious situation today, perhaps because people are more mature about religious matters, or perhaps because religion no longer matters very much to most people. After all, people are still quite willing to fight and kill for things that really matter to them: political ideology or material resources, for example.

In contrast to religion's way of knowing, science's way allows the discussion and resolution of differences. If they were older, Brad and Dan might discuss their criteria for judging the mayor. Of course, they still might disagree; Brad might value the city's financial state, while Dan might rate city services the best measure of the mayor's ability. However, each would be able to rationally discuss, to have a give and take, and, most importantly, to change their mind if they decided they were wrong. In contrast, the younger Brad and Dan can only cling to the "faith of their fathers."

Science bases its claims on demonstrated fact; it accepts genius but doesn't demand belief in supernatural events; it's open to disagreement, discussion, and improvement. In this respect, science's way is the better way of knowing.

Science and Consistency and Truthfulness
Another claim of the revelational way of knowing is that scripture is consistent and truthful. Are science's beliefs more consistent and truthful than religious beliefs based on scripture? Yes.

Science's theories - unlike religion's - are truthful to the known facts. When something is discovered false, science acknowledges it as false, even if a famous person once declared it true and millions of people believed it for hundreds of years. Special Relativity and Quantum Mechanics are excellent examples, as we'll see.

By 1900, scientists had found in the theories of Newton and his successors an unparalleled understanding and mastery of the natural world. Yet, the orbit of the planet Mercury disagreed with the predictions of Newton's theories. Slightly. Nonetheless, the orbit disagreed. Einstein introduced a new theory, the theory of Relativity,

56

that better explained (i.e., predicted) Mercury's orbit. But Relativity fundamentally disagreed with Newton's theories. That is, it contradicted what science had accepted as true for over two centuries. Science eventually acknowledged the superior truth of Einstein's ideas.

In contrast, it's instructive to imagine what might have happened if science acted like religion. Had Newton been considered a saint or divine Incarnation, had his theories been considered Eternal Law, Einstein might have been ignored, banned, perhaps even tortured and put to death. But science values consistence and truth more than religion does. It values them above any historical person and above any fixed set of beliefs. Therefore, Einstein's theories were eventually acknowledged to be true. A small disagreement of Mercury's orbit with Newton's theory led, not to a "prophetic reinterpretation" of the orbit, but to a revision of the theory. A simple regard for the truth led to a superior truth. And this superior truth - Einstein's theories - opened the way to undreamt-of power, the power of the atom and atomic energy.

The story of Quantum Mechanics - although not as well-known as Relativity - also shows science at its best. The are many good books that tell that story, the story of scientists groping for a truth they couldn't fully understand (and don't fully understand even today). Again, scientists refused to bend the truth to their beliefs, but rather modified their beliefs to conform to the truth.

We find another example of science's truthfulness (but not, unfortunately, consistency) when Relativity and Quantum Mechanics are compared.

A scientific theory describes a certain part of the universe. It must be self-consistent. Ideally, it should also be consistent with other scientific theories that describe other parts of the universe. Unfortunately, Relativity and Quantum theory do not fully agree with each other. As Stephen Hawking observes:

> Today scientists describe the universe in terms of two basic partial theories - the general theory of relativity and quantum mechanics. . . . Unfortunately . . . these two theories are known to be inconsistent with each other - they cannot both be correct. ([H02],11).

Could the two theories be brought into perfect agreement if scientists allowed truth to be bent a bit, if they ignored certain facts, facts that are, perhaps, "insignificant"? No. To science no fact is insignificant when it contradicts belief. In every case, belief must make way for fact. Science does not intentionally ignore facts. Relativity and Quantum Mechanics disagree. Scientists acknowledge that simple fact, instead of ignoring and hiding it, like a scandal. Science's search for a *genuine* "unified" theory that describes the entire universe is still in progress.

In its expression of "Truth," religion has sometimes done violence to the simple and humble truth. Science's devotion to truth is higher. In its single-minded devotion to the plain and simple truth, science's way of knowing is superior to the revelational way.

Science and Completeness and Finality

A third claim of the revelational way of knowing is that scripture is final and complete. Are science's beliefs also final and complete? No. In fact, scientists explicitly deny that their laws and theories are final. For example, one writer has:

> To propound one's beliefs in a scientific spirit is to acknowledge that they may turn out wrong under continued examination, that they may fail to sustain themselves critically in an enlarged experience. ([S03],1),

and

> [i]f I put forward a hypothesis in scientific spirit, I suppose from the outset that I may be wrong, by independent tests to which I am prepared to submit my proposal. ([S03],10).

Therefore, Stephen Hawking writes:

> Any physical theory is always provisional, in the sense that it is only a hypothesis: you can never prove it. ([H02],10).

Instead of being the last and final word, theory is science's best current approximation to truth, always open to revision and improvement. As a consequence, science's world view remains open to challenge and criticism, to revision and improvement. Science's openness to more and better truth is the opposite of religion's fixed, final and closed-minded world view.

> More properly, perhaps, "the scientific method" should be called "the scientific spirit." Its antithesis is the sort

of closed philosophical system which caused the
Church to forbid the great Galileo to argue that the
earth moves around the sun. Its wellspring is . . .
unswerving dedication to truth . . . curiosity . . . open-
mindedness and a skepticism which refuses to accept
as truth anything which cannot be demonstrated.
([L02],31).

Science's openness greatly contributes to its consistency and truthfulness. Because it sees itself as an ongoing, imperfect, human approach to truth, it avoids the closed, fixed nature of revelation. While religious truth has remained stagnant for centuries, science has provided ever-increasing insight into, and control of, the natural world.

Because science is open to, and indeed eager for, new discoveries and truths, it's superior to the revelational way of knowing.

Science and Salvation, Enlightenment, Liberation

The last claim we'll discuss is that revelation is required for salvation, enlightenment or liberation. How does science measure up to this claim? Science doesn't consider its truths necessary for salvation, enlightenment, or liberation. In this, science's way of knowing is again the superior way. For science refuses to condemn most of past and present humanity to eternal torment simply because they were unacquainted with some scripture, or refused to believe it.

Yet, science entirely ignores questions of salvation, enlightenment, or liberation, as well as questions about our place in the universe, how we should live our lives, and what happens after death. In this ignorance, this voluntary limitation, science is inferior - vastly inferior - to religion.

Strange. Ask what happens in an electrical circuit, a supernova, or the heart of an atom and science has much to say. Ask why we were born, how we should live, what happens after death, and science says "I don't know" or even "I don't know and I don't care. That's none of my business."

How can a discipline that seeks to understand us, our world, and our place in it say such questions are none of its business? Couldn't an endeavor whose goal is an explanation of the "whole working of the universe" address questions such as "What is my place in this universe?", "Is there an optimum way to live my life?" and "What happens after death?" Couldn't science make such questions its business?

Of course, many people believe such "supernatural" questions are beyond the reach of natural philosophy, i.e. of science and its way of knowing. Their opinion, however, isn't shared by everyone. In fact, Dampier writes

> . . . philosophers are coming to see that, in a
> metaphysical study of reality, the methods and results
> of science are the best available evidence. . .
> ([D01],vii).

At the very least, science could study religion in a descriptive way. It could describe and classify various religious beliefs. But such a study would only produce a descriptive science that was a second-hand account of beliefs in existing religions. It wouldn't yield a new, living religion which was, in addition, a science.

To create a living religion which was also a science, science would have to explore the religious domain actively, directly, first-hand. It would have to apply its way of knowing to religious issues. If it did, the laws and theories it found would be both religion and science - science because they are a product of science's way of knowing; religion because they deal with questions in religion's domain. In addition to being a descriptive science, such a scientific religion would be experimental and explanatory. Perhaps it would even be predictive or exact.

Summary

In this and the last chapter, we investigated the religious and scientific way of knowing. We found science's way of knowing superior. Because of this superiority, scientific knowledge is often of a higher quality than religious knowledge. That is, scientific knowledge is often true when religious "knowledge" is not. Why this is so follows directly from science's way of knowing.

First, science's knowledge is based on experimental evidence that others may repeat and test, not on hearsay reports and ancient records that may or may not be true.

Second, scientific theories are tested, not blindly accepted on faith. A theory must prove itself by answering criticism and challenge before science accepts it as true. Science has no "sacred cows," no beliefs above question and criticism. It accepts no theory merely on the authority and prestige of some scientist.

Third, as far as possible, scientific knowledge is consistent; where physics and chemistry intersect, for example, they agree. And it's universal; quantum mechanics and relativity apply equally well in China or Spain.

Lastly, scientific truth is always open to revision and improvement. It isn't frozen and final, forced to ignore new knowledge that doesn't fit its theories. Science bends its beliefs to fit the facts rather than bending the facts to accommodate its beliefs.

Not only is science's way of knowing superior, but the two different way of knowing are what truly separate science and religion. Science and religion fundamentally differ in *how* they know. Moreover, the scientific way of knowing is what makes a science a science. If a religion adopted the scientific way of knowing, it too would become a science.

But don't science and religion also fundamentally differ in *what* they know? After all, religion talks about God; science doesn't. So how could a religion adopt the scientific way of knowing, even if it wanted to? Isn't it ridiculous to think science could test religious claims like "There is no God but Allah" or "There are Three Persons in One God"? Yes, it is. But there's no need to test any and all religious claims. For many people it would be enough if science addressed questions such as *Who am I? Why am I here?* and *What happens when I die?* As we'll see, science can address these questions and many more. To understand how, we must first discuss the domain - the *what* - of science and of religion. We must investigate how these two domains differ and, more importantly, what they have in common.

The next chapter discusses the scientific domain of knowing, with emphasis on a portion which borders on the philosophical and metaphysical. The subsequent chapter discusses the religious domain of knowing, also with emphasis on a portion which borders on the philosophical and metaphysical. Then, after a chapter about knowers, we discuss the application of the scientific way of knowing to religious questions.

3

- Science's Domain Of Knowing -

Chapter Summary: This chapter explores what science studies. It finds that energy is the basis of everything science studies in the physical world. After discussing if the universe is eternal, it examines the term "the eternal substance" and explores ideas that are fundamental to the rest of the book.

Ways of knowing was the topic of the previous two chapters. Domains of knowing is the topic of this and the next chapter. Two questions arise immediately: What are domains of knowing? and Why discuss them?

The idea of domain of knowing is simple, as an illustration will show. Suppose you ask a medical doctor how to treat back pain. Because the question is in medicine's "domain of knowing" the doctor can answer, speaking as a doctor. If you ask the doctor to recommend a good mystery novel, however, the doctor can't answer as a doctor, only as an individual. Why? Because back pain is in medicine's domain of knowing, but good mystery novels are not. So, a field's domain of knowing is simply its subject matter. Religion's domain of knowing includes theology and morals. Biology's domain of knowing is living things. Geology's domain of knowing is the earth.

What's science's domain of knowing? We can describe it (somewhat abstractly) in terms of an idea we've already seen, the scope of science's way of knowing. In the last chapter, we met the idea of scope twice: we saw that theories have scope and that science's way of knowing has scope. What is the scope of science's way of knowing? We said it's all the fields science studies (physics, chemistry, but not yet theology). But all the fields science studies is also its domain of knowing. Therefore, the scope of science's way of knowing is identical to science's domain of knowing.

But why discuss domains of knowing? The object of these first six chapters is to apply science's way of knowing to the field of ultimate questions. Applying science's way of knowing to a new field enlarges science's domain of knowing, because it enlarges the sum total of what science studies. A builder first studies a house before enlarging it, checking the foundation, the walls, the roof before adding a new room. Similarly, before attempting to enlarge science's domain of knowing we'll explore it, as it is today. And the builder examines the land the new addition will rest upon. Similarly, we'll explore religion's domain before attempting to bring part of it into science's domain.

Therefore, this chapter discusses science's domain of knowing, and the next discusses religion's domain of knowing. Both chapters emphasize a common area upon which a scientific religion can be built. The fifth chapter discusses how a scientific religion could provide itself with observational and experimental data. The sixth builds on all that has gone before when it discusses the application of science's way of knowing to ultimate questions.

What Science Studies

What is science's domain of knowing? That is, what does science study? Science studies the flow of blood in the body, the flow of thoughts in the mind, the flow of water in a river, the flow of the planets in space, the flow of electrons in a semiconductor - and more. Its domain of knowing is vast and includes a number of branches. In 1964 there were ([M04],75) 620 recognized branches; today there probably are more.

There are ways of classifying, of bringing order, to science's many branches. One way is by dividing them into social science, biological science, physical science, and mathematical science, though mathematics is sometimes thought to be an independent discipline, not a science itself but the "queen of sciences." Like mathematics, philosophy is also thought to be something different from science, though it originally gave birth to science. In fact, an early term for science was "natural philosophy" - the philosophy of the natural world as distinct from the type of philosophy that treats religious, theological, or metaphysical issues. Perhaps philosophy should be called the "mother of sciences."

Another way to classify science's many branches is by the size of the phenomena they investigate. On the largest scale, cosmology investigates the entire universe, its origin, evolution, and future. On a smaller scale, astronomy deals with galaxies, stars, and planets. Geology's focus is the earth, its continents, mountains, and rivers. There's a group of sciences - medicine, psychology, sociology, etc. - which study human beings. Biology's concern is living entities in general. Chemistry deals with phenomena on the molecular scale. And, finally, nuclear physics probes the heart of matter, atomic and sub-atomic phenomena. (*Powers of Ten* [M17] vividly illustrates these different scales.)

Not all scientific fields can be classified by size, however. The study of nuclear reactions in stars, for instance, concerns two widely differing scales. And some disciplines investigate phenomena that have no scale, that is, phenomena which don't exist on the physical level. Psychology studies emotions and mental states; mathematics investigates concepts and ideas, as does philosophy. This suggest yet another way of classifying scientific disciplines: by the type of phenomena they investigate - physical, emotional, or intellectual.

Basis of the Physical Universe
Do science's many branches have anything in common? One commonality is that in each field it studies science strives to uncover and understand the objectively true.

> [S]cience articulates, in a self-conscious and
> methodologically explicit manner, the demands of
> objectivity over a staggering range of issues of natural
> fact, subjecting these issues continuously to the joint
> tests of theoretical coherence and observational
> fidelity. ([S03],3).

But this just says each field uses science's way of knowing. Do science's many branches have anything in common aside from their way of knowing? Are they actually branches of a single tree? Or are they unconnected fields of study?

To see what science's branches all share, let's begin with what physical objects all share. What can one physical object have in common with another?

Shape is one possibility. For example, imagine two chairs. Though they're different from one another, they have enough in common (their shape) to be recognizable as chairs. Though they may differ in composition (one wood, the other metal), color, height, weight, etc., they're still recognizable as chairs. Substance is another possibility. Imagine a desk, a table, and a chair, all made of wood. Though their shape differs, they are recognizable as distinct manifestations of a single substance, wood.

If *all* physical objects are ultimately made from the same substance, then science's many branches (at least, the ones that study physical objects) have something in common, aside from their use of science's way of knowing: they all study various manifestations of one and the same substance.

Are all physical objects ultimately made of one substance? The question is ancient. Centuries before science existed, philosophers debated the idea that

> . . . there is only one basic substance or principle as the ground of reality . . . that reality consists of a single element. ([R01],862).

This belief is called "monism", from the root "mono" which means one. Philosophers in ancient Greece, and perhaps before, searched for the one single substance that composes all things. In fact, physics takes its name from a Greek word that indicates a single, universal substance.

> In the ancient Greek schools in Ionia and Elia, the essential essence of all things . . . ([W10],148)

was called

> . . . the "physis," from which our word "physics" is derived . . . ([W10],148).

While some Greek philosophers hypothesized a single, universal substance, Aristotle taught that all things are made out of not one, but four, basic elements: fire, earth, water, and air. His four-element theory was wrong, but was nonetheless believed for centuries. It's remarkable that the "elements" correspond to energy and the three states of matter - solid, liquid, and gas.

The search for the *physis* continued in 19th century Europe, where chemists discovered that all physical objects are composed of atoms. Eventually, about 92 different kinds of atoms were discovered and arranged in the "periodic table," which many high-school students study. Physicists soon discovered that atoms themselves are composed of sub-atomic particles such as protons, neutrons, and electrons. True to their name, they searched deeper for *physis*, the one ultimate basis of matter, the essential essence of all things. Physicists hunted for a smaller particle that composes all the sub-atomic particles. Their search was not successful.

Yet even if physicists had found an ultimate particle - an infinitesimal unit that makes up that sub-atomic particles that compose the atoms that constitute wood and all other matter - they would not have found the basis of everything. For even 19th century physicists knew of something that isn't in the periodic table or composed of sub-atomic particles. Energy. In the 19th century, there were two laws of conservation: one said matter couldn't be created or destroyed; the other said energy couldn't be created or destroyed. It seemed matter and energy were two forever separate and distinct entities, and "never the twain would meet." Science had replaced Aristotle's four elements with two irreducible elements: matter and energy.

In the early 20th century, however, Albert Einstein showed in his famous equation, $E=mc^2$, that matter is a form of energy. As Nigel Calder writes in *Einstein's Universe*:

> Einstein's formula $E=mc^2$ expresses the equivalence of mass and energy. In it E is energy, m is mass and c^2 is the square of the speed of light. The c^2 comes in only because of the conventional ways in which physicists reckon energy and mass. You could just as well, and more simply, write $E = m$ and adjust your units of measurement to suit. ([C02],31).

Schrodinger is a bit more emphatic. (As always, italics are in the original.)

> . . . *particles* have now turned out to be *quanta of energy*, because - as Einstein discovered in 1905 - *mass and energy are the same thing.* ([S06],54-5).

So there is, in fact, a single basis for the entire physical world. In the words ([J01],176) of physicist James Jeans, energy is "the fundamental entity of the universe". The age-old search for the ultimate basis of the physical universe has ended. Though there's much more to be discovered and understood, physics has finally fulfilled its original mission. It's finally discovered the essential essence of all things, an essence which underlies everything science studies. And it's answered an ancient philosophical question: monism accurately describes the physical universe.

The Table Illustration

The things we see around us all have a common basis. They are all different manifestations of the same "fundamental entity." Yet they all seem very different from one another. Wood seems very unlike water; metal seems very unlike oil. They certainly don't appear to be different manifestations of a single substance. How can wood and water, metal and oil, all be manifestations of a single universal essence? We'll illustrate by taking a simple physical object, a wooded table, and following the path to its essential essence.

Imagine a wooden table.

What is it? - A table.

How old is it? - Let's suppose you recall the date the table came into existence. Suppose you made the table on your last birthday. So the number of days the table has been in existence can be counted.

What's the table made of? - "Wood" is one answer. In one sense the table is made of wood. In another sense, it's made of its parts, its components - a top and four legs. But more than components are needed to make a table. The components must have the proper relation to each other. Disassemble the table - cut the top in half and pile on the legs - and the components no longer form a table. They merely form a pile of wood. True, the pile contains the pieces of a table. But the pile is not itself a table. The parts are still wood, however.

What is it even if it's rearranged? - Wood.

So what is it? - A table made of wood.

How old is it? - Let's suppose the tree was cut, the lumber created, a month before your last birthday. As a table, it has existed since your last birthday. As wood it has existed a month longer.

What is wood made of? - Once

> . . . it was thought that the *content* of a solid was what determined its characteristics: what made diamonds hard, leather tough, iron magnetic and copper conductive. . . . Today we know that many of the properties of a solid are determined by its *structure*: by the way the material's basic building blocks - its atoms - are ordered, and by the way they join together. ([L02],99).

Like the table, wood also consists of parts: various atoms (such as carbon, hydrogen, and oxygen) form wood molecules. And, like the table, the atoms must have a particular relation to each other to make a wood molecule.

Things like the table that are created by a particular arrangement of components are called "component things" or "component objects." The table is a component object because it's formed by a particular arrangement of its components or parts. Wood molecules are component objects, too. In fact, words themselves are component objects. The word "are", for instance, is a component object because it's created by a particular arrangement of components, its letters. Rearrange the letters and we get "era", an entirely different word. Similarly, rearrange the atoms in wood and we get a different molecule. Furthermore, a component object ceases to exist when its components are rearranged. Rearrange its letters and the word "are" ceases to exist. Rearrange its atoms and a wood molecule ceases to exist. Yet the atoms continue to exist.

What is it even if it's rearranged? - Atoms.

So what is it? - A table made of wood made of atoms.

How old is it? - Most of the atoms have been in existence for millions or billions of years. Most atoms were created in supernova explosions a long time ago, and have existed as atoms ever since.

What are atoms made of? - Atoms are made of components, various subatomic particles, such as neutrons, protons, and electrons.

> Every atom has the same basic internal structure, and
> the protons, neutrons and electrons in any one atom
> are identical to those in any other atom. ([L02],10).

Again, the components - the protons, electrons, neutrons, etc. - must be in proper relation to each other to form a particular kind of atom. If the atom's subatomic particles are rearranged, then that atom ceases to exist and other atoms are formed. For example, ([L02],170) if the subatomic components of an uranium atom are suitably rearranged, the uranium atom ceases to exist and two barium atoms come into existence.

What is it even if it's rearranged? - Subatomic particles.

So what is it? - A table made of wood made of atoms made of subatomic particles.

How old is it? - While it's possible to create protons and neutrons, most of the table's neutrons, protons, electrons, and other subatomic particles are billions of years old.

What are subatomic particles made of? - Neutrons and protons are composed of fundamental particles; the electron is itself a fundamental particle. In 1990, there were three known families of fundamental particles. Moreover, just as family members may be male or female, fundamental particles in any of these three families may be classed as quark or lepton.

What is it even if it's rearranged? - Fundamental particles.

So what is it? - A table made of wood made of atoms made of subatomic particles made of fundamental particles.

How old is it? - Perhaps billions of years.

What are fundamental particles made of? - Energy. The E of Einstein's famous $E = mc^2$ composes fundamental particles. If a particle's energy is "rearranged", then the particle ceases to exist and other particles are created.

What is it even if it's rearranged? - Energy.

So what is it? - A table made of wood made of atoms made of subatomic particles made of fundamental particles made of energy.

How old is it? - Isaac Asimov, a well-known science writer:

> In a way, of course, we might argue that the energy of the universe (including matter, as one form of energy) has always existed and always will exist since, as far as we know, it is impossible to create energy out of nothing or destroy it in nothing. This implies, we can conclude, that the substance of the universe - and therefore the universe itself - is eternal. ([A11],184).

What is energy made of? - As far as we know, energy has no components, no parts which compose it. It's not made of any other thing. It's made of itself.

What is it even if it's rearranged? - Energy, as far as we know, cannot be disassembled or rearranged.

So what is it? - Energy.

How old is it? - It's eternal.

Creation?

The table illustration demonstrates what science's many different branches have in common, other than their way of knowing. They all study various manifestations of a single entity, energy. After all, we can do for wood and water, metal and oil, what we did for the table. We can take each and follow the path to its essential and eternal essence, energy. Therefore, science's branches - at least the ones that study physical objects - all study different manifestations of the same thing. What about scientific disciplines which study emotions and thoughts? Are emotions and thoughts also manifestations of the single essential essence? For now, we'll assume they are. A later chapter discusses the question in more detailed.

The illustration also raises the question of the age of the universe. Asimov makes an unusual claim: the universe is eternal. Other scientists say the universe was created in the big bang. Perhaps that's why Asimov hedges. He begins "In a way, of course, we might argue" and continues:

> That, however, is not what we really mean. We are
> concerned with more than the mere substance of the
> Universe. ([A11],184).

Is the universe eternal? Or did it begin during the big bang? First of all, what is the big bang?

Ten to eighteen thousand million years ago, scientists currently believe, the universe was in an extremely dense, hot, unstable state called the "cosmic egg" or "primeval fireball." A tremendous explosion, the big bang, started the universe expanding. It continues to expand, even today. Will the expansion ever stop? Some scientific theories say no. Others predict the universe will eventually stop expanding and start to contract. They predict the universe will someday collapse upon itself, form another cosmic egg, and undergo another big bang. If such predictions are true, then the big bang isn't unique. The universe regularly expands and contracts, as if it's breathing.

But was the universe created in the big bang? A college astronomy text claims it was.

> Some unique explosive event must have occurred 10
> to 18 billion years ago, apparently creating matter and
> sending it flying out on its expanding journey. Many
> astronomers refer to this time as the **age of the
> universe**, marking the creation of the universe as we
> know it. ([H01],428).

Of course, the universe was hardly "as we know it" after the big bang. It took millions of years for stars and planets, minerals and life to develop. Perhaps the writer meant "marking the creation of the universe that we know" because nothing is known about any universes that may have existed before the big bang. Physical laws seem to break down in the cosmic egg. The text continues:

> What existed before the big bang? No one is sure that
> this question has meaning or that any observations
> could reveal an answer. . . . [T]he mysterious
> explosive event at the beginning is *assumed* to have
> mixed matter and radiation in a primordial soup of
> nearly-infinite density, erasing any possible evidence
> of earlier environments. ([H01],434).

Science knows nothing about what form and shape the universe may have taken before the big bang, but about its substance science knows this: it's eternal.

So, does the big bang create the universe, or not? In one sense, it does; in another, it does not. The big bang creates the universe just as joining pieces of wood together creates a table. The big bang creates matter just as packing snow together creates a snowman, or packing sand together creates a sand castle. The table parts exist before the table is created, the snow and sand exist before the snowman and sand castle are created. In each case the basic "stuff" isn't created. It already exists. Similarly, energy existed before the big bang created matter. Therefore, in another sense the big bang didn't create the universe because - as far as science knows - the big bang didn't create energy. Energy existed before the big bang. So, as Asimov claims, the universe is eternal. It was never created.

Turning to Philosophy

In one sense the universe is created during the (last?) big bang. In another sense it is not, because its basis is eternal. Both answers are correct. Yet scientists favor the first and ignore the second. They speak of the big bang as the time of creation and don't often acknowledge the universe's eternal essence. Calder, Schrodinger, Jeans, and Asimov are exceptions, though Asimov mentions it only in passing and even calls it the "mere" substance of the universe. Other scientists and science writers fail to mention it at all.

Why is the universe's ultimate essence so rarely acknowledged by scientists and science writers? And how can they ignore such a fundamental entity, even if they wish to? We'll explore those two questions after we've discussed the ultimate source a bit more.

71

Because the universe's fundamental entity will prove fundamental to coming chapters, too, because it's the foundation upon which coming chapters are built, we'll need to understand it in as many ways as possible. We've seen what science says about it. Now, we'll turn to philosophy and see what it has to say about the universe's ultimate essence. The next chapter explores what religion has to say about the eternal substance.

In our philosophical exploration we'll meet many names and ideas that apply to energy. Our purpose in examining them is to lay a foundation for coming chapters, where we'll meet the pronouncements of saints, sages and mystics who experienced the essential essence. Not being scientists, these men and women described their experience in the language of philosophy and religion rather than science. Discussing philosophical and religious terms for energy will help us better understand what they said.

So, let's turn to philosophy and see what it has to say about the essential essence. We'll begin with the phrase "energy, the eternal substance." The phrase expresses three ideas about energy. One, that it's unique, "*the* eternal substance" not "*an* eternal substance". We'll discuss uniqueness last. Second, the phrase says that energy is eternal. Lastly, it says energy is a substance. Are these claims true?

Is energy actually eternal? As we've seen, energy can neither be created nor destroyed. Obviously, something which can't be created has no beginning and therefore has always existed. Similarly, something that can't be destroyed has no end and will therefore last forever. Because the word "eternal" means

> . . . lasting forever; without beginning or end; always existing ([R01],453),

it follows that energy is eternal.

Is energy actually the substance of the physical universe? The word "substance" comes from the Latin "sub stantia" which translates "that which stands under." Since the table is made of wood, wood "stands under" the table, that is, the table's substance is wood. Similarly, wood is made of wood molecules, which therefore "stand under" wood. Molecules are made of atoms which are made of subatomic particles which are made of fundamental particles which are made of energy. Therefore, the table's substance is wood; wood's substance is molecules of atoms; etc. But neither table, wood, molecule, atom, or particle is eternal. Energy, however, is eternal. Therefore, energy is an eternal substance.

But is energy *the* eternal substance? It seems to be. Modern science, as far as I know, says only energy is an eternal substance. Therefore, energy is the eternal substance rather than one eternal substance among many. Energy is the unique eternal substance.

Philosophy has many other names and ideas that apply to the eternal substance. We need to briefly examine some. Many ideas fall into one of the three categories we just discussed. That is, many express either uniqueness or eternal existence or substantive existence. We'll begin with names and ideas that imply eternal existence.

The *Eternal* Substance

Ideas that express eternal existence include uncreated, unoriginated, unborn, and unformed. Let's see how each applies to energy.

Since energy is eternal, it was never created, originated, or born at anytime in the past. Since it was never created, originated, or born, it's uncreated, unoriginated, and unborn. The term "unborn," of course, has another meaning: that which is not yet born. However, I'll only use "unborn" to mean something that exists but was never born, originated, or created at anytime in the past.

Energy is also unformed. Why? A meaning of "formed" is created, originated. An idea is created when it's formed in the mind; a business corporation is formed when a few people come together and create it. Therefore, energy is unformed because there was never a time when it was created. It's unformed in another sense, as well, because "form" has another meaning:

> 1. The shape or contour of something as distinguished
> from its substance . . . ([F08],523).

Energy is "unformed" because it has no shape or contour. Form is also the opposite of substance, so it's appropriate that the eternal substance is unformed.

The terms we just discussed - uncreated, unoriginated, unborn and unformed - concern the past. What about the future? Ideas that express future eternal existence include indestructible, enduring, lasting, persevering, permanent, and undying. Let's discuss them

Something that's eternal will never cease to exist in the future. Therefore, it can't be destroyed. It's indestructible. As the only eternal, indestructible entity known to science, energy is "the indestructible." In a similar sense, it may be called "the enduring," "the lasting," "the persevering," "the permanent," and "the undying."

The next set of ideas we'll discuss refer both to past and future eternal existence.

That which is eternal is unaffected by time, either past or future. Though mountains wash to the sea and continents change their shape, the eternal remains, untouched by time, beyond time, and, in this sense, "the timeless." Moreover, it's not finite in time, not bounded or limited by time. Therefore, it's infinite, unbounded, and unlimited with respect to time.

The Eternal *Substance*

Philosophy has other ideas that express not eternal existence but substantive existence. "Ultimate ground of existence" is an important one. What does it mean? Let's begin with the simpler idea "ground of existence."

The table exists but doesn't exist independently. Rather its existence depends on wood. Therefore, the table is based on and "grounded" in wood. Wood is the table's ground of existence. If the wood ceases to exist then the table stops existing, too. Though wood is the table's ground of existence, the inverse it not true: the table is not wood's ground of existence because if the table ceases to exist, the wood can still exist. If the table's top and legs are cut into pieces, for example, the table - as a table - ceases to exist. But the wood still exists. But if the table is not wood's ground of existence, what is? Wood's existence is grounded in wood molecules, which in turn have their existence grounded in atoms, which in turn are grounded in subatomic particles, etc. The process of taking "the ground of existence of the ground of existence of the . . ." doesn't go on forever. When it ends we've arrived at the table's ultimate ground of existence - energy. Wood, molecules, atoms, etc. are all grounds of existence for the table, but only energy is the table's ultimate ground of existence.

By the way, the process of taking the ground of existence indicates dependence. The table depends on wood for its existence, which depends on wood molecules for its existence, etc. Energy, however, as ultimate ground of existence, depends on nothing else for its existence. It has independent, self-sufficient existence. We'll return to the idea of independent, self-sufficient existence soon.

Other philosophic terms which express the idea of ultimate ground of existence include core, center, root, source, primal, fundamental, the first, and essence. Let's examine them.

As we look deeper into the table - to wood, to molecules, to atoms, etc. - as we get closer to its ultimate ground of existence, we get closer to its "core" and "center." Therefore, energy is the table's center and core. But the table isn't unique. For any other physical entity we'd arrive at the same core and center. We'd find the same ultimate ground. Therefore, energy is unique; it's "*the* core" and "*the* center."

That which is the core and center is also "the root." Because an entity derives its existence from energy, it's "rooted" in energy, just as a snowman is rooted in snow, a sand castle in sand. As root, energy is also "the source" since the table "flows" from its root and source, down through fundamental, subatomic, and atomic particles, molecules, and wood to finally become a table. As the root and source, energy is also primal and fundamental, the first. Finally, it's the essence because essence is

    1. the basic or necessary constituent of a thing
      ([R01],451),

and that which is root and source is obviously a basic and necessary constituent.

The terms we've seen suggest inanimate objects: apples have cores, circles have centers, plants have roots, and rivers have sources. Philosophy has many other names and ideas that express substantive existence. Some - such as mother, father, creator, identity and self - express almost person-like qualities. Others - such as isness and suchness - deny all qualities, person-like or not. We'll discuss both types.

The root and source of all that exists can be pictured in a personal way, as the "Father," the "Mother," the "Creator" of everything. Moreover, regarding energy in this way makes the universe appear as a child, the son of an eternal Father, the daughter of an eternal Mother, and a creation of the Creator

What of identity and self? A son or father, a daughter or mother, is a person with an identity, a self. Does the idea of identity and self also apply to energy?

First of all, what do the words mean? A dictionary defines "identity" as

> 1. the state or fact of remaining the same one or ones, as under varying aspects or conditions . . .
> ([R01],659).

What of "self"? One definition is

> 1. An individual known or considered as the subject of his own consciousness. ([F08],1218).

A later chapter discusses awareness of one's own consciousness. A better definition of "self" for what we're discussing now is

> 5. being the same throughout ([R01],1193).

Let's see how these definitions apply to people.

Imagine an actor in costume. The question "Who is that?" has two senses. The first sense refers to appearance. "Who are they playing today? Macbeth? Caesar?" The second refers to the underlying identity and self. "Who is the actor? Is that Joe or John?"

So, identity and self refer to the person who "stands under" various appearances. Our clothes change and, over the years, our body changes. That which remains the same in all those different appearances - which "stands under" the appearances - is our identity and self.

Now, let's generalize the idea of identity and self to inanimate objects. Imagine (once again!) a table. The question "What is that?" also has two senses. In the sense of appearance, the question means "What is the wood 'playing' now? A table? Yesterday it was just a pile of wood. Today, however, I see that someone has fashioned it into a table." In the sense of standing under, however, the question refers to identity and self. "What is the enduring reality underneath the appearance of what was yesterday a pile of wood and is today a table?" We can even ask, "What is 'playing' the table?"

The answer is that energy, the ultimate ground of existence, is "playing" the table. Energy is the enduring reality behind the table's appearance. It's the identity and self of the table and every other entity.

Mother, father, identity and self express personal qualities. Isness and suchness, on the other hand, express an absence of all qualities, personal or not. Nonetheless, isness and suchness are loosely related to the ideas of identity and self. Let's see how.

Something that remains the same under different appearances and conditions is independent of appearances and conditions. Appearances come and go, conditions change, but the identity and self remains. But does something that's independent of particular appearances and conditions have any appearances and conditions to call its own? Does it have any qualities of its own? For example, a person's identity and self is independent of particular appearances and conditions. A person is born with a tiny body that later increases to ten or more times its original weight. A person today may be cheerful whereas yesterday they were sad. Yet we feel the same person exists underneath the different bodies and moods. Therefore, the person's identity or self must be independent of appearances, such as body, and conditions, such as mood. If it is, can identity and self have distinctive qualities of its own? We return to that question in a later chapter which discusses the question of personal identity.

What of energy, the identity and self of physical objects? Does it have any distinctive qualities? A persuasive philosophical argument says no. It's as follows.

As we go toward center, physical characteristics are lost. A table may have a smooth finish but the idea of "smooth finish" doesn't exist on the atomic level. A bell may have a nice tone, but tone (nice or otherwise) has no meaning on the subatomic level. Diamonds are hard, but the carbon atoms which compose them are no harder or softer than the carbon atoms that compose soot. Therefore, as we get closer to center we lose distinctive qualities. If we take this process to the limit, we reach pure energy, which therefore has no specific qualities. We reach undifferentiated existence, existence itself. If this argument is true, then energy is "pure isness" or "pure suchness" - pure existence, devoid of particular qualities.

Some writers put it differently: they say that pure existence has *all* qualities in a latent state. Just as sunlight has all colors of the spectrum (as passing it through a prism demonstrates), they say that pure suchness contains all qualities in an undifferentiated state.

We've just examined philosophical concepts that express eternal existence and substantive existence. We'll discuss a few more philosophical concepts after we pause to consider a question we raised earlier.

Historical Bounds

Decades ago, science finally discovered the universe's single, eternal basis. Physics at last found the essential essence of all things. Strangely, not very much is made of this discovery. One reads innumerable popular science books and rarely sees it expressed. Why not? Why does science so often ignore or overlook what seems a significant find? Historically, science has ignored the universe's ultimate basis, along with questions of ultimate purpose, for two reasons: fear, and a desire to avoid useless discussions.

We've already discussed the well-known difficulties early science experienced with religion about physical facts, for instance, the earth's path around the sun. In those times, "Eternal Substance" and similar phrases concerned the supernatural world - they were applied to God. If scientists had been so foolish (or daring) as to include the Ultimate Ground of Existence (God) in their discussions, they might have suffered much more drastic persecution from religious authorities. Science might not have survived. So science from its earliest days has included only the natural world in its scope. As early as 1663, Robert Hooke, curator of the Royal Society, one of England's foremost scientific societies, wrote:

> The business and design of the Royal Society is - To improve the knowledge of naturall things, and all useful Arts, Manufactures, Mechanick practices, Engynes, and Inventions by Experiments - (not meddling with Divinity, Metaphysics, Moralls, Politicks, Grammar, Rhetorick or Logick). ([G05],122).

And it's probably no accident science went by the name *natural philosophy* back then. Newton, for instance, labeled his landmark work *Philosophia Naturalis Principia Mathematica* (The Mathematical Principles of Natural Philosophy). When religion could have crushed science, scientists adopted the attitude: "We are only trying to understand the *natural* world. We have no intention of intruding on religious questions. So please leave us in peace." Science's domain was limited by prudence, if not self-preservation. By fear.

The second reason is more positive: a desire to avoid fruitless discussions. In medieval Europe, the explanation of the universe and our place in it fell within the domain of religion and much (perhaps too much) had already been said about it. Scholasticism, a philosophical system based on theology, had examined such questions in detail. As Nobel physicist Werner Heisenberg wrote:

> So much had already been said about the larger
> scheme of things by philosophers and theologians that
> there was no longer much new to say about it;
> scholasticism had produced weariness of thought. But
> the details of natural processes had as yet been
> scarcely looked into. ([H03],216).

So, in certain scientific circles

> . . . it therefore became an absolute principle that only
> observed details should be discussed, not the larger
> connection of the whole. ([H03],215-6).

Heisenberg believed that even though such a restricted scope risked

> . . . the danger of losing sight of the totality, the
> interconnected unity of the whole. ([H03],215),

it was, at the same time,

> . . . precisely the reason for the abundant fruitfulness
> of the new natural science. ([H03],215).

Even today, scientific disciplines carve out a fixed portion of the natural universe as their domain of study and ignore any relation to the larger whole. Because science intentionally ignores connections to the larger whole, most scientists, speaking as scientists, can say nothing about our place in the universe and our relation to the eternal basis of everything. For the two reasons we've seen, and perhaps others, science ignores such questions.

But the historical conditions which fostered this tradition no longer exist. True, scientists are still opposed by religious people who dispute the theory of evolution or some other scientific idea. But such disagreements no longer threaten science's survival. If it wished, science could investigate our place in the universe and other questions of ultimate meaning. In a few chapters we'll discuss how it might do so.

Independent, Uncontingent, Unconditional, Uncaused

Let's return now to discussing philosophical terms we'll see in coming chapters. We've already discussed names and ideas which express energy's eternal and subsistent existence. Next, we'll discuss independent, self-sufficient existence, an idea that's related to ultimate ground of existence. All the terms we'll explore in this section mean almost the same thing. We discuss them because we'll see them in coming chapters. Let's begin with the terms self-sufficient, self-existent, independent, and sovereign.

We saw earlier that taking "the ground of existence of the ground of existence of the . . ." indicates dependence. Because wood is the table's ground of existence, the table can't exist without the wood. But the wood can exist without the table. Saw the table into pieces and the wood still exists. Wood's existence, however, depends on wood molecules, whose existence, in turn, depend on atoms, whose existence, ultimately, depends on energy. Therefore, the table depends on each of its grounds of existence for its own existence. If the wood, the molecules, the atoms, or energy cease existing the table ceases existing, too. In general, an entity that has a ground of existence in something other than itself depends on that other thing. The ultimate ground of existence is its own ground of existence, however, therefore it exists independently and depends on nothing else for its existence.

At one time, an educated person might have imagined entities that were their own grounds of existence. For instance, they might have believed that Aristotle's four elements, earth, water, air, and fire, really are independent elements, existing in and of themselves, independent of any lower-level entity. Today, however, modern science knows that each and every material entity is ultimately based on energy. Today, science recognizes only one self-sufficient entity, existing independent of anything else. Energy is its own ground of existence and depends on nothing else for its existence. Therefore, it's self-sufficient, self-existent, independent, and sovereign.

Uncontingent and unconditioned are two more terms we'll see in coming chapters. They, too, express independent existence. To see how let's begin with their opposites, contingent and conditioned existence, taking contingent existence first. What is contingent existence? It's a kind of existence an entity has when it

. . . must be considered as dependent for its actual
existence on some being *other* than itself. ([C11],47).

For example, a snowman depends on snow for its own existence - it
can't exist unless snow exists. So, the snowman exists contingently. It
has contingent existence.

Something which has dependent, contingent existence also has
conditioned existence. Why? Because a contingency or dependency is
also a condition. Because the snowman depends on snow for its
existence, it exists contingent upon snow. So, it exists on the
condition that snow exist. It has dependent, contingent, and
conditioned existence.

Now, let's discuss the opposite of contingent, conditioned
existence: uncontingent, unconditioned existence. Something which
exists independently, without any contingency or conditions, has
uncontingent, unconditioned existence. Energy has such existence.
And, as the only known entity with such existence, it's "*the*
unconditioned" and "*the* uncontingent."

Cause is the last concept we'll discuss in this section. A cause is
similar to a dependency, a contingency, and a condition. So,
dependent existence, existence with contingencies or conditions, is
also caused existence. Let's discuss an example.

What causes a snowman to exist? It's initial existence is caused by
snow existing and by the person who builds it. It's continued existence
is caused by snow's continued existence. If any cause were absent, the
snowman would not exist. For a snowman, snow is a dependency,
contingency, and condition. But it's also a cause. Snow causes the
snowman to continue existing. In general, something's dependency,
contingency or condition is its cause, as well.

But if something was never created then it has no initial causes.
And if it exists independently, if it is its own ground of existence, then
it has no cause which keeps it in existence. Therefore, it's "uncaused"
because nothing causes it. Or it may be thought of as causing itself, as
"self-caused." And if it's the cause of anything else, it's an "uncaused
cause" and "first cause" as well. Because energy is free of dependency
it's free of cause. Therefore, it's all of the above: uncaused, self-
caused, an uncaused cause, and the first cause.

We have yet to discuss philosophical names and ideas which express energy's uniqueness. We'll do so after pausing to consider another question.

The Cyclic Method

We've seen some historical reasons why sciences ignore the eternal substance and questions of ultimate meaning. But if the root and source is as central and important as it seems, *how* can it be overlooked, even if a science chooses to do so? Physicist Sir Arthur Eddington described one method, which he calls "the cyclic method." In ([E01]) *The Nature of the Physical World*, Eddington presents a lengthy example of the cyclic method where Einstein's potentials are defined in terms of intervals are defined in terms of . . . are defined in terms of potentials. Here's another, shorter example.

> Electric force is defined as something which causes motion of an electric charge; and electric charge is something which exerts electric force. ([E01],264).

Electrical force is defined in terms of electrical charge; electrical charge is defined in term of electrical force. Wood is what the table is made of; the table is made of wood. The definitions are cyclic: they refer to each other. Therefore, there's no need to discuss anything deeper. Eddington concludes:

> And you can see how by the ingenious device of the cycle[,] physics secures for itself a self-contained domain for study with no loose ends projecting into the unknown. All other physical definitions have the same kind of interlocking. ([E01],264).

There is, however, one science for which the cyclic method fails to work, sub-atomic physics. This science looks deep down into the very heart of matter. It can't avoid reference to the universe's ultimate basis. Its investigations have uncovered the universe's eternal source, and proven Einstein's claim that matter and energy are one.

The Eternal Substance

We've finally reached our last group of philosophical terms that apply to energy. We've examined terms which express its eternal, substantive, self-sufficient existence. Now, let's examine names and ideas which express its uniqueness. Some of the terms we'll discuss are the one, the pure, the unmixed, the unadulterated, and the simple.

Because energy is the single, unique root and source of all that exists, it alone exists on the ultimate level. On the level of everyday objects there are many different things. On the atomic level there are only about ninety-two different kinds of atoms. But on the ultimate level nothing exists but energy. On that level, energy is "the one." Moreover, because nothing else exists on that level, there is no possibility of mixture. There is nothing else there for energy to mix with. Therefore, energy is pure, unmixed, and unadulterated.

Energy is simple because it's composed of only one substance or element, itself. Entities which aren't simple are compounded of two or more elements. Water, for example, is composed of two components, hydrogen and oxygen. If energy wasn't simple, if it was composed of elements, then we would have go at least one level deeper to arrive at the ultimate level. In this case energy wouldn't be the ultimate ground of existence because its existence would be grounded in its elements. But, in fact, energy is its own ground of existence. Therefore, it's simple. Moreover, it's "the all" because all things and everything is ultimately energy. All it's also "the one". Therefore, it's "the all and the one."

As the ultimate substance of everything that exists, energy pervades every corner of the cosmos, as matter, electromagnetic radiation, etc. Therefore, it's present everywhere, it's omnipresent, because where anything exists, energy - as ultimate ground of existence - also exists. That which is omnipresent is also "unbounded" and "beyond space" in the sense that space does not limit or confine it to any particular place. It's not finite or limited in space. It is therefore infinite and unlimited with respect to space.

But is energy actually omnipresent? After all, it's usually considered different from space. Wouldn't a region of space in perfect vacuum, free of electromagnetic radiation, also be free of energy? To be omnipresent, the "all and the one," shouldn't energy somehow include space, too? And what about time? Energy is eternal, it knows no time. But if energy is the "all" shouldn't it include time?

The two questions aren't different if relativity theory is correct, because that theory says space and time aren't two distinct things. Rather, they're two sides of one entity, space-time. Energy is infinite, unbounded, and unlimited with respect to space-time. But is it also the ultimate ground of existence of space-time?

It may be. In quantum theory, "empty space" is known to be neither. For example, the *ABC's of Quantum Mechanics*, in the section *There is No Emptiness!*, has

> . . . vacuum or void or emptiness is generally
> nonexistent. Only matter and fields fill all of space.
> ([R09],248).

And another writer, explaining the discoveries of the famous physicist Paul Dirac, says:

> What we call "empty space" is actually a sea of
> negative energy electrons! ([G04],125).

So space-time may be a manifestation of energy.

Or it may not. Space-time may be entirely different and independent of energy. If it is, the world view this book presents will need some revision. No problem.

If this book's world view claimed to be absolutely true and revealed by God, then we'd have to insist that energy is somehow the ultimate basis of space-time. We'd have to treat our world view like religious dogma, demanding acceptance regardless of any evidence to the contrary. But what we're discussing is a tentative world view that, I hope, is compatible with science and, I'm sure, can be corrected and improved. It's a living, growing set of ideas, capable of change, adaptation, improvement, and correction - not the last word, the final, frozen, ultimate and absolute truth. If it needs some revision then it will be revised. No problem.

We've finally reached this chapter's last two philosophical ideas. Both ideas express energy's uniqueness and completeness. They are "absolute" and "perfect." Let's take "absolute" first. What does it mean? It has

> . . . two chief uses: as a adjective it is used in contrast
> with relative, comparative or conditioned; as a noun it
> is used by philosophers to denote the universe
> conceived as a single whole or system. ([N04],v1,49).

In subsequent chapters, we'll see "absolute" used as an adjective, often as the opposite of relative or conditioned. Now, we'll discuss its use as a noun, as "the absolute." What is "the absolute"? The

> . . . ultimate whole is the Absolute . . . Only the
> Absolute is fully real . . . It is timeless or eternal . . . It
> is *causa sui*, self-caused; for there can be no cause or
> ground outside itself. ([N04],v1,50).

It's

> . . . all-comprehensive; there could not, even in theory,
> be anything outside it . . . ([N04],v1,50).

Is energy the absolute? We've already seen it's timeless, eternal, self-caused, the all and the one. In a later chapter, we'll see how only energy is fully real. If we assume for now that only energy is fully real, then it deserves to be called "the real," the eternal and the ultimate reality. And if "all-comprehensive" is taken in the sense of "all that ultimately exists," then, on the ultimate level, energy is all-comprehensive, too. Thus, energy is the absolute.

However, the phrase "all-comprehensive" may be taken to mean *all* levels of existence. Besides the ultimate level, there's the sub-atomic level; the emotional and intellectual levels where we experience feelings and thoughts; and the level of sense experience, the level we live on much of the time, sensing people and objects. Words which suggest this sense of all-comprehensive are "cosmos" and "universe." In this sense, energy isn't the absolute. I'll use absolute in the first sense, as a synonym for energy, and use "cosmos" or "universe" to indicate all levels of existence.

Our final concept is "perfect." The word has many meanings, among them are

> 2. excellent or complete . . . 7. pure or unmixed . . . 8.
> unqualified; absolute. ([R01],986).

We've already seen energy is the absolute, the pure, and the unconditioned. Since unqualified has a meaning similar to unconditioned, we may say energy is "the perfect."

Summary

In this chapter, we surveyed what science studies. We saw that science recognizes energy as "the fundamental entity of the universe." So, science's many branches (at least, the ones that study physical phenomena) actually study various manifestations of energy. Therefore, energy, the universe's ultimate ground of existence, is itself a valid object of scientific study, because it's the basis of everything science studies.

Then, we turned to philosophy for a deeper understanding of the universe's fundamental entity. We examined philosophical names and ideas that apply to energy. We saw that many express either uniqueness or eternal existence or substantive existence. In coming chapters, saints, sages, and mystics will use those terms to express their visions and insights about the universe's fundamental entity.

But saints and seers use more than philosophical ideas to express their intuitions and visions. They use the language of religion as well. Therefore, we'll now turn to religion for a yet deeper understanding of the universe's uncaused cause. We'll examine religious names and ideas that apply to the eternal substance. Of the religious terms we'll discuss, one stands out above all others. It's a simpler, less philosophical, and more emotional idea that the ones this chapter discussed. Children learn this idea on their mother's knee. Yet saints and the wise ponder its meaning throughout their lives. It's an idea that has moved men and women to heroic acts of mercy and self-sacrifice. It has also moved them to inhuman acts of violence and torture. The idea is expressed with a simple, three-letter word. God.

4

- Religion's Domain Of Knowing -

Chapter Summary: This chapter discusses what religion studies and claims knowledge of - God. It examines two ideas of God, God as a Person and God as a non-personal entity, and presents examples of the second idea of God as it occurs in various religions. Then the two ideas of God are compared.

What does religion know? What is its domain of knowing? Religion claims knowledge of many different fields. Some religions claim cosmological knowledge, that is, knowledge about how the universe came into existence. Some claim historical knowledge (the Tower of Babel, Noah's ark, etc.). Most claim moral knowledge, knowledge about how a person should conduct their life. Yet, most religions are only secondarily interested in cosmology, history, and even morals. Their primary concern is God. As a rule, religion - first and foremost - claims knowledge of God.

God. Does any other word evoke more contradictory images? Universal Father or mythical parent? Pure Truth or wish fulfillment? Our Creator or our creation? Source of all that's good or source of religious wars and inquisitions? People certainly have many different ideas about God. Yet most people, whether they believe in God or not, think of God as a Person. Christians believe in the three Persons of the Trinity. Jewish people worship Yahweh, the God of Israel. Islamic nations proclaim Allah. In these traditions, the popular idea of God is a God who is a Person, not some impersonal, inanimate stuff that makes up tables and cars, you and I. Yet the world's religious literature sometimes describes God as Eternal Substance, Ultimate Ground of Existence, and all the other terms of the previous chapter. Are these instances coincidental? Are they merely sloppy, inexact writing? Or are the writers trying to convey a picture of God very different from the popular image of God as a Person? Why would they? What's wrong with thinking of God as a Person?

Personhood and God

Is God actually a person? Richard McBrien, a Chairman of Notre Dame University's Department of Theology, discusses this question.

> Is God "a person"? We are not asking here the
> question of Trinity, whether there are three Persons in
> the one Godhead. We raise instead the question
> whether God is a separate Being among beings. . . .
> [T]he answer is "Of course not. God is not a person
> because God is not any one thing or being." But if the
> noun *person* is taken analogically, the answer has to
> be different. Does the reality we name "God" have
> qualities which we also attribute to persons? Yes,
> insofar as we understand persons as centers of
> intelligence, love, compassion, graciousness, fidelity,
> and the like. What we mean by . . . *God* certainly must
> comprehend such qualities as these. In other words,
> it's better to attribute "personality" to God than to deny
> it entirely and to look upon God as some impersonal,
> unconscious cosmic law. And yet the attribution is
> always analogical; i.e., God is *like* a person, but God
> is also very much *unlike* a person. ([M07],333).

McBrien asks "Is God a person?" and immediately rephrases "Is God a separate Being among other beings?" Why? Does being a person necessarily imply an entity separate from the rest of creation? It seems it does. An introduction to one of the works of "Dionysius," translated by C. E. Rolt, discusses this point.

> Now an individual person is one who distinguishes
> himself from the rest of the world. I am a person
> because I can say: "I am I and I am not you."
> Personality thus consists in the faculty of knowing
> oneself to be one individual among others. And thus,
> by its very nature, Personality is (on one side of its
> being, at least) a finite thing. The very essence of my
> personal state lies in the fact that I am not the whole
> universe but a member thereof. ([D08],4).

McBrien and Rolt seem to agree that a God who is an actual Person is too limited, too finite. To be a person one must be able to say "I stop here. Everything outside of this is not me." Rolt rejects this idea of God.

> God, on the other hand, is Supra-Personal because
> He is infinite. He is not one Being among others, but in
> His ultimate nature dwells on a plane where there is
> nothing whatever besides Himself. ([D08],4).

Notice, as we travel from table to wood to atom to subatomic particle, we reach energy which dwells on a plane where there is nothing whatever besides itself.

Monotheism and Religious Monism

Thinking of God as a Person may not be entirely satisfactory, but do the concepts of the previous chapter work any better? Does it make sense to regard the "mere" substance of the universe as God, not as some God who is a Person, but as God nonetheless? Many people, no doubt, would immediately answer "No!" because the concept of God as a Person is so pervasive and ingrained. Most people grow up with it. And, whether they believe in God or not, they usually think of God as a Person, as Jesus or Krishna, Allah or Jehovah. On the other hand, the idea of God as Ultimate Ground of Existence is a new concept for many people. Indeed, many have never even heard of such an idea of God.

Why? Why is God so often pictured as a Person, even by people who don't believe in God? And why is God so rarely thought of as the Eternal Substance, even by people who are religious?

One possible answer is that monotheistic religion is much more common that monistic religion. Monotheistic religion teaches that God is a Person, separate from creation. The three major Western religions - Christianity, Islam, and Judaism - are primarily monotheistic. And monotheistic ideas also occur in Eastern religions such as Hinduism, Buddhism and Taoism. On the other hand, monistic ideas occur much more rarely. And they're found in Western religion even more rarely than in Eastern religion.

But the answer doesn't get to the heart of the matter because it doesn't explain why monotheistic religion is so much more common than monistic religion. Why is religious monotheism so much more common than religious monism? We'll return to that question in a later chapter, devoted to exploring the idea of Gods who are Persons. For now, let's try to better understand the monist idea of God by exploring the idea of God as Ultimate Ground of Existence, Eternal

89

Substance, and all the other ideas of the previous chapter. Before we do, however, we'll need to say something more about monism itself.

The type of monism we're now discussing - religious monism - differs from science's type of monism. Scientists today accept a monist description of the physical universe. They believe every physical phenomena is a manifestation of one entity, energy. But they don't regard energy religiously and they don't call it God. Therefore, science's type of monism isn't religious, rather it's natural and secular. Religion's type of monism, on the other hand, does regard the single Ultimate Ground of all Existence as God, not necessarily as a God who is a Person, but as God nonetheless. The religious monist believes that God is the one entity which stands under all other entities, that all phenomena, natural or otherwise, are manifestations of God, just as ocean, snow, and steam are all manifestations of water.

Religious monism is often unknown or ignored. For example, a college dictionary ignores religious monism when it attempts to define monotheism as the

> . . . doctrine or belief that there is but one God.
> ([F08],877).

The definition is faulty. Why? Because a proper definition describes the thing in question and rules out everything else. If it fails to exclude, it's only a description, not a definition. For example, "a whole number between eleven and thirteen" could be considered a definition of twelve, while "a whole number between eleven and fifteen" could not, because it fails to rule out fourteen. So, the dictionary "definition" is faulty - it's merely a description - because it describes monotheism *and* religious monism. It applies to either belief because both accept one God. Monotheists believe in one God who is a Person; religious monists believe in one God that is not a Person, but the unique, eternal, ultimate substance.

We'll try to be clearer by always keeping in mind there are two very different ideas of God in question. To avoid confusing the two, I'll use different terms. For the monotheistic idea of God - the idea of God as a distinct Person, separate from creation - I'll use the phrase "God who is a Person" or "God as a Person." And for the monistic idea of God, I'll use the phrase "the God which is not a Person." Other authors use the term "Godhead" instead.

90

Let's now examine instances of monist ideas in the world's religions. We'll meet the statements of religious people who definitely thought of God as Eternal Substance, Ultimate Ground of Existence, and many of the other terms of the previous chapter. And we'll meet the statements of religious people who *may* have thought of God in that manner. Because Eastern religions are more monistic than Western religions, we'll begin with an Eastern religion, one of largest, Hinduism.

Hinduism

Monotheism teaches that God is a Person, a separate entity, distinct from the rest of creation. Religious monism teaches a God that is not a Person, but is the Ultimate Ground of Existence. In a word, monotheism says there is only one God; monism (religious or natural) says there is only One - period!

India has both conceptions of God. It has religions that worship Gods who are Persons such as Krishna, Kali and Shiva. And it has Vedanta, a religion that worships "Brahman," the God which is not a Person. What is Brahman? The world of people and objects we see around us has an ultimate ground of existence. Science calls it energy. Vedanta calls it Brahman. In 1919 Swami Paramananda, a Vedanta monk, wrote that

> Brahman is the vast ocean of being, on which rise
> numberless ripples and waves of manifestation.
> ([U03],107),

the manifestations, of course, being chairs and tables, you and I. He says

> . . . nothing in the created world can exist independent
> of Brahman, who is the basis of all existence.
> ([U03],87),

and

> . . . all created things have their origin in Him. He is
> the foundation of the universe. There is nothing
> beyond Him. ([U03],86).

But don't the last two quotes refer to Brahman as a Person? Don't "*who* is the basis" and "nothing beyond *Him*" show Brahman is a Person? No. Vedanta specifically says Brahman is not a Person. Why, then, would the Swami refer to the God that is not a Person as if It

were a Person? This question is addressed in the chapter that
discusses Gods who are Persons.

The Swami was a monk of India's Ramakrishna order, founded by
the 19th century saint, Sri Ramakrishna, who India considers one of
its greatest holy men. Sri Ramakrishna teaches

> God alone is real, the Eternal Substance; all else is
> unreal, that is, impermanent. . . . God is the only
> Eternal Substance. ([G03],81-2),

and

> God alone is the real and permanent Substance; all
> else is illusory and impermanent. ([G03],179).

Ramakrishna, in turn, echoes one of his teachers, Totapuri, who says

> Brahman . . . is the only Reality, ever-pure, ever-
> illumined, ever-free, beyond the limits of time, space,
> and causation. Though apparently divided by names
> and forms . . . Brahman is really one and undivided.
> ([L07],159).

Though relatively recent, these statements about Brahman could
have been made at any time in India's religious history. For example,
Shankara, India's great 8th century religious reformer, writes of
Brahman:

> It is the ground upon which this manifold universe . . .
> appears to rest. It is its own support . . . eternal . . .
> eternally free and indivisible . . . Though one, it is the
> cause of the many. . . . It is the one and only cause . .
> . It has no cause but itself. . . . It is unchangeable,
> infinite, imperishable. . . . It . . . appears . . . as a
> manifold universe of names and forms and changes.
> ([S11],71-2).

Even more ancient than Shankara's writings are India's oldest and
most authoritative scriptures, the Vedas, written by ancient seers. The
Upanishads are the philosophic part of Vedas. In the Taittiriya
Upanishad, we find reference to an even older writing:

> Concerning which truth it is written: Before creation
> came into existence, Brahman existed as the
> Unmanifest. From the Unmanifest he created the
> Manifest. From himself he brought forth himself.
> Hence he is known as the Self-Existent. ([U04],56).

And in another upanishad, the Svetasvatara Upanishad, we read of
Brahman:

> He is the substance, all else the shadow. He is the
> imperishable. The knowers of Brahman know him as
> the one reality behind all that seems. ([U04],119).

At this point, I suppose it's entirely obvious what also deserves to be called "the imperishable" and "the one reality behind all that seems." How ancient seers, such as the upanishad authors, discovered the eternal Basis of everything, thousands of years before science, is the subject of the next chapter.

Brahman is the Eternal Basis of the world we see outside. As such, it's a religious analogue to what science calls energy. What about the world inside? Inside we see a world of emotion and thought. Is energy its ultimate basis, too? Science doesn't say but Hindu Vedanta answers yes. It says "Atman" is the inner world's Ultimate Ground of Existence and teaches that brahman and atman are identical; they are two words for the same Thing, seen from two different perspectives, from within and from without. Other religions also identify the inner world's Eternal Substance with the Eternal Substance of the physical world. For example, some Buddhists call the universe's Eternal Substance "mind".

Buddhism

Buddhism is another major Eastern religion. Its founder, Buddha, is often said to have avoided any teaching about God, gods, and Ultimate Reality. For example:

> The Buddha departed from the main lines of traditional
> Indian thought in not asserting an essential or ultimate
> reality in things. ([N05],v3,375).

Although Buddha does not speak of God or gods, he does speak of what we are calling the God which is not a Person. For in the Udana scripture (VIII.3), a very old Buddhist writing, he declares:

> There is an Unborn, Unoriginated, Uncreated,
> Unformed. If [there] were not this Unborn, this
> Unoriginated, this Uncreated, this Unformed, escape
> from the world of the born, the originated, the created,
> the formed, would not be possible. But since there is
> an Unborn, Unoriginated, Uncreated, Unformed,
> therefore is escape possible from the world of the
> born, the originated, the created, the formed.
> ([B14],32-3).

93

He also declares in another scripture that:

> All things are made of one essence . . . as if a potter
> made different vessels out of the same clay. . . . There
> is no diversity in the clay used . . . ([C04],163).

Moreover, Buddha describes the religious quest as a search for that one essence.

> [S]omeone, being liable to birth . . . seeks the unborn .
> . . being liable to ageing . . . seeks the unageing . . .
> being liable to decay . . . seeks the undecaying . . .
> being liable to dying . . . seeks the undying . . .
> ([C13],206).

Centuries after Buddha, *The Awakening of Faith* was written and attributed to Asvaghosha, a philosopher and poet. In an introduction, we read that

> . . . the all-inclusive Reality, the unconditional
> Absolute, is called Suchness. . . . What is real is
> Suchness alone; all else is unreal, a mere appearance
> only, because it is relative, being devoid of
> independent self-nature or own-being. ([A15],12,15).

Science, of course, doesn't teach things are mere appearances. But it finds them lacking independent existence because all physical entities ultimately depend for their existence on energy, suchness, the unconditional absolute, what Hindu's call Brahman.

As we've noted, some Buddhists call the ultimate ground "mind". For example, *The Tibetan Book of the Great Liberation* declares that mind is

> . . . non-created and self-radiant . . . Reality . . . ever-
> existing . . . partaking of the Uncreated . . . self-born . .
> . ([T06],212,3,4).

This text also teaches that the external world is rooted in mind. Thus, we read

> . . . matter is derived from mind or consciousness, and
> not mind or consciousness from matter. ([T06],213).

Is not matter derived from energy, which is non-created, ever-existing, uncreated, self-born?

In this century, Buddhadasa, a Thai Buddhist monk, divides things into two classes:

> . . . conditioned, impermanent objects produced by
> causes . . . ([B13],34);

and things

94

> . . . free from the process of coming into being and
> ceasing . . . things uncreated by causes. ([B13],34).

He also mentions a state "free from coming into being," that is, a state
that's unoriginated and unborn.

> When . . . conditioned things are dissolved, an
> unconditioned, indestructible, self-existing state
> remains . . . ([B13],34).

Perhaps this describes a meditative state he has experienced. A later
chapter discusses meditation.

Taoism

Taoism is an Eastern religion indigenous to China. Huston Smith, in
his *The Religions of Man*, distinguishes three senses of the word
"Tao" and names the first "ultimate Tao."

> Ineffable and transcendent, this ultimate *Tao* is the
> ground of all existence. It is behind all and beneath all,
> the womb from which all life springs and to which it
> again returns . . . *Tao* in the first and basic sense can
> be known, but only through mystical insight . . .
> ([S15],199).

One translation of the *Tao Te Ching*, the most authoritative Taoist
book, speaks of ultimate Tao when is says:

> Ultimate reality is all-pervasive; it is immanent
> everywhere. All things owe their existence to it . . .
> (XXXIV, [T01],36).

And in another translation we read:

> There was something formless yet complete,
> That existed before heaven and earth;
> Without sound, without substance,
> Dependent on nothing, unchanging,
> All pervading, unfailing.
> One may think of it as the mother of all things under
> heaven. (XXV, [W01],174),

and

> . . . return to the root is called Quietness;
> Quietness is called submission to Fate;
> What has submitted to Fate has become part of the
> always-so
> To know the always-so is to be Illumined . . .
> (Ch.XVI,[W01],162).

Like Buddhism, Taoism also speaks of meditative states. A Taoist story tells that Confucius, upon finding the Taoist Lao Tan in a deep meditative state, said:

> Just now you appeared to me to be a mere lifeless block, stark as a log of wood. It was as though you had no consciousness of any outside thing and were somewhere all by yourself. ([W01],117).

Lao Tan replied:

> True. I was wandering in the Beginning of Things. ([W01],117).

Sikhism

The Sikh religion is the last Eastern religion we'll discuss. It was founded by Guru Nanak, who was born in 1469 C.E. and died seventy years later. His teachings about God have been summarized thus:

> This Being is One. He is eternal. He is immanent in all things and the Sustainer of all things. He is the Creator of all things. He is immanent in His creation. ([M09],163).

And one of Nanak's hymns, the Japji, opens

> There is One God
> His Name is Truth.
> He is the Creator,
> He is without fear and without hate.
> He is beyond time Immortal,
> His Spirit pervades the universe.
> He is not born,
> Nor does He die to be born again,
> He is self-existent. ([H13],43).

Christianity

In the last chapter we saw that energy is the self-existent, the eternal, the ultimate, the uncreated. In this chapter we gave energy another name: the God which is not a Person. For a religion like Vedanta, which accepts the idea of God, this name may be acceptable. For a religion like Buddhism, which does not use the idea of God, this name may be unacceptable; Buddhism might prefer "the supreme reality which is not a person" instead. Each religion speaks of what we are calling the God which is not a Person in its own words, using its own terms.

96

In Western religion, the idea of God is pervasive. Therefore, our term "the God which is not a Person" is probably more fitting, even though the idea itself occurs less frequently in the West. Yet it does occur. We'll begin with a few examples from Christianity.

The Carthusian order, founded in 1086 by Saint Bruno, is one of Roman Catholicism's most austere and conservative orders. Its monks remain anonymous and rarely publish. Yet one has written:

> God is subsistent being itself. The word 'being' applies strictly only to God . . . For all other things, ourselves included, compared to that pure and perfect Substance, are not even shadows. That is why God gave his name when speaking to Moses as *He who is*. ([A06],101).

We've seen Energy is the pure and perfect Substance. It's also "subsistent existence" since It's the Existence underlying and subsisting under all existence.

The anonymous monk isn't the only Roman Catholic who believes God is Subsistent Existence Itself. One of the greatest teachers of the Catholic Church, Saint Thomas Aquinas, the famous 13th century theology professor, writes:

> Now existence is more intimately and profoundly interior to things than anything else . . . So God must exist and exist intimately in everything. (Ia,8,2, [A07],v2,113),

and

> God exists in all things by substance . . . exists in everything as causing their existence. (Ia,8,4, [A07],v2,121),

and lastly

> God is sheer existence subsisting of his very nature. (Ia,44,1, [A07],v8,7).

To the religious Thomas Aquinas "God is sheer existence." To the scientist, is not Energy "sheer existence"? Does not the Self-Existent subsist of its very nature, independent of any other entity?

Of course, for many Christians the definitive question would be: is there any evidence Jesus thought of God the Father as an impersonal entity? There is. Although some modern translations have Jesus using the personal pronoun "who" to refer to the Father, the King James version has many instances where Jesus uses the impersonal pronoun "which." The following are all taken from King James.

> Let your light so shine before men, that they may see your good works, and glorify your Father which is in heaven. ([H08],Mt5:16);
>
> That ye may be the children of your Father which is in heaven: for he maketh his sun to rise on the evil and on the good, and sendeth rain on the just and on the unjust. ([H08],Mt5:45);
>
> But thou, when thou prayest, enter into thy closet, and when thou hast shut thy door, pray to thy Father which is in secret; and thy Father which seeth in secret shall reward thee openly. ([H08],Mt6:6);
>
> After this manner therefore pray ye: Our Father which are in heaven, Hallowed be thy name. ([H08],Mt6:9).

Other instances are Mt 6:18, 7:11, 16:17, 18:10, 23:9, Mk 11:25, 11:26, and Lk 11:2.

Which pronoun is correct? *The New Testament Study Bible, Mark* has ([N10],317) Mark's original Greek side by side with an English translation. But the Greek seems to have neither "who" or "which." I have no knowledge of ancient Greek, so I don't know which interpolation is more justified.

Judaism

Other references to the God which is not a Person occur in scriptures holy to Jews and Christians alike. For example, the Biblical book of Exodus records a conversation between Moses and God. From a burning bush, God commands Moses to lead the Israelites out of Egypt.

> And Moses said unto God, Behold, *when* I come unto the children of Israel, and shall say unto them, the God of your fathers hath sent me unto you; and they shall say to me, What *is* his name? what shall I say unto them?
>
> And God said unto Moses, I AM THAT I AM: and he said, Thus shalt thou say unto the children of Israel, I AM hath sent me unto you. ([H08],Ex3:13-14).

"I am that I am" is a curious phrase which, I'm sure, has puzzled many a religious student. What does it mean? Solomon Nigosian discusses this question.

'*Ehyeh-Asher-Ehyeh*' is usually translated as 'I Am
Who I Am', or 'I AM What I Am' . . . The word *Ehyeh*
derives from the Hebrew root word *hayah,* which
means 'life', 'being'. The same root is also the
antecedent of the word, or name, YHWH - the four
letters of the ineffable name of God, never
pronounced by Jews. Basically then, the Self definition
of God as '*Ehyeh-Asher-Ehyeh*' is understood to mean
that God is a BEING - an Absolute, Immutable Being,
but beyond human comprehension. ([N13],19).

The Christian interpretation is similar. W. F. Albright, a
"distinguished biblical archaeologist," suggests it means "He who
brings into being whatever comes into being" ([M07],285). If we
paraphrase using impersonal pronouns we get "That which brings into
existence whatever comes into existence" - a description of the First
Cause and Ultimate Ground of Existence. In much the same spirit of
the Carthusian monk who wrote "all other beings . . . compared to that
pure and perfect Substance, are not even shadows," the Jewish teacher
Schneur Zalman teaches:

The fact that all created things seem to have existence
and being in their own right is because we can neither
conceive nor see, with our physical eyes, the Force of
God which is in the created world. . . . There is really
nothing in existence besides God. ([W08],56-7).

Might not a scientist say the same of energy?

Other hints of the God which is not a Person occur in the writing
of various Jewish saints and mystics. A Jewish mystic called "The
Besht," for example, has a teaching similar to Schneur Zalman.

. . . [T]here is nothing in the whole universe except
God himself, who fills the whole world with his glory.
The essential part of this contemplation is that a man
should think of himself as absolutely nothing, for he is
really only the soul within him, which is a part of the
divine itself. Thus the whole of reality is only the one
God himself. ([W08],48).

And Dov Baer of Lubavitch, Schneur Zalman's son, wrote

There is no reality in created things . . . [F]rom the
point of view of the divine vitality which sustains us,
we have no existence . . . From which it follows that
there is no other existence whatsoever apart from His

existence . . . (From Introduction, Dov Baer of
Lubavitch, *Tract on Ecstasy*, 44-45 in [E04],136-7).

Obviously, nothing has existence apart from That which is the Ground
of all Existence. On the other hand, if God is a Person separate from
creation, then things exists apart from God, regardless of how they
were originally made.

If Energy is the Root and Source of everything which exists, then
things may be thought of as emanations of It. Just as light emanates
from a fire, the world we see emanates from God.

> How beautiful is the mystical conception of the divine
> emanation as the source of all existence, all life, all
> beauty, all power, all justice, all good, all order, all
> progress . . . The basis for the formation of higher,
> holy, mighty, and pure souls is embodied in it.
>
> The divine emanation, by its being, engenders everything.
> ([K06],165).

Rabbi Abraham Isaac Kook (1865-1935), chief rabbi of what was
then Palestine, penned the last quotation. He also writes:

> The forms that robe reality are precious and holy to
> us, and especially to all who are limited in their
> spiritual perception. But always, when we approach a
> life of enlightenment, we must not swerve from the
> perspective that light flows from the incomprehensible
> to the comprehensible, by way of emanation, from the
> light of the En Sof. ([K05],208).

A footnote describes En Sof as "The Infinite, a mystical term for
God." "The forms that robe reality" is a poetic image that well
expresses science's understanding of physical entities: they are forms
which embody but at the same time conceal their eternal basis.

One last observation concerning Judaism. Solomon Nigosian
writes that the famous Jewish writer Moses Maimonides, in his *Guide
of the Perplexed*, teaches

> God's existence is absolute . . . It includes no
> composition . . . ([N13],20).

The dictionary defines "composition" as

> 1. putting together of parts, ingredients, etc., to form a
> whole ([F08],277).

Something produced by combining parts, elements, or components is
created by composition and is therefore a "component entity" or
"component object." As we've seen, the Real is pure in the sense of

unmixed and simple in the sense of composed of only one substance or element. It has no parts and includes no composition.

Islam

Islam is the world's youngest major faith. Like the God of the Judeo-Christian scriptures, Allah is described with personal attributes in the Koran (sometimes spelt Quran), the holy book of Islam. Many of the 99 Islamic names of Allah, such as The Forgiver, The Knower, The Just, and The Wise, refer to a God who is a Person. Yet, as in the Judeo-Christian tradition, Islam has references to the God which is not a Person, too. For example, other names of Allah such as The Eternal, The Self-Sufficient, The Real, The One, and The Light, may be taken as applying to that God. Moreover, Allah is unborn: "Allah is One, the Eternal God. He begot none, nore was He begotten." (Sura 112, [K07],265). And Sura 57:24 describes Allah as "the Absolute" ([M10],389) in one translation and "self-sufficient" ([K07],109) in another. Allah is also described as "the Living One, the Ever-Existent" (Sura 20:111, [K07],231) in one translation, and as "the Living, the Eternal" ([M10],233) in another. Finally, being everywhere, omnipresent - "He is with you wherever you are." (Sura 57:4, [K07],107) - Allah isn't separate from creation.

We've seen that Energy is the unique absolute Reality on which the existence of everything else depends. So is Allah.

> The central concept around which all Quranic teaching revolves is that of *tawhid,* the unity or oneness of Allah, the Quranic name for God. Such a concept emphasizes a rigorous monotheism, stating Allah to be a unique absolute Reality . . . Thus in Islam Allah is the sole reality on whom the existence of everything else depends. ([B16],316).

The idea that Allah is the unique absolute Reality seems more monist than monotheistic.

Islam has many other instances of a monist conception of Allah. For instance, Imam Ali Zain al-Abedin (680-712 C.E.) writes

> O my Lord! Thou art the One. Thou art the Omnipotent. Thou art the Everlasting. All else perishes . . . Thou art and Thou alone! Thou art the High! Thou art superior to all. There is none else, save Thee, O my Lord! ([R04],47).

And Shah Wali Allah (1703-1763 C.E.) claims
> . . . Reality is independent of any creator. As such, it is
> the Source of all existence and must exist before
> every other existence. This Existence is all-embracing.
> Anything outside it is non-existent . . . This Existence
> is, therefore, the Very Person of God. All that exists in
> the universe exists because of His Existence.
> ([R04],337).

Indeed, even the phrase "I am that I am" is applied to Allah. The
Islamic seer Dhun-Nun writes Allah must be addressed as
> Thou art who Thou art, eternally, in eternity.
> ([S04],44).

And the famous mystic Ghazzali sees that
> He is everything, nay, HE IS THAT HE IS . . .
> ([A03],111).

He writes:
> Real Being is Allah . . . [T]here is nothing in existence
> save Allah alone . . . the Prime Reality . . . [T]here is
> no Existent except God . . . All existence is,
> exclusively, His Aspect. ([A03],103-5).

For Ghazzali, Allah is the "One Reality" ([A03],128), the "Real
Existence" ([A03],172). Thus,
> Every time you indicate anything, your indication is, in
> reality, to Him . . . ([A03],112).

Finally, Ibn al-'Arabi, considered "undeniably one of the greatest
figures in the Muslim mystical tradition and probably one of the
world's greatest mystics ([I01],22)," realizes, as did Aquinas, that
> He is Being Itself, the Essence of Being . . .
> ([I01],135).

Though many Africans are Islamic, others have a variety of
religious beliefs. Even small, indigenous religions have references to
the God which is not a Person, however.
> A number of African peoples think of God as self-
> existent and pre-eminent. ([M06],43).

For example, the Zulu name of God, U-zivelele, translates
> He Who came of Himself into being; He who is of
> himself ([A02],109).

And other Africans ([M06],43) believe God is uncreated and self-
originating.

Zoroastrianism

Zoroastrianism is the last religion we'll discuss. It's an ancient religion founded in northeastern Iran about six centuries B.C.E., and is sometimes forgotten when Western and Eastern religions are mentioned. Today, it's a small religion, with followers in India and Iran. Once, however, it was the predominate religion of the Persian state. Greek, Jewish, Christian, and Islamic belief show its influence. The Zoroastrian phrase for God is "Ahura Mazda." In a catechism written in 1869 we find:

> Q. What is the meaning of the word <u>khuda</u>?
> A. It is a Persian word for God, and it means self-created, i.e. one who has no creator. The Avestan word is Ahura Mazda. ([K08],4).

So Ahura Mazda is Self-Existent. *Zoroastrian Theology* ([D06]) elaborates, describing Ahura Mazda as

> . . . the supreme godhead of Zoroastrianism . . . the Being par excellence . . . He is not begotten, nor is there one like unto him. Beyond him, apart from him, and without him, nothing exists. He is the Supreme Being through whom everything exists . . . He is the most perfect being. He is changeless. He is the same now and for ever. He was, he is, and he will be the same transcendent being, moving all, yet moved by none. In the midst of the manifold changes wrought by him in the universe, the Lord God remains changeless and unaffected . . . ([D06],19-20).

Notice, the author uses the pronoun "he" to refer to Ahura Mazda, but he also writes:

> Ahura Mazda is a spirit. There is no anthropomorphic trait in the lofty conception of Ahura Mazda, for he is devoid of all human imperfections. ([D06],20).

Another author writes of the Zoroastrians

> . . . the name by which God is known in their scriptures is Ahura Mazda, which may be translated Eternal Light, or, by a figure of speech, Abiding Wisdom . . . ([D05],18).

As we'll see later, the Ultimate Ground is often experienced as "eternal Light."

Being, the Great Chain of Being, Absolute Being

Hinduism, Buddhism, Taoism, Sikhism, Christianity, Judaism, Islam, and Zorastrianism - these religions speak of what we are calling the God which is not a Person; sometimes explicitly, sometimes obscurely. Of course, they also speak of another kind of God, a God who is a Person. Christianity speaks of Jesus and God the Father, Islam speaks of Allah, Hinduism speaks of Krishna. Are the two ideas of God - God as some Person and God as Eternal Substance - complementary, or contradictory? Is it necessary to choose between one or the other? Or are both ideas true? Are they two different but equally good ways of thinking about God? Perhaps they are. Perhaps both ideas point to the same Reality. Nonetheless, the two ideas themselves shouldn't be confused. They really are different.

Not only are the ideas different, some writers say the realities are different, too. They say that Gods who are Persons are quite different from the God which is not a Person. For example, Meister Eckhart, a famous Christian mystic, says

God and Godhead are as distinct as heaven and earth. ([M12],114).

P. Coffey and Ken Wilber also believe the God which is not a Person can't be equated with any God who is a Person. Coffey was an Irish professor of metaphysics in the centuries old Scholastic tradition; Ken Wilber is a contemporary writer. Before we see what they have to say, we'll need to examine three concepts they use: being, the Great Chain of Being, and the Absolute Being. After we discuss the concepts, we'll see how Coffey and Wilber use them.

Let's begin with "being." What does it mean? As we'll see, Coffey and Wilber use that word with two different meanings, even in the same sentence. Their first meaning of "being" is entity, that is, a particular person, thing, place, idea, etc. The second meaning is essence, ground of existence. Coming chapters avoid such ambiguity by generally not using "being" as a noun. Instead, the chapters use "entity" for being's first meaning (person, place, etc.) and "existence" for its second.

Both meanings of "being" apply to our second concept, the Great Chain of Being. What is the Great Chain of Being? It's a way of arranging everything that exists into a hierarchy by ranking entities that have more qualities higher than those with less. For example, rocks and birds exist, but rocks have only existence while birds have another quality, life. Therefore, birds are higher in the Great Chain of Being than rocks. Similarly, birds and people have existence and life, but people have reason, too. Therefore, people rank higher than birds (and rocks). In general, entities with more qualities are placed higher in the hierarchy than those with less.

People have qualities other than reason. They have knowledge, goodness, justice, power, truth and beauty, all in varying - but limited - amounts. Some people have more knowledge than others. Some have more goodness. But all have only a limited amount of knowledge, goodness, and other qualities. Imagine now the top of the Great Chain of Being, where there is unlimited life, knowledge, power, truth, goodness, reason, and other qualities. Can more than one (not necessarily human) person exist there? No. For example, two persons can't have unlimited power because each would limit the other's power. So, there can be at most one person at the top of the Great Chain of Being, a person who has unlimited power, life, knowledge, and other qualities.

Is there a person at the top of the Great Chain of Being? Coffey believes there is, and calls that entity the "Absolute Being," which brings us to our third and last concept. For Coffey, "Absolute Being" (with capitalization) means the person at the top of the Great Chain of Being, the Person with an infinite amount of goodness, truth, beauty, power, justice, knowledge, and other qualities. Traditionally, the Absolute Being has been identified with some God who is a Person, a Supreme Person. But Coffey uses the same phrase, "absolute being" (without capitalization, however) to indicate something different. For Coffey, "absolute being" is absolute existence, the self-existent and self-caused essence of an entity. With this meaning, the phrases "absolute being," "ultimate ground of being" and "ultimate ground of existence" are all synonymous. Wilber does something similar: he uses "Spirit" (with a capital S) for what Coffey calls absolute being; and he uses "spirit" (with a lower case s) for something similar to what Coffey calls Absolute Being. Confused? Perhaps things will be clearer after we've seen what Coffey and Wilber have to say.

Two Ideas of Gods
Now that we've discussed being, the Great Chain of Being and
Absolute Being, let's see what Coffey and Wilber say about our two
ideas of God. Coffey believes the Absolute Being (person) is the
opposite of absolute being (the eternal substance). He writes:

> [T]he being which realizes in all its fullness the reality
> of *being* is the Absolute Being in the highest possible
> sense of this term. This concept of Absolute Being is
> the richest and most comprehensive of all possible
> concepts: it is the very antithesis of that other concept
> of "being in general" which is common to everything
> and distinguished only from nothingness. ([C11],49).

Notice, in a single sentence Coffey uses "being" with two
different meanings. He says the being (entity, here a person) which
realizes the reality of being (existence) is the Absolute Being
(Absolute Person, some God who is a Person). Notice, too, that even
though Coffey's "Absolute Being" is a God who is a Person, he uses
the impersonal pronoun - " the being *which* realizes . . . "

Coffey's concept of the Absolute Being brings a problem to mind.
Goodness, truth, beauty, power, etc. are positive qualities. One might
wonder why the Absolute Being doesn't possess, as well, an
exceedingly rich, comprehensive set of negative qualities, such as
evil, falsehood, ugliness, feebleness, injustice, ignorance, etc.
Following Augustine, one answer is that negative qualities don't really
exist; they don't have positive existence. In this view, evil doesn't
really exist but is merely an absence of good, just as darkness doesn't
have its own positive existence but is merely a lack of light. So
possessing light in an infinite amount would naturally rule out
possessing any darkness.

Ken Wilber also uses the Great Chain of Being to draw a sharp
line between our two ideas of God. In *Quantum Questions*, Wilber
uses Spirit (with a capital S) for the Eternal Substance, which he
describes as ([Q01],18) "the Ground of Reality of *all* levels." Wilber
says Spirit has

> . . . no specific qualities or attributes itself, other than
> being the "isness" . . . or "suchness" or "thatness" . . .
> of all possible and actual realms - in other words, the
> unqualifiable Being of all beings . . . ([Q01],18).

106

Notice that, like Coffey, Wilber uses "being" in two different ways. By "the unqualifiable Being of all beings" he means the unqualifiable existence or essence of all entities.

Wilber uses Spirit (with a capital S) to indicate the Ultimate Ground of Existence, but he uses "spirit" (with a small s) to indicate something entirely different, the ([Q01],17) "highest dimension or summit of being," that is, the top of the Great Chain of Being. Wilber emphasizes that Spirit and spirit are different, and claims that many an

> . . . outrageous philosophical sleigh-of-hand has been perpetrated. . . ([Q01],19)

by those people who confuse the two. Coffey also condemns people who confuse the two ideas of God.

> Hegel and his followers have involved themselves in a pantheistic philosophy by neglecting to distinguish between those two totally different concepts. A similar error has also resulted from failure to distinguish between the various modes in which being that is relative may be dependent on being that is absolute. God is the Absolute Being; creatures are relative. So too is substance absolute being, compared with accidents as inhering and existing in substance. But God is not therefore to be conceived as the one all-pervading substance, of which all finite things, all phenomena, would be only accidental manifestations. ([C11],49-50).

Perhaps people would be more successful distinguishing two "totally different concepts" if less similar phrases were used to refer to those concepts. After all, it's very easy to confuse "absolute being" and "Absolute Being", or "spirit" and "Spirit". Though our two phrases are also similar, the reader will probably find it easier distinguishing "the God which is not a Person" and "a God who is a Person."

Pantheism

Coffey says Hegel and his followers commit the "error" of pantheistic philosophy. What is pantheism? A dictionary defines it as

> 1. the doctrine that God is the transcendent reality of which the material universe and man are only manifestations. ([R01],961).

107

Is pantheism wrong? Coffey evidently thinks so because his last quote has "God is not . . . to be conceived as the one all-pervading substance. . ." Why not? The quote comes at the end of a chapter, and Coffey doesn't explain why, in his opinion, God may not be thought of as the one all-pervading substance of which all finite things, all phenomena, all particular entities, are manifestations. After all, isn't that exactly the God which is not a Person?

Is Coffey right? Is it wrong to think of God as the Ultimate Ground of Existence and Eternal Substance? The answer depends on which idea of God is in question. Recall the "work" illustration, where we saw there were two different meanings of "work" in question. Rather than use the word "work" we decided to use "everyday work" or "physicist work" depending on which of the two very different ideas of work we had in mind. The situation with the word "God" is similar. When most people say "God" they have in mind some God who is a Person. For such a God, pantheism doesn't make sense. Pantheism is wrong when it's applied to Gods like Allah, Jehovah, and Krishna. These Gods are not the one all-pervading Substance. Rather, they are Persons, separate from the rest of creation. On the other hand, when religious monists speak of God they have in mind the God which is not a Person. Pantheism describes this God very well. The God which is not a Person certainly is the one all-pervading substance of which all finite things are a manifestation.

So, there are two different concepts in question. Therefore, coming chapters use two different phrases to avoid any "outrageous philosophical sleigh-of-hand." As long as one idea of God is clearly distinguished from the other, no confusion is possible.

Summary
Let's pause now to see where we've been and where we are going. The first two chapters discuss the different ways of knowing used by religion and science. The second chapter finds the scientific way of knowing superior. This and the last chapter discuss the different domains of religion and science, and find a common ground: the Ultimate Ground of Existence.

For centuries, religions have spoken about the God which is not a Person (as well as Gods who are Persons, of course.) Holy men and women in the world's religious traditions have described God as

Existence itself, the Eternal Substance of all, the Self-Existent, the Blessed and Uncreated Light. Today, science also speaks of an eternal substance, a self-existent entity underlying everything. Science has finally arrived at an insight long known to seers and mystics - that the physical universe is a manifestation of a single, Eternal, Self-Existent Ground of Existence. But even with its superior way of knowing, science's insight falls short of religion's in a few ways.

First, participation in science isn't available to everyone because understanding and practicing science demand certain intellectual, analytical, and mathematical abilities. Many people cannot participate. In addition, understanding and practicing science requires education, which depends on a standard of living not enjoyed by a large portion of humanity. To those struggling to survive, education is impossible. Such people cannot participate in science, even if they possess the necessary intellectual gifts.

Second, science doesn't address the whole human being. In general, science offers much for the mind and body (science and technology are the basis of many industries), but little for the heart. Science gives us much to know and do, but little to love. It admits the existence of an eternal source, but places little value on it. The scientist's relation to energy is cold and impersonal. How different is the relation of the religious devotee to the Father, the Mother, the Ultimate Ground of Existence!

Third, science fails to address some important questions. Why are we here? What is our place in the universe and our fate after death? Science is silent.

Lastly, science theorizes a root and source but offers no way to directly experience it. The ground of existence remains a mental concept rather than an actual experience. Science, as physicist James Jeans admits, has not yet reached contact with the Center. Jeans writes:

> Many would hold that, from the broad philosophical standpoint, the outstanding achievement of twentieth-century physics is not the theory of relativity with its welding together of space and time, or the theory of quanta with its present apparent negation of the laws of causation . . . [I]t is the general recognition that we are not yet in contact with ultimate reality. ([J01],150-1).

Will science ever attain "contact with ultimate reality"? Jeans doesn't think so. He believes that for science

> . . . "the real essence of substances" is for ever unknowable. ([J01],155).

But if the real essence of any and all entities is unknowable, then my own real essence must be forever unknowable, too. So, there's a part of me - in fact, my most vital part, my real essence - that I can never know. Strange. How can I *avoid* knowing my real self? How can "I" never know "I"? Or rather does Jeans mean that I can know my own real essence, but not scientifically? that I can somehow reach contact with ultimate reality in some private way, inaccessible to science?

Whatever the answer, it seems natural someone would want to know their own real essence, and, if possible, the essence of other entities, even of the entire universe. In fact Albert Einstein, who differentiates three stages of religion, believes that the person whose religion is of the highest type

> . . . feels the futility of human desires and aims and the sublimity and marvelous order which reveals themselves both in nature and in the world of thought. ([E03],38).

To this kind of person, continues Einstein, individual existence seems to be

> . . . a sort of prison and he wants to experience the universe as a single significant whole. ([E03],38).

In its present state, science offers no way to experience the universe as a whole, no way to experience the universe's Root, Essence and Ultimate Reality, no way even to experience our own ultimate Essence. Yet without science, seers and mystics, some poor men and woman with no formal education, much less scientific education, claimed to have experienced the universe and themselves in this way. And their descriptions of their experience are often descriptions of energy. They discovered a truth that science found after hundreds of years of search and experimentation. How did they do it? How do seers and mystics know? Perhaps by *seeing*.

5

- Knowers -

Chapter Summary: The chapter discusses types of knowledge with emphasis on direct knowledge, particularly of God. Discusses people who claim such knowledge, mystics. Direct knowledge of the God which is not a Person is discussed, along with mystics who claim such knowledge. Lastly, various levels of direct knowledge of God are explored.

How did they know?

Was the anonymous Carthusian monk only stating dogma when he wrote "God is subsistent being itself"? Was Buddha just voicing an opinion when he said "There is an Unborn, Unoriginated, Uncreated, Unformed"? And was Ramakrishna merely repeating something he once heard when he claimed "God alone is real, the Eternal Substance"? How did they come to know? How did they acquire knowledge of God?

If scriptures are true, a person can gain knowledge of God through religion's way of knowing. If the "fundamental entity of the universe" is identified with the God which is not a Person, a person can gain knowledge of God through science's way of knowing. But in each case, the type of knowledge is "second-hand." Ramakrishna, Buddha, and perhaps the Carthusian monk, seem to have had a different kind of knowledge, "first-hand" knowledge. We'll discuss these two types of knowledge before we discuss the knowers themselves.

Types of Knowledge

What is "second-hand" knowledge? It's knowledge derived from books or another person. It's also called ([T07],212) "knowledge by description." A court of law would call it "hearsay" knowledge.

How is second-hand knowledge acquired? For example, how can I acquire second-hand knowledge of New York City? I might talk to people who have been there, or read travel books, historical books, and descriptions of the city's history, politics, peoples and neighborhoods, streets and transit system, etc. Similarly, to acquire second-hand knowledge of God I might listen to the accounts of holy people or read sacred scriptures and other holy books.

First-hand knowledge, also called ([T07],212) "knowledge by acquaintance," is a different kind of knowledge. A court of law would call it "eye-witness" knowledge.

How is first-hand knowledge acquired? I can acquire first-hand knowledge of New York city only by traveling there and seeing it myself. Similarly, only actual, direct experience of God qualifies as first-hand knowledge.

First-hand knowledge - direct experience - of God is the domain of mysticism. From an encyclopedia:

> [M]ysticism . . . the doctrine that a person can experience direct awareness of ultimate reality . . . ([M13],v13,41).

First-hand experience is also called "gnosis", an ancient Greek philosophical term that means

> . . . direct knowledge of God . . . not the result of any mental process . . . ([N11],71).

More broadly, it means knowing and occurs in the English words "agnostic" and "diagnostic."

"Gnosis" has another meaning we should discuss. Some early Christians who claimed first-hand knowledge of God were labeled "gnostic." *The Gnostic Gospels* ([P01]) presents a vivid historical and philosophical picture of those ancient Christians. Throughout this book, however, "gnosis" doesn't refer to those Christians or their doctrines. Rather, it means mystical knowledge: direct experience of and even union with the Eternal.

So, mysticism and gnosis concern direct experience of the Real.

> Mysticism . . . an experience of union with divine or ultimate reality . . . ([C14],v17,114).

The mystic's first goal is direct experience of the Center. The mystic's ultimate goal is union with It.

> The goal of mysticism is union with the divine or sacred . . . [M]ysticism will always be a part of the way of return to the source of being . . . ([N06],v12,786).

Buddha and Ramakrishna, and perhaps the anonymous Carthusian monk, acquired knowledge of the Eternal through direct experience. They "saw" It.

Direct experience of and union with the Ultimate Ground of Existence is the topic of this chapter. We'll examine the writings of those who claim first-hand knowledge - direct experience - of the God which is not a Person.

What They Don't Know

Let's begin by discussing what mystics don't know. Often, mystics who know the All are loosely spoken of as "knowing everything." Do they really know everything? In one sense they do, because they know the Root and Source of everything. Yet, as one writer asks, what ancient mystic knew

> . . . that Hydrogen and Oxygen combined to produce water? . . . [W]hy could they not know that man can fly like birds in the air with the help of machines? ([S01],586).

None, probably.

Therefore, in the literal sense mystics don't "know everything." Mystical experience doesn't give full and complete knowledge of mathematics, history, physics, finance, music, the past and future, etc. Attaining direct knowledge of the All is not the same as attaining all knowledge.

Yet devotees sometimes think of mystics as "knowing everything" in the literal sense. For example, one mystic, Therese Neumann, conversed from a state of "elevated calm" where

> [s]he would sit up straight and almost immobile, her eyes closed, her face happy and relaxed . . . Her responses were lively, clear, and so tactful that many people . . . were deeply moved and even shared her serenity. ([S22],34).

People who wished to build low-income housing ([S22],36-7) took her praise as a guarantee of success. They went ahead before all the necessary legal permits were approved. When the projected failed, they sustained heavy financial losses.

Therese was in fact a genuine mystic; we'll see more about her later. However, her mystical attainment didn't make her a financial advisor, political advisor, or expert on the local bureaucracy.

Mystics don't speak infallibly, even about religious questions. For example, many mystics have predicted the world's imminent end, a belief which is dogma in some religions. So, even though mystics sometimes accept and preach their religion's dogmas, those dogmas aren't necessarily true. Indeed, other mystics often disagree with some dogma or other. Rather than follow the religious way of knowing and uncritically accept the claims and insights of the mystics, we'll follow the scientific way of knowing and critically evaluate the mystics and their claims.

Experience of Uncreated Light

Mystics don't have knowledge of any and all things. They do, however, claim direct knowledge, that is, direct experience of God.

How can direct experience of God be described? If the God in question is a God who is a Person, scripture often has an answer. Moses on Mount Sinai, the transfiguration of Jesus, and Arjuna's experience of Krishna are descriptions of such experience. So, a mystic who experiences some God who is a Person may have a vision of Jesus or Krishna, or feel the presence of Yahweh.

But what is an experience of the God which is not a Person like? Since most of our experience is sense experience, we might expect a direct experience of the Uncreated to be in some way like sense experience. In fact, mystics sometimes do "hear" ([U01],77) the "Divine Harmony." More commonly, however, the mystic raises

> . . . the eye of the soul to the universal light which
> lightens all things . . . (*Republic* 540, [D07],VII,406)

and "sees" the Uncreated Light. Such a mystic might take a liturgical phrase such as

> Light of light, enlighten me ([H12],71)

quite literally.

A mystic who has seen the Uncreated Light has been "illuminated" if only for a short while. Such mystics

> . . . assure us that its apparently symbolic name is
> really descriptive; that they do experience a kind of
> radiance, a flooding of the personality with a new light.
> . . . Over and over again they return to light-imagery in
> this connection. . . . [T]hey report an actual and
> overpowering consciousness of radiant light, ineffable
> in its splendour. . . ([U01],249).

Mystics sometimes experience the God which is not a Person as Uncreated Light. Sometimes, their words accurately describe their experience. Other times, their statements are inaccurate, or are misunderstood or misinterpreted. Why? First, the word "light" is misinterpreted as intellectual understanding. Phrases such as "a light dawned" or "seeing the light" demonstrate this meaning. Second, "light" is misunderstood as a reference to some God who is a Person. Perhaps, such misunderstanding is why "light" is so often applied to Gods who are Persons. For example, the following are from a Presbyterian hymnal.

> Christ, the true, the only Light . . . Fill me, Radiancy divine ([H12],52),

and

> O God of Light, Thy Word, a lamp unfailing, Shines through the darkness of our earthly way . . . Guiding our steps to Thine eternal day. ([H12],217).

Lastly, the mystic misunderstands the experience. Imagine someone who experiences the Uncreated Light. They intuitively know It is holy, so they naturally try to understand It in the context of their religion. They call It the Light of Christ or the Light of Krishna. Thus, what was actually an experience of Uncreated Light is inaccurately described as an experience of Christ or Krishna.

A Modern, Secular Seer

Many of the mystics we'll meet are religious figures: some were monks and nuns who lived hundreds, even thousands, of years ago. But it would be wrong to think that mystical experience only happens to those who are religious, or that it's a thing of the past. It's still going on today. Let's look at a contemporary instance.

The World Was Flooded with Light, subtitled *A Mystical Experience Remembered* was published in 1985. It recounts the spontaneous mystical experience of a woman who was raised Protestant and had become a Jungian psychiatrist. She had read about mysticism in college as an English major, in her study of the Catholic Mass and the mystical hymns of the 14th, 15th, and 16th centuries, but she had found it

> . . . all very far away and long ago and not to be taken seriously except as an object of literary study. ([F05],33).

In the spring of 1945, however, when she was 42 years old, a mystical experience forever changed her world view. For five days,

> [t]here was light everywhere. . . . [T]he world was
> flooded with light, the supernal light that so many of
> the mystics describe . . . [T]he experience was so
> overwhelmingly good that I couldn't mistrust it. . . .
> [G]lory blazing all around me. . . . I realized that some
> of the medieval poems I had been so innocently
> handling were written to invoke just such an
> experience as I had had. (That stuff is still alive, I tell
> you.) ([F05],43-4).

In 1985 at age 82, she says her experience was

> . . . so far from anything that I had thought in the realm
> of the possible, that it has taken me the rest of my life
> to come to terms with it. ([F05],36).

Western Christian Seers

The preceding account describes a mystical experience that was accompanied by a supernatural light and had a life-long effect. Can we identify that light with the Uncaused Cause? Might there be a purely natural cause? In some cases, there probably are. Or might the light be the radiance and glory of some God who is a Person? Perhaps. In others cases, however, it seems direct experience of the Ultimate Ground of Existence is the best explanation. Some of the accounts we'll see only suggest such experience of the Uncreated Light. Others have explicit descriptions of It. Let's turn to one now.

The Confessions of Saint Augustine was written by one of Christianity's most influential figures. It describes an experience of an Unchangeable Light that is God.

> And being thence admonished to return to myself, I
> entered even into my inward self, Thou being my
> Guide: and able I was, for Thou wert become my
> Helper. And I entered and beheld with the eye of my
> soul (such as it was), above the same eye of my soul,
> above my mind, the Light Unchangeable. Not this
> ordinary light, which all flesh may look upon, nor as it
> were a greater of the same kind, as though the
> brightness of this should be manifold brighter, and with
> its greatness take up all space. Not such was this
> light, but other, yea, far other from all these. Nor was it

> above my soul, as oil is above water, nor yet as
> heaven above earth: but above to my soul, because It
> made me; and I below It, because I was made by It.
> He that knows the Truth, knows what that Light is; and
> he that knows It, knows eternity. Love knoweth it. O
> Truth Who are Eternity! and Love Who art Truth! and
> Eternity Who art Love! Thou art my God . . .
> ([A14],126-7, Bk. VII,Ch.X).

Augustine saw an Unchangeable Light, said It was not like ordinary light, and called It God. He also said It's above him because It made him. In what sense did It make him? In what sense does It make us? In the sense that It created us, brought us into existence, caused our birth? Or in the sense that It composes us at this very moment? That our existence depends upon It's Existence even now?

Like Augustine, George Fox saw the Light. Unlike Augustine, he identified it with a God who is a Person. Fox writes in his journal that

> . . . there did a pure fire appear in me . . . The divine
> light of Christ manifesteth all things . . .
> (Ch.1,[J05],14).

Had Fox been born in India would he have called the Light the divine light of Krishna? I don't know. But he certainly emphasized the experience was an experience of Light. Fox's emphasis on Inner Light pervades the religious group he founded, the Society of Friends, also known as the Quakers.

> Early Friends made constant reference to the Light
> Within. ([F01],7)

In fact, Fox's early followers called themselves ([F01],1) "Children of the Light."

But how did Fox come to know about the Light? A contemporary Friends' pamphlet, *What is Quakerism?*, asks and answers this question.

> How did Fox know? "Experimentally" . . . "know" had a
> somewhat different meaning before the scientific
> revolution. Today we almost unconsciously think of
> knowing as a subject-object relationship, the way a
> scientist observes facts . . . In the seventeenth century
> knowing still had its Biblical (e.g. Jeremiah 1:-10)
> dimension of the union of knower and known. To Fox
> his experience was a transforming power that
> changed his daily behavior. ([P08],10-11).

Fox was a mystic; he knew by direct experience.

Even today, Quakers know that the Light isn't merely a thought, an intriguing idea. Rather, it's something that can be directly experienced and change a person's life. Some Friends even seek something more than direct experience. They seek union. They seek not only to experience the Light, but to unite with It.

> Early Friends often emphasized the experience of obedience to the Inner Light; while some modern Friends . . . speak of the total immersion of the individual will and identity in the divine One. ([P08],10).

We'll discuss union with Uncreated Light later.

Pascal also seems to have had a mystical experience of Light. Blaise Pascal was one of the great mathematicians of the seventeenth century and with Pierre Fermat discovered the theory of probability. When he died, a servant discovered a piece of parchment sown up in his jacket. On this parchment, the *Memorial of Pascal*, Pascal recorded the memory of a mystical experience, written around the figure of a flaming cross. Translated from the French, part of the text is as follows:

> From about half past ten in the evening until half past twelve – FIRE. God of Abraham, God of Isaac, God of Jacob, not of the philosophers and savants. Certitude. Certitude. Feeling. Joy. Peace. ([C12],137-8).

Pascal, too, seems to have experienced Light and Fire, and associated It with a God who is a Person.

Of Pascal's Memorial, Evelyn Underhill in her classic work *Mysticism: A Study in the Nature and Development of Man's Spiritual Consciousness* writes:

> He seems always to have worn it upon his person: a perpetual memorial of the supernal experience, the initiation into Reality, which it describes. ([U01],188).

She believes the experience concluded

> . . . a long period of spiritual stress, in which indifference to his ordinary interests was counterbalanced by an utter inability to feel the attractive force of that Divine Reality which his great mind discerned as the only adequate object of desire. ([U01],189).

Underhill mentions other Christian mystics whose experience of Light parallels the experience of Augustine, Fox, and Pascal.

> LIGHT, ineffable and uncreated, the perfect symbol of pure undifferentiated Being: above the intellect, as St. Augustine reminds us, but known to him who loves. This Uncreated Light is the "deep yet dazzling darkness" of the Dionysian school, "dark from its surpassing brightness . . . as the shining of the sun on his course is as darkness to weak eyes." It is St. Hildegarde's *lux vivens*, Dante's *somma luce*, wherein he saw multiplicity in unity, the ingathered leaves of all the universe: The Eternal Father, or Fount of Things. "For well we know," says Ruysbroeck, "that the bosom of the Father is our ground and origin, wherein our life and being is begun." ([U01],115).

Is not the Ultimate Ground of Existence our ground and origin, root and source as well? Is not Energy pure Isness, pure Suchness, pure undifferentiated Existence?

Eastern Christian Seers

Of course, other religions speak of direct experience of the Eternal Light. Hesychasm, a mystical tradition of the Eastern Orthodox Christian Church, is particularly explicit. The Eastern Orthodox and the Roman Catholic churches were once branches of a single Christian church. About 1054, the two divided. Hesychasm is the monastic tradition of the Eastern Orthodox church that expresses its ([M02],106) "central mystical doctrine." Hesychast monks claim direct experience of God in the form of Uncreated light.

One of the greatest Hesychast saints lived about a thousand years ago. His name is Symeon. He's called "the New Theologian" to indicate he ranks second ([S26],37) only to theologian "par excellence," Gregory of Nazianzus. In his *Third Theological Discourse*, Symeon writes:

> God is light, a light infinite and incomprehensible . . . one single light . . simple, non-composite, timeless, eternal . . . The light is life. The light is immortality. The light is the source of life. . . . the door of the kingdom of heaven. The light is the very kingdom itself. ([S25],138).

We've seen how Energy is simple and non-composite (not composed of parts), timeless and eternal.

Symeon emphasized it's possible to experience the Light which is God.

> Our mind is pure and simple, so when it is stripped of every alien thought, it enters the pure, simple, Divine light . . . God is light - the highest light. ([W11],132),

and

> For if nothing interferes with its contemplation, the mind - the eye of the soul - sees God purely in a pure light. ([W11],137).

How did Symeon know that God is a simple, non-composite, eternal Light which can be seen? He claimed his knowledge was first-hand.

> I have often seen the light, sometimes it has appeared to me within myself, when my soul possessed peace and silence. . . ([L09],118-9).

Symeon's relation to the Light was anything but cold and impersonal. In his twenty-fifth hymn he writes:

> - But, Oh, what intoxication of light, Oh, what movements of fire!
> Oh, what swirlings of the flame in me . . . coming from You and Your glory! . . .
> You granted me to see the light of Your countenance that is unbearable to all. . . .
> You appeared as light, illuminating me completely from Your total light. . . .
> O awesome wonder which I see doubly, with my two sets of eyes, of the body and of the soul! ([S26],24-5).

And he left no doubt he considered this Light God.

> It illuminates us, this light that never sets, without change, unalterable, never eclipsed; it speaks, it acts, it lives and vivifies, it transforms into light those whom it illumines. God is light, and those whom he deems worthy of seeing him see him as light; . . . Those who have not seen this light have not seen God, for God is light. ([L09],121).

As might be expected, the experience of the Light which is God can be quite intense.

> If a man who possesses within him the light of the
> Holy Spirit is unable to bear its radiance, he falls
> prostrate on the ground and cries out in great fear and
> terror, as one who sees and experiences something
> beyond nature, above words or reason. He is then like
> a man whose entrails have been set on fire and,
> unable to bear the scorching flame, he is utterly
> devastated by it . . . ([W11],113).

Some years after Symeon, the orthodoxy and validity of the Hesychast experience of God as Light was questioned. Many rose to Hesychasm's defense. Gregory Palamas (1296-1359) is perhaps the most famous. Palamas gave the Hesychast experience of Uncreated Energy a philosophical basis acceptable to Orthodox Christianity.

> Gregory argued that the divine energies manifest the
> Godhead in an effulgence of light, which it is possible
> for humans beings to see, God willing. The light that
> the apostles saw on the mount of the transfiguration
> was uncreated light, not a created effulgence.
> ([P15],69).

Like Symeon, Palamas insists Uncreated Light is an actual experience, not a symbol or a metaphor for intellectual understanding.

> Palamas affirms the utter reality of the saints' vision of
> God, constantly repeating that the grace that reveals
> God, like the light that illumined the disciples on Mount
> Tabor, is *uncreated.* ([M14],120).

Palamas's defense of the Hesychast experience was successful. The Orthodox Church ([N04],v11,465) accepts his teachings and numbers him among its saints.

Uncreated Light

Hesychast monks almost always describe their experience of God as an experience of Uncreated Light. They associate that Light with the light of a scriptural incident.

> Take for example the term "Taborite light," with which
> hesychasts always describe their experience of God.
> For they identify the divine reality that reveals itself to
> the saints with the light that appeared to the Lord's
> disciples at His Transfiguration on Mount Tabor. Such
> an identification seems to them justified not merely as
> a symbol but as something very real. ([M14],116).

But why is the Light called Uncreated? The biblical account doesn't use that word. It says while Jesus was praying

> . . . his face changed its appearance, and his clothes
> became dazzling white. . . . Moses and Elijah . . .
> appeared in heavenly glory . . . his companions . . .
> saw Jesus' glory . . . ([G02],Lk 9:29-32).

But Luke doesn't call the light "uncreated" and neither does Matthew (17:1-8) or Mark (9:2-8). Matthew uses light imagery when he describes Jesus' face (Mt 17:2,[G02],18) as "shining like the sun." And Mark says (Mk 9:3,[G02],42) Jesus' clothes "became shining white - whiter than anyone in the world could wash them." But why is the light called Uncreated? Perhaps, the monks' experience of the Light which is Uncreated came first, and identification with the light which shone on Mount Tabor second.

Imagine monks see a reality within themselves. The reality is a kind of Light. They realize that the reality is God, that seeing the Light is an experience of God. So they naturally identify It with an incident in their own scriptures, the Light that shone at the transfiguration of Jesus Christ on Mount Tabor. But if the same Light was experienced by a Hindu mystic, would it be called the light of Krishna? Might not a Buddhist identify such a light with the pure Essence of Mind or the Clear Light of the Void?

Could it be that mystics of all times and cultures have had vision of the same Eternal Light? Was the Islamic mystic, Sumnun, speaking of that Light when he wrote

> I have separated my heart from this world -
> My heart and Thou are not separate.
> And when slumber closes my eyes,
> I find Thee between the eye and the lid. ([S04],62)?

And was Angelus Silesius, a Christian mystic of the 17th century whose simple, clear verses have a Zen-like quality, speaking of that same Light when he wrote:

> A heart awakened has eyes:
> perceives
> the Light
> in dark of night. ([B05],109)?

And how might a scientist who happened to see that Light speak of It? As the E in $E=mc^2$? Probably not. If they spoke of It at all, it would probably be in a religious context.

Essence and Energies

We'll see other mystics' records of the Light, but let's pause to consider a question. Of God, St. Paul writes:

> He alone is immortal; he lives in the light that no one can approach. No one has ever seen him; no one can ever see him. ([G02],1Tm 6:16).

If Paul was right, then the mystics are wrong. They didn't see God at all. Who is wrong, Paul or them? Paul, I believe.

It is probably worth re-emphasizing that this book's world view agrees with some beliefs of established religions and disagrees with others. Since religions disagree among themselves, no world view could possibly agree with all beliefs of all religions. So I usually don't remark when a point I'm making disagrees with some religion or another, except when the disagreement, like the one we're discussing now, brings up an interesting point.

Can God actually be directly experienced. Can God be seen? Vladimir Lossky in *The Mystical Theology of the Eastern Church* writes:

> It would be possible to draw up two sets of texts taken from the Bible and the Fathers, contradictory to one another; the first to show the inaccessible character of the divine nature, the second asserting that God does communicate Himself, can be known experimentally, and can really be attained to in union . . . ([L08],68).

(Lossky, in fact, draws up such a list in the second chapter of *The Vision of God* ([L09]), a book which treats the above contradiction in great detail.) He continues:

> The question of the possibility of any real union with God, and, indeed, of mystical experience in general, thus poses for Christian theology the antinomy of the accessibility of the inaccessible nature. ([L08],69).

The Eastern Orthodox Church resolves this scriptural contradiction by distinguishing between the essence of God, which is inaccessible, and the "energies" of God, which are

> . . . forces proper to and inseparable from God's essence, in which He goes forth from Himself, manifests, communicates, and gives Himself. ([L08],70).

The famous philosopher Immanuel Kant has a similar idea: he defines

> . . . noumena, the things in themselves, which we can
> never know, and the phenomena, the appearances,
> which are all that our senses can tell us about.
> ([D04],329).

An analogy to both ideas might be this: No one has ever really
experienced fire's essence; they've only experienced fire's energies,
that is, seen its light, felt its heat, or heard the sound of its burning.
Similarly, the God which is not a Person may be considered (compare
[U01],109) transcendent, inaccessible, and unknowable. From this
viewpoint, Energy is not identical with the God which is not a Person
but rather is that God's first emanation, the primary manifestation
upon which the entire universe is based. This distinction may be
applied to the Christian Trinity so that the Father is transcendent
existence, and the Son or Logos is the Father's first-born, the first
manifestation through which the universe is made.

> God has real existence in the world insofar as He
> *creates* the world, i.e., gives it existence by giving it a
> share in His own real existence in and through the
> *energies*. ([M03],72).

Therefore, the Logos is the Uncreated Light considered as the
Root of the universe, exterior to one's self. Paul pictures Jesus, the
Logos, in this way:

> Who is the image of the invisible God, the firstborn of
> every creature; For by him were all things created . . .
> And he is before all things, and by him all things
> consist. ([H08],Col,1:15-17).

And what is the Spirit? The same Root and Source seen interiorly, as
one's own Ultimate Ground of Existence.

Jewish Seers

Let's now turn to other religions and examine what Jewish, Islamic,
Hindu, Sikh and Buddhist seers say about the Light which is God.

Jewish records I've found aren't as explicit as Hesychast
descriptions of Uncreated Light. Yet, some Jewish mystics do speak
of divine Light. In fact, Rabbi Kook expressed the mystic's goal in
terms of Light.

> The divine light sustains all life, is to be found in
> everything that exists, and is also the goal of all
> creation . . . [T]he mystic's goal is to perceive and

experience this divine light and to be united with the universe. ([C16],30).

Was Kook, a man who undoubtedly believed in a God who is a Person, speaking of the God which is not a Person? Perhaps. It's hard to see how perceiving the radiance of some God who is a Person, separate from creation, would unite a mystic with the universe. But perceiving the universe's ultimate Substance unites a mystic with the universe in an obvious and intimate way.

In any case, Kook identifies light with God. He writes of ([K05],221) "the light of En Sof, the light of the living God" and says that

> holy men, those of pure thought and contemplation, join themselves, in their inner sensibilities, with the spiritual that pervades all. Everything that is revealed to them is an emergence of light, a disclosure of the divine . . . ([K05],208),

and also writes of ([K05],225) the "light of eternity . . . in which the temporal and the eternal merge in one whole." Might not a vision of Eternal Light merge the temporal and eternal - for instance, the table and its Eternal Basis - into one whole?

Islamic Seers

The next religion we'll discuss is Islam. The Koran, Islam's scripture, speaks of Allah's Light.

> Allah is the light of the heavens and the earth. His light may be compared to a niche that enshrines a lamp, the lamp within a crystal of star-like brilliance. . . . Light upon light; Allah guides to His light whom He will. (Sura 24:35, [K07],217).

And the Sufis, who are Islam's mystics, call themselves ([N11],1) "the followers of the Real", and speak of "a pillar of light formed from the souls of . . . saints" and "the preexistent light of Muhammad" ([S04],56), as well as the "light of God" ([S04],60) which guides the mystic.

Ghazzali is one of the most famous Sufis. His *The Niche for Lights* ([A03]) "shows a highly developed light metaphysics - God is the Light" ([S04],96). Moreover, Ghazzali believed mystics can see God. He writes:

> . . . Allah's gnostics rise from metaphors to realities . . .
> and at the end of their Ascent see, as with the direct
> sight of eye-witness, that there is nothing in existence
> save Allah alone. ([A03],103-4).

Thus,

> [t]hese gnostics, on their return from their Ascent into
> the heaven of Reality, confess with one voice that they
> saw nought existent there save the One Real.
> ([A03],106).

If someone could see down to each and every entity's ultimate level,
the level of its Ultimate Ground of Existence, would they not see
there that nothing is in existence but That?

Other Sufi mystics also describe their experience of God as
experience of Light. Abu Yazid al-Bestami, for instance, writes

> I gazed upon God with the eye of certainty . . . He had
> advanced me to the degree of independence from all
> creatures, and illumined me with His light, revealing to
> me the wonders of His secrets and manifesting to me
> the grandeur of His He-ness. . . . I saw my being by
> God's light. ([M18],105-6),

while Abu 'l-Hosian al-Nuri says

> I saw a light gleaming in the Unseen. . . . I gazed at it
> continually, until the time came when I had wholly
> become that light. ([M18],226).

And al-Shebli uses light imagery to describe a meditative experience.

> It was a soul severed from all connections, passed
> away from all carnal corruption. It was a soul come to
> the end of its tether that could endure no longer,
> visited successively inwardly by the importunate
> envoys of the Presence Divine. A lighting-flash of the
> beauty of the contemplation of this visitation leaped
> upon the very core of his soul. ([M18],283).

Hindu Seers

Eastern religions also speak of God as Light. From the Hindu
tradition, I'll present the writings of the contemporary mystic, Swami
Paramananda, a monk of the Ramakrishna order. He addresses the
Uncreated Light when he writes:

> O Thou Effulgent Light,

> Thou who are ever unchanging, without beginning or
> end,
> Make Thy effulgence felt in my heart.
> Fill my whole being with it, leaving room for naught
> else. ([P04],137),

and

> I know in my inmost depth
> That no mortal light
> Can reveal Thy immortal face;
> Thou art seen only in thine own effulgence.
> ([P04],112).

Did Paramananda directly experience the Light? He writes:

> Out of the deep darkness of night
> A light burst upon my soul,
> Filling me with serene gladness.
> All my inner chambers
> Are opened at its touch;
> All my inmost being
> Is flooded by its radiance. ([P04],108).

A Sikh Seer

I know of no first-hand account of the experience of Guru Nanak,
who founded the Sikh religion. But in his hymns we find:

> Blend your light with the Light Eternal
> Mingle your consciousness with His . . . ([H13],71),

and

> Men of God see the Divine Light. . . .
> In every heart shines the Light Eternal . . . ([H13],71),

and finally

> God is hidden in all things
> Those that serve Him attain the gates of paradise
> Those whose bodies and souls
> Are permeated with the divine Word
> Blend their light with the Light of God.
> Evil-doers perish, are reborn only to die
> Men of God merge with the True One. ([H13],177).

Did Guru Nanak believe experience of God was possible? W.
McLeod in *Guru Nanak and the Sikh Religion* writes:

> The ultimate essence of God is beyond all human
> categories, far transcending all powers of human

expression. Only in experience can He be truly known.
([M09],165).

Buddhist Seers

The last religion we'll discuss is Buddhism. Buddha not only said the Uncreated exists, but made Its attainment his goal, a goal expressed in terms of light: to achieve the Uncreated is to become enlightened.

> The Master described his Enlightenment: . . . Being
> liable to birth because of self, to age and sorrow and
> death, I sought the unborn and undecaying and
> undying. I attained this in the last watch of the night
> and won the stainless, the freedom from bondage,
> Nirvana. Knowledge and vision came to me.
> ([W07],26).

And another Buddhist scripture has

> Is not Amitabha, the infinite light of revelation, the
> source of innumerable miracles? . . . Amitabha, the
> unbounded light, is the source of wisdom, of virtue, of
> Buddhahood . . . Amitabha, the immeasurable light
> which makes him who receives it a Buddha . . .
> ([C04],172-3,175).

So, direct experience of Uncreated Light is what transformed the man, Siddhartha Guatama, into the man who was called Buddha. "Buddha" means the "awakened or enlightened one," and Siddhartha Guatama deserved that title only after attaining enduring vision of the Unborn, the Undecaying, the Undying.

Yet another Buddhist scripture, *The Tibetan Book of the Great Liberation*, also mentions a

> self-originated Clear Light, eternally unborn . . . non-
> created . . . Total Reality ([T06],218-219).

It says:

> Although the Clear Light of Reality shines within one's
> own mind, the multitude look for it elsewhere.
> ([T06],220).

Degrees of Intimacy

Can science's way of knowing be applied to religious questions? Can science address ultimate questions? What might a scientific religion be like? We've posed those questions more than once; now we've covered all the material we'll need to answer them. Before we do, however, we'll finish our discussion of knowers by discussing some mystics who claim to have gone beyond first-hand experience of God to a more intimate relationship, a claim that shocks many people. How to judge those mystics and their claims leads us to the application of the scientific way of knowing to religion, in the next chapter.

We've discussed second-hand knowledge of God and a more direct type of knowledge, first-hand knowledge. Is second-hand knowledge of God actually possible? The atheist would say no; God doesn't exist therefore no true knowledge of God is possible. But most religious people would agree second-hand knowledge of God is possible.

Is first-hand knowledge of God - knowledge by acquaintance - actually possible? Here, even a religious person might answer "No. Not in this life. It may have occurred to Moses or Mohammed, but it's certainly not possible for an ordinary person." Yet, as we've seen, some people have had such experience. The Jungian analyst and Pascal, for example, seem to have had at least one first-hand experience of God. Though the experience faded, each life was deeply affected.

But is a single, brief experience of God the most intimate kind of relationship possible between a human being and God? First-hand knowledge is also called knowledge by acquaintance, so the question can be re-phrased "Is acquaintance the most intimate relationship possible between a human being and God?" Asked that way, the question almost answers itself. There are many levels of intimacy beyond acquaintance. For one, there's friendship. And, indeed, mystics who are blessed with numerous experiences of the Eternal, as was Symeon, sometimes view God as more than an acquaintance, as Friend.

Of course, it could be said that God, for each and every person, is much more than acquaintance and friend: God is our Father and Mother. However, we are discussing what relationship a person directly experiences for themselves. Even if God actually is everyone's ultimate Parent, some people would say they have had no experience of God at all, others would claim an occasional experience, while a few would claim frequent first-hand experience. Such people might be called strangers, acquaintances and friends of God, respectively.

So, by stranger we mean someone who has never "met" God, in the sense of a direct experience. By acquaintance, we mean someone, Pascal, for example, who has had at least one meeting with God, one direct experience. And by friend, we mean someone, like Symeon, who regularly experiences God first-hand.

But is friendship the highest type of relationship possible between a human being and God? Or is a relationship even more intimate than friendship possible? What is the highest degree of intimacy with God possible to an ordinary person? What is the most intimate knowledge of God possible to a human being? The remainder of this chapter discusses those questions.

Symbols of Intimacy

Can a person enjoy a relationship with God even more intimate than friendship? Some mystics say "Yes" and see in marriage a symbol of their relationship with their God, who is usually a Person. In fact, wedding symbolism is common in some religions. For example:

> To women mystics of the Catholic Church, familiar with the . . . metaphor which called every cloistered nun the Bride of Christ, that crisis in their spiritual history in which they definitely vowed themselves to the service of Transcendent Reality seemed, naturally enough, the veritable betrothal of the soul. . . . [T]he constant sustaining presence of a Divine Companion, became, by an extension of the original simile, Spiritual Marriage. ([U01],138).

The imagery of marriage is appropriate for several reasons. First, the stereotype in many cultures of women as passive suggests the helplessness of a soul, often pictured as female, even a bride, waiting for the male Lord to initiate union.

> '*Let Him kiss me with the kisses of His mouth.*' Who is it speaks these words? It is the Bride. Who is the Bride? It is the Soul thirsting for God. (St. Bernard, *Cantica Canticorum*, Sermon vii, in [U01],137).

Second, the desire of husband and wife for each other suggests the intense desire of the mystic for God. Third, the delights of sexual union suggest the rapture of first-hand experience.

> There is no point at all in blinking the fact that the raptures of the theistic mystic are closely akin to the transports of sexual union, the soul playing the part of the female and God appearing as the male. There is nothing surprising in this, for if man is made in the image of God, then it would be natural that God's love would be reflected in human love, and that the love of man for woman should reflect the love of God for the soul. . . . To drive home the close parallel between the sexual act and the mystical union with God may seem blasphemous today. Yet the blasphemy is not in the comparison, but in the degrading of the one act of which man is capable that makes him like God both in the intensity of his union with his partner and in the fact that by this union he is a co-creator with God. All the higher religions recognize the sexual act as something holy: hence . . . adultery and fornication . . . are forbidden because they are a desecration of a holy thing . . . ([Z02],151-2).

Lastly, though the image of marriage expresses closeness and intimacy, it also maintains the eternal distinction and separation of human being and God. Even if two married people enjoy such an intimate relationship that they feel united and one, they nonetheless retain their separate identities. Like two strands woven into one cord, husband and wife unite together but remain separate persons. Therefore, marriage is an apt symbol for the relationship some mystics feel between themselves and a God who they regard as an actual, separate Person. It expresses the feeling of intimacy and union as well as the reality of eternal difference between them and God.

Of course, marriage isn't the only symbol of intimate relation with God. Another is iron in fire. The Sufi Hujwiri ([N11],159) uses that image, as does John Ruysbroeck, a great medieval Christian mystic, when he writes:

> [A] piece of iron is penetrated by fire, so that with the fire it does the work of fire, burning and giving light just as fire does . . . and yet each retains its own nature - for the fire does not become iron, nor the iron fire, but

> the union is without intermediary, for the iron is within
> the fire and the fire within the iron . . . ([R07],259).

We'll soon return to Ruysbroeck's idea of union with and without intermediary.

The Many Meanings of Union

We've examined four degrees of intimacy with God: the stranger, who has had no first-hand experience of God; the acquaintance, who has had a rare, perhaps single, experience; the friend, who has regular experiences; and the spouse, who has frequent, intimate experiences. Of the four, the spouse enjoys the most intimate relationship with God.

Is a degree of intimacy even above spiritual marriage possible between mystic and God? Is full and complete union with God possible? The question is difficult to answer, partly because "union with God" has so many different meanings. So, before we discuss degrees of intimacy beyond spiritual marriage, we'll need to examine the many different things "union with God" can mean.

Let's begin with the plainest, simplest meaning. Just as a raindrop merges with the ocean until it loses its own identity and becomes the ocean, union with God (in the plainest, simplest sense) occurs when a person loses their own identity and merges with God, becomes God. Of course, whether "union with God" in this sense is possible, whether it's impossible, whether the very idea is blasphemy, are entirely different questions (which this chapter discusses). But the plain-and-simple meaning of the phrase is union, merging until only God remains.

But is this plain and simple sense what most mystics mean by "union with God"? No. Many mystics claim "union with God." A few of them mean union in the sense just described, but many don't. They don't mean they have become God, and would be shocked to hear anyone claim identity with God.

Similarly, many religions speak of "union with God." As Geoffrey Parrinder in *Mysticism in the World's Religions* writes:

> The religious experience of the ordinary believer is
> often spoken of as 'communion' with God, and this is
> one of the commonest Christian expressions . . .
> 'Comm-union' means 'union with' . . . That the ordinary
> believer seeks communion with God is witnessed by

> countless hymns and devotional writings, Christian
> and Hindu, Muslim and Buddhist. ([P05],191).

A few of those religions mean union in our sense, but many don't. In fact, many religions explicitly deny that union in the raindrop-and-ocean sense is possible.

What, then, do mystics and religious people mean by "union with God"? They may mean almost any type of experience of God. They may mean second-hand experience. Or first-hand experience. Or a type of experience even more intimate than first-hand experience.

For example, Ruysbroeck seems to use "union" for all three types of experience. Ruysbroeck describes someone who is "united" with God through a virtuous life, through grace, or through the Church's sacraments as ([R07],253) united "through an intermediary". Such union seems to be second-hand experience of God, or perhaps a weaker type of first-hand experience. He also describes another type of contact with God, which he ([R07],259) calls "union without intermediary." Such "union," which he symbolizes by iron in fire, seems to be first-hand experience. Ruysbroeck also speaks of a third, more intimate type of union, "union without difference," a

> . . . state of beatitude . . . so simple and so modeless
> that . . . every distinction of creatures pass away, for
> all exalted spirits melt away and come to nought by
> reason of the blissful enjoyment they experience in
> God's essential being, which is the superessential
> being of all beings. ([R07],265).

"Union without difference." "Every distinction of creatures pass away." "Melt away and come to nought." The phrases suggest a more intimate relation than spiritual marriage. They suggest the raindrop-and-ocean union of human and God until only God remains. Is this what Ruysbroeck means?

Perhaps not, because he also insists there's ([R07],265) "an essential distinction between the soul's being and God's being." And he says that union in the sense we've discussed is impossible.

> [N]o created being can be one with God's being and
> have its own being perish. If that happened, the
> creature would become God, and this is impossible ...
> ([R07],265-6).

In fact, he even denies such union to the soul of Christ.

> [N]o creature can become or be so holy that it loses its
> creatureliness and becomes God - not even the soul
> of our Lord Jesus Christ, which will eternally remain
> something created and different from God.
> ([R07],252).

The logic is clear. Christ has a human soul. Human souls are created
things, forever different from God, who is uncreated. Therefore,
Christ's human soul will remain forever different from God. Of
course, as an orthodox Catholic, Ruysbroeck admits Christ also has a
divine nature which is God. Our concern, however, is not the dogma
of Christ's dual nature, but various types of union with God.

Participatory Union and Illusory Union

Ruysbroeck denies that any person can achieve union with God in the
raindrop-and-ocean sense. His denial accords with traditional
Christian dogma, which rejects the idea of utter transformation of
human into God until only God remains. Like many other religions,
Christianity believes that because God is a separate Person, human
and God are eternally different and distinct, so no literal union of
human and God is possible. It sees two forever separate persons, God
and mystic, so no union like raindrop and ocean can occur.

Nonetheless, Ruysbroeck's "union without difference" seems to
describe just that type of union. Augustine and Athanasius speak even
more explicitly of such union. Augustine writes of a voice which
says:

> I am the food of grown men: grow, and thou shalt feed
> upon Me; nor shalt thou convert Me, like the food of
> thy flesh into thee, but thou shalt be converted into Me
> . . . verily, I AM that I AM. ([A14],127, Bk.VII,Ch.X).

And Athanasius writes:

> The Word was made flesh in order to offer up this
> body for all, and that we, partaking of His Spirit, might
> be made gods. ([S08],88),

and

> He was not man, and then became God, but He was
> God, and then became man, and that to make us
> gods. ([S08],88).

"Thou shalt be converted into Me." "Might be made gods."
Augustine, Athanasius, and, at one point, Ruysbroeck, appear to

disagree with the Christian dogma that says a person can never unite with God in the sense of merging, of becoming God. Yet all three men are Christians-in-good-standing; Ruysbroeck is called blessed, while Augustine and Athanasius are saints. It seems we've uncovered another contradiction, reminiscent of "seeing an unseeable God." Now, the contradiction is that Western Christianity says a person cannot unite with God in our raindrop-and-ocean sense, but some of its representatives say such union is possible.

A similar contradiction appears in Eastern Christianity. The Eastern Orthodox Church also affirms the basic distinction between human and God.

> In Palamite terminology, as in that of the Greek
> Fathers, God is essentially apart from other beings by
> His uncreated nature. The proper condition of these
> beings is the created state . . . ([M14],120).

Yet, it also speaks of union with God - for when created entities

> . . . transcend their own domain by communication
> with God, they participate in uncreated life . . . [W]hat
> the Christian seeks, what God grants him in
> sacramental grace, is uncreated divine life, deification.
> Knowledge of God, then, according to Palamas, is not
> a knowledge that necessarily demands that the
> knowing subject be exterior to the object known, but a
> union in uncreated light. ([M14],120).

God is "essentially apart from other beings." Yet, a mystic may achieve "union in uncreated light" and even "deification." A contradiction? Of course. The solution? The mystic transcends their own domain (the domain of normal, separate human existence, apparently) and participates in uncreated life.

What is participation in uncreated life? Uncreated life apparently indicates God. Ruysbroeck's iron-and-fire analogy illustrates the idea of participation.

Iron in fire experiences fire's heat and light. As the experience continues, the iron begins to radiate heat and light, and becomes more and more like fire. By participating in fire's "life," the iron in some measure transcends its own nature and takes on the nature of fire. Yet, it always remains iron and never becomes fire.

Similarly, the mystic in God experiences God's energies, God's awareness and love. As the experience continues, the mystic begins to radiate awareness and love, becoming more and more like God. By participating in God's "life," the mystic in some measure transcends their own nature and takes on the nature of God. Yet, the mystic always remains their self and never becomes God. We'll call such first-hand experience of God "participatory" union.

Islam has a similar contradiction and another solution. Like Christianity, Islam believes God is a Person. Yet, some accepted Islamic mystics speak of union with God in the raindrop-and ocean sense. Ghazzali, who we've met before, uses the idea of "illusory" union to resolve the contradiction.

Ghazzali describes mystics whose experience of God is so intense that they *feel* they've become God, because they feel they've lost their individual self. Those mystics reach such a level of what Ghazzali ([A03],108) calls "unification," "identity," "extinction," (and, might we add, deification?) that they feel like the famous Sufi mystic Hallaj, also known as Mansur, who said:

> I am He whom I love and He whom I love is I.
> ([A03],107),

and

> I am The ONE REAL! ([A03],106).

However, says Ghazzali, when such mystics regain their senses, they realize they have not achieved

> . . . actual Identity, but only something resembling Identity . . . ([A03],107).

According to Ghazzali, some Sufis experience the illusion of union but don't actually experience union itself. Glowing white hot, participating in fire's life, the iron (he seems to say) may forget itself and believe it has become fire, when in fact it has not. We'll call such first-hand experience of God "illusory" union. Illusory and participatory union are the most intimate types of union possible in a religion that believes God is a Person.

Literal Union

We've discussed many types of union, from Ruysbroeck's three kinds of union, to the union of spiritual marriage, to participatory and illusory union. None is union in the plain, simple sense. None is like the union of raindrop and ocean. Now, we'll discuss union with God in the sense of raindrop and ocean. For brevity, we'll refer it as simply "union." So, from now on "union" means the joining of human and God, until only God remains.

Various mystics suggest such union is possible, and even claim it for themselves. For example, Meister Eckhart, one of the leading figures of medieval German mysticism, writes:

> [W]e are not wholly blessed, even though we are looking at divine truth; for while we are still looking at it, we are not in it. As long as a man has an object under consideration, he is not one with it. ([M11],200),

and

> [W]hen I cease projecting myself into any image, when no image is represented any longer in me . . . then I am ready to be transported into the naked being of God, the pure being of the Spirit. . . . I am translated into God and I become one with him - one sole substance, one being, and one nature: the Son of God. ([M12],134).

Eckhart's last statement and others like it - for example,

> The ground of the mind and the ground of God are one sole essential being. ([M12],107)

- aroused the interest of the Inquisition. He was brought before religious authorities but escaped with his life. Hallaj, as we'll see, was not so fortunate.

Even in Islam some mystics claim union, Ghazzali's explanation notwithstanding. For example:

> Some one came to the cell of Bayazid and asked, "Is Bayazid here?" He answered, "Is any one here but God?" ([N11],159).

And Jalalu 'd Din, said:

> With Thy Sweet Soul, this soul of mine
> Hath mixed as Water doth with Wine.
> Who can the Wine and Water part,
> Or me and Thee when we combine?
> Thou art become my greater self;
> Small bounds no more can me confine.
> Thou hast my being taken on,
> And shall not I now take on Thine? ([U01],426)

Lastly, we have Hallaj who once declared:

> Thy Spirit is mingled in my spirit even as wine is mingled with pure water.

> When anything touches Thee, it touches me. Lo, in
> every case Thou are I! ([N11],151),

In another instance, knocking at a door Hallaj was asked "Who is there?" He answered:

> I am the Absolute . . . the True Reality . . . ([S04],66).

For such statements, the Islamic orthodox convicted Hallaj of blasphemy, cut off his hands and feet, and sent him to the gallows.

Understanding Union

Can a human being actually unite with God, become God? Many people find the idea blasphemous or absurd. Historically, the mystic who claimed such union risked torture and execution. Today, that mystic risks the same treatment in some countries. In others, they risk being committed to a psychiatric institution. But are all mystics who claim union deluded? Or are some speaking the simple truth?

Union (in the sense we've now discussing) seems incredible when the God in question is some God who is a Person. It means the mystic has become God. It means the mystic who claims union with Jehovah has become the Person who created the world in six days. It means the mystic who claims they've united with Allah gave the Koran to Mohammed. It's hard to see how such a claim could be true.

On the other hand, when the God in question is the God which is not a Person, the claim of union might well be true. How? One way of understanding such union is as a kind of dissolving back into the ocean of Uncreated Light in which we live. Movies offer a helpful analogy. (Free-standing, three-dimensional holograms would be a better analogy.)

Consider a movie. All that exist on the movie screen is actually light. Because of the way the light "dances" on the screen, we see people, animals, trees, sky, and buildings on the movie screen. Yet all that exists, all we ever see, is light.

Similarly, Energy, the Uncreated and Unchangeable Light, composes us and the rest of creation. All that exists in our world is that Ultimate Reality. Because of the way It "dances" (whatever that might mean; more later), we see people, animals, trees, sky, and buildings.

Understood in this sense, Hallaj and Bayazid speak the simple truth. They, you, I, and every other entity is a manifestation of the one

Eternal Entity. We are images of the Eternal Light which is God. So the Christian mystic Angelus Silesius speaks the simple truth when he says:

> I am not outside God.
> He is not outside me.
> I am His radiance;
> my light is He. ([B05],109).

And even Eckhart speaks the truth when he says

> God's being is my being . . . ([M12],87),

if he's understood as saying that he, we, and any God who is a Person all share the same Ultimate Ground of Existence.

The movie analogy also applies to second- and first-hand experience of God. Second-hand experience is like a movie-screen person reading a book about the movie light which is their root, source, and ultimate ground of existence. First-hand experience of Uncreated Light - also called illumination, as we'll see - is like a movie-screen person experiencing the movie light as something other than and external to themselves, though it is in fact their root, source, and ultimate ground of existence.

Finally, union with Uncreated Light is like a movie-screen person experiencing the movie-screen light as their own ultimate ground of existence, realizing it's not something different from them, but rather is their own, true, inmost self. Such experience isn't second- or even first-hand experience. Rather, it's a third type of experience, unitive experience.

Knowledge which derives from unitive experience is called unitive knowledge. For unitive experience and knowledge, the analogy of wood and fire applies better than iron and fire, because wood not only gets hot and glows like fire, it not only participates in fire's life, but it eventually is consumed and transformed into fire itself.

Transcending the Triad of Knower, Knowing, Known
The movie analogy and the wood-and-fire analogy are two ways of understanding unitive experience. Another way, which occurs often in mystical literature, involves the idea of the triad of knower, knowing, known. Unitive experience and knowledge is said to occur when separation dissolves, when the knower unites with the Known, transcending the triad of knower, knowing, and known.

But what is triad of knower, knowing, and known?

Second-hand knowledge has three elements: the knower, the act of knowing, and the known. For example, when someone acquires second-hand knowledge of God through scripture, the "known" is the scripture, the knowing involves reading and understanding, and the person reading the scripture is the knower.

First-hand knowledge has the same three elements: knower, knowing, known. If New York city is known, then seeing and experiencing directly is the knowing, and some person is the knower. If God is known, then the knower has direct experience of God and therefore is a mystic.

In second-hand knowledge of God, the knower's separate identity remains. Even in first-hand experience of God, also called "illumination," the knower's separate identity remains intact.

> All pleasurable and exalted states of mystic
> consciousness in which the sense of I-hood persists,
> in which there is a loving and joyous relation between
> the Absolute as object and the self as subject, fall
> under the head of Illumination. ([U01],234).

In both types of knowledge, the knower remains distinct and separate from the Known. I-hood, the separate personality of the knower, remains. Even in the most intimate degrees of union - spiritual marriage, participatory and illusory union - the triad of knower, knowing, and known remain.

Yet, many mystical traditions speak of transcending the triad. Nicholson speaks of it when he writes the Sufi's path leads to the realization that

> . . . knowledge, knower, and known are One.
> ([N11],29).

The Hindu sage Ramana Maharshi also speaks of it when he says that:

> (d) The one displaces the triads such as knower,
> knowledge and known. The triads are only
> appearances in time and space, whereas the Reality
> lies beyond and behind them. ([T03],173-4),

a conclusion he bases on three insights - that Reality is:

> (a) Existence without beginning or end - eternal.
> (b) Existence everywhere, endless - infinite.
> (c) Existence underlying all forms, all changes, all
> forces, all matter and all spirit. . . . ([T03],173).

Lastly, *The Awakening of Faith* speaks of transcending the triad when it describes how Buddhist knowers

. . . realize Suchness. We speak of it as an object . . .
but in fact there is no object in this realization . . .
There is only the insight into Suchness (transcending
both the seer and the seen) . . . ([A15],87).

With unitive knowledge, our New York City illustration fails. There is no way for a person, in any sense, to actually transcend the triad and unite with, become one with, the city - though a person might feel united with it or feel they participate in its life. There is, however, an apt Hindu analogy. The mystic who aspires to first-hand knowledge of God is like someone who wants to taste sugar. To enjoy the taste of sugar, the taster must remain distinct from sugar. The mystic who seeks union with God, on the other hand, is like someone who wants to become sugar. That mystic seeks actual and literal union, until only the One remains.

The Unreal Self

But aren't we already united with our Ultimate Ground of Existence? Yes. Every human being is *already* united with the God which is not a Person. Every human being is already one with their Ultimate Ground of Existence. We all have union with that God. But only a few of us are conscious of it.

Such consciousness is what mystics like Hallaj claim. They claim that they have directly experienced what is for others a theoretical identity. So, instead of claiming identity with some God who is a Person, mystics like Eckhart and Hallaj claim conscious union with the God which is not a Person, of which everything, including Gods who are Persons, are a manifestation.

Therefore, a mystic like Hallaj may not see himself as fundamentally different from anyone else. In fact, far from feeling his personal self is exalted, he may feel individual personhood is an illusion.

Why? Imagine a person who experiences their Ultimate Substance, their real Self, which is God, the God which is not a Person. They see, perhaps for the first time, their real, eternal Self. In contrast, what they have previously considered their self, their ego, may seem insignificant. Writes Underhill:

> All its life that self has been measuring its candlelight
> by other candles. Now for the first time it is out in the
> open air and sees the sun. ([U01],200).

Therefore, the person who experiences the Ultimate Ground of Existence as their own real Self, may well come to consider their individual, distinct personal self as an unimportant illusion. The Christian saint Catherine of Siena may have come to this opinion after she heard a voice which said:

> In self knowledge, then, thou wilt humble thyself;
> seeing that, in thyself, thou dost not even exist."
> ([U01],200).

In any event, mystics who aspire to union with the God which is not a Person often desire

> [n]ot to become *like* God or *personally* to participate in
> the divine nature . . . but to escape from the bondage
> of . . . unreal selfhood and thereby to be reunited with
> the One infinite Being. ([N11],83).

And, in fact, such mystics sometimes reach the state of

> . . . passing-away in the divine essence . . . He
> contemplates the essence of God and finds it identical
> with his own. . . . [H]e becomes the very Light.
> ([N11],155-6).

The mystic who reaches such a state of union may claim identity with God because they can find no other Self to call their own.

> As the rain-drop ceases to exist individually, so the
> disembodied soul becomes indistinguishable from the
> universal Deity. ([N11],167)

having achieved

> . . . self-annihilation in the ocean of the Godhead.
> ([N11],168).

Of such union, Nicholson writes:

> Where is the lover[?] . . . Nowhere and everywhere:
> his individuality has passed away from him.
> ([N11],119).

Ramakrishna likened ([G03],103) unitive experience of the Eternal to a salt doll dissolving in the ocean. Ramakrishna himself ([L07],153) experienced this state, which is called "Nirvikalpa Samadhi" in Hindu religious literature.

The Tragedy of Hallaj
Many scriptures hint at the identity of human self and the God which is not a Person, or even state it outright. For example, words in the very first chapter of the Bible.

> So God created man in his *own* image, in the image of
> God created he him; male and female created he
> them. ([H08],Gn1:27).

Of course, someone who believes in a God who is a Person might think these words mean that God has two arms, two legs, a torso, and a head with two eyes, ears, one nose, etc., and so created us in His own image. But someone who believes in the God which is not a Person would probably interpret them as saying that we are literally, at this very moment, an image of the Uncreated Light, the Absolute Existence and Ultimate Ground of Existence. Like fish in an ocean of water, we live in an ocean of Uncreated Light. From this viewpoint, words from a Christian hymn are literally true:

> Lord, Thou hast been our dwelling place
> Through all the ages of our race ([H12],84),

as are words the Christian Saint Paul spoke to the people of Athens before their high tribunal, the Areopagus,

> For in him we live, and move, and have our being . . .
> ([H08],Acts,17:28).

Sometimes, followers of the God which is not a Person who have realized that, in the ultimate, ontological sense, they already are the Ultimate Ground of Existence, that their inmost self is identical with that God - sometimes they celebrate their intimacy, union, and identity in word and song. In some religions, their exaltations are understood. More commonly, however, they are misunderstood, often because our two ideas of God are confused.

For example, returning to Hallaj, it's probable that he and the Islamic orthodoxy who executed him had two very different ideas of God in mind. Reynold Nicholson in *The Mystics of Islam* describes these two ideas, as they occur in Islam.

> Both Moslem and Sufi declare that God is One . . .
> The Moslem means that God is unique in His
> essence, qualities, and acts; that He is absolutely
> unlike all other beings. The Sufi means that God is the
> One Real Being which underlies all phenomena. . . .

the whole universe, including man, is essentially one
with God. . . ([N11],79-80).

Nicholson even implies that orthodox religion and mystic habitually
see God in these two different ways.

Religion sees things from the aspect of plurality, but
gnosis regards the all-embracing Unity. ([N11],74).

Therefore, Hallaj probably meant he had achieved unity with the
God which is not a Person. In fact, a few other mystics understood his
claim in exactly that way.

"I am the Absolute Truth," or, as it was translated later,
"I am God," led many mystics to believe that Hallaj
was a pantheist, conscious of the unity of being.
([S04],72).

Tragically, the Islamic orthodox did not.

Summary

Many chapters ago we posed a question: Can the scientific way of
knowing be applied to religious questions? that is, can it be applied to
the religious domain of knowing? Before we could answer this
question we had to investigate the domains of science and religion.
The third and fourth chapters showed the Ultimate Ground of
Existence lies within both domains. This common ground makes the
task of constructing a scientific religion much easier. After all, the
difference between religious monism and natural monism is one of
attitude. One regards the Eternal Substance religiously, the other does
not.

Religions obtain much of their knowledge of the Ultimate from
writings which they suppose perfectly true. Science does not allow
itself such blind belief. Science has obtained whatever knowledge it
has of the Eternal through its investigations into physical phenomena.
But such investigations, though valuable, have failed to address life's
really important questions.

We saw in this chapter that human beings can have, and in fact
have had, direct experience of the God which is not a Person, the
Ultimate Ground of Existence.

For people who've experience It, the Real is a concrete reality;
indeed, sometimes the only Reality. For others, however, the Eternal
is a concept, an abstraction, a theoretical construct. We've seen that
scientific theories often contain theoretical constructs, things whose

existence is provisionally accepted until their existence is either proved or disproved, experimentally. In this book, the Root is an theoretical construct. Anyone interested in experimentally proving or disproving It's existence for themselves is free to adopt some form of mystical life.

We've also seen experience of the Ultimate Ground of Existence - either first-hand or unitive - comes under the heading of mysticism or gnosis; that the pinnacle of mystic experience is union with Ultimate Reality; and that mystics who achieve such union sometimes describe the experience as one of "deification."

Deification. At first sight, the word may seem shocking, absurd, or blasphemous. And even when its meaning is understood - the transformation of human self until only God remains - it's still easy to feel uncomfortable. The term seems too liable to be misunderstood, misused, or exploited.

When the God in question is the God which is not a Person, "deification" implies only becoming conscious of what already is. Our union with the Eternal is already a fact; the deified mystic is someone who has realized this in a vivid and concrete manner.

On the other hand, when the God in question is some God who is a Person, the idea of deification has a great potential for misuse and exploitation, especially since it's often accompanied by the following reasoning: Since a God who is a Person isn't bound by religious observances (this reasoning goes), and since everything such a God does is moral, even if it seems otherwise, the same must apply to one who has achieved deification and is united with that God. Therefore, even if a deified person steals, rapes, or murders they commit no crime, for they are above all earthly laws and understanding.

The dangers inherent in such reasoning are obvious. As Rufus Jones writes:

> These doctrines - that the universe is a Divine Emanation, that God is being incarnated in man, that each person may rise to a substantial union with God, that external law is abolished and ceremonial practices outdated, that the final revelation of God is being made through man himself - these doctrines are loaded with dangerous possibilities as soon as they receive popular interpretation. ([JO3],188-9).

Rolt comments on some similar ideas in the writings of "Dionysius":

His doctrines are certainly dangerous. Perhaps that is a mark of their truth. For the Ultimate Truth of things is so self-contradictory that it is bound to be full of peril to minds like ours which can only apprehend one side of Reality at the time. Therefore it is not perhaps to be altogether desired that such doctrines should be very popular. They can only be spiritually discerned, though the intensest spiritual effort. Without this they will only too readily lead to blasphemous arrogance and selfish sloth. ([D08],47)

The next chapter presents instances of the inappropriate, unhealthy application of some mystical ideas. The need for some way of evaluating mystical claims is presented. The revelational way of knowing is shown to have been the way such claims were evaluated in the past. It's suggested the scientific way of knowing should be used to evaluate such claims today. Lastly, details of the application of the scientific way of knowing to the claims of the mystics are considered.

6

- Science Plus Religion -

Chapter Summary: This chapter shows how mysticism can be misused, and then discusses the need for a way of judging supposed mystical statements. Judging with science's way of knowing would yield a scientific religion. The chapter discusses how the four elements of science's way would apply to mystical statements. After restating some fundamental premises, the chapter closes with an allegory.

Mysticism: Proper, Improper and Counterfeit

We've seen that mysticism means direct experience of either the Ultimate Ground of Existence or a God who is a Person. "Mysticism", however, is also used - or rather, misused - to mean the psychic, the magical, the occult, and even the demonic. A popular publisher, for example, offers a *Mysteries of the Unknown* series whose topics include UFOs, astral projection, clairvoyance, channeling, and ([M19]) *Mystic Places*. Shops which sell love potions, good luck charms, and charms that ward off the "evil eye" also sell "mystic" dream interpretations books. And even the prestigious *Scientific American* magazine published the following.

> Superstitions, cults and mysticism appear with surprising consistency during a social crisis. Today it is ESP and UFOs, astrology and clairvoyance, mystic cults and mesmeric healers. . . . [A]t the same time the fortunes of these mystics have risen, the number of popular science magazines and television programs has declined markedly. ([K01],34).

Throughout this book, "mysticism" is used in its proper sense, to indicate direct experience of the Ultimate, not UFOs or ESP. But it's worth investigating how the other meanings arose.

Imagine a person with a newly-found interest in things which seem outside the natural world. Since mysticism, magic, the psychic, the occult, and the demonic all concern such things, a person at first might see them as similar, and mistake one for the other. Or a person interested in one might naturally wonder what the others were about, too. It's not surprising therefore that

> . . . in every period of true mystical activity we find an outbreak of occultism, illuminism, or other perverted spirituality . . . ([U01],149).

Mysticism, magic, the psychic, the occult, and the demonic are indeed similar in that they all treat things beyond the normal, everyday world. But they are fundamentally different, too. (The seventh chapter of [U01], by the way, describes some of the differences between mysticism and these other types of activity.)

In time, anyone who progresses in mysticism (or in the other activities) will realize their own genuine interests. They'll learn to distinguish the mystical from the magical, the psychic from the occult, and may even come to see them as different as day and night. The unaware person, however, may still lump them all together. Thus, to the uninformed "mystical," "magical," and "occult" may suggest more or less the same thing.

This book is not about the psychic, the magical, the occult, or the demonic. Such phenomena don't concern us, even when they're inappropriately labeled "mysticism," as long as it's understood that they aren't genuine mysticism. Unfortunately, even when all such phenomena are ignored, there remains a large number of ridiculous, perverted, and even insane acts and beliefs that seem based on genuine, though perhaps inferior and misinterpreted, mystical experiences and ideas. As Underhill observes, great indeed are the errors

> . . . into which men have been led by a feeble, a deformed, or an arrogant mystical sense. The number of these mistakes is countless; their wildness almost inconceivable to those who have not been forced to study them. ([U01],149).

We'll examine a few instances, specifically the Nazis, the Ranters, some deviant Sufis, and Bhagwan Shree Rajneesh.

While it's well-known that the Nazis murdered millions of people during the Second World War, it's often forgotten that their ideology was based, in part, on mystical and scientific ideas. Mystical ideas, particularly those of the medieval mystic Meister Eckhart, formed a part of Nazi belief. A forward in a book published in 1941 about Eckhart has:

> According to the Nazis, Eckhart is a member of their party in good standing. ([M11],xv).

However, science (or, at least, what was thought science) contributed much more to Nazi ideology than mysticism. In an illuminating article, *Biological Science and the Roots of Nazism*, George Stein writes:

> Hitler's views are rather straightforward German social Darwinism of a type widely known and accepted throughout Germany and which, more importantly, was considered by most Germans, scientists included, to be scientifically true. . . . Hitler did not invent national socialist biopolicy. . . [A]lmost every element of Nazi biopolicy was already well established in the German political culture in both a vulgar, man-in-the-street sense, and, more importantly, among the educated elite who took their views from the representative science of the day. ([S21],51),

and:

> There really was very little left for national socialism to invent. The foundations of a biopolicy of ethnocentrism, racism, and xenophobic nationalism had already been established within German life and culture by many of the leading scientists of Germany well before World War I. ([S21],57).

Given the disastrous consequences of Nazi beliefs, it might certainly be argued that the mystics are best left ignored, and that science should stay within its present boundaries and not meddle with the values, ethics, and morals that have traditionally been the concern of religion. Yet if this reasoning is accepted, wouldn't similar reasoning about the Inquisition and other religiously motivated atrocities prove that religion also fails as a sure guide to right and wrong? If the Nazi episode demonstrates a fatal flaw in the scientific way of knowing, then doesn't the Inquisition expose a similar flaw in the religious way of knowing? Perhaps, they do. Perhaps, both ways of knowing are imperfect. But I believe the scientific way is the better of the two.

The Nazis were primarily a political movement; perhaps it's not surprising that removing mystical concepts from their spiritual context and forcing them to serve political, nationalistic, or imperialistic causes had disastrous results. Of course, the results aren't always unfortunate. Democracy and the idea of the equality, I've read, derive from mystical ideas. Certainly, equality isn't a common sense idea. After all, even some children (generally one's own) are brighter, stronger, more agreeable, more attractive, better behaved, etc., than others.

Sadly, mystical ideas are subject to abuse and misapplication even in a religious context. For example, about the time the Society of Friends - the Quakers - was being born in England, another group, the Ranters, was also growing. The Ranters seem to have had some appreciation of the Ultimate Ground of Existence.

> The *central idea* of Ranterism was the doctrine that
> God is essentially in every creature. ([J03],467).

One Ranter claimed

> . . . that he was *not the* God, but he was God, because
> God was in him and in every creature in the world . . .
> ([J03],475).

"He was not the God, but he was God" expresses an idea that understood, would have saved many a mystic from torture and death. It denies the Ranter is Jehovah or Jesus, but says he's identical with the Eternal Substance, his ground of existence.

Unfortunately, some Ranters fell into the mistakes Jones and Rolt mentioned at the end of the last chapter. They decided that because they "were God" moral or legal rules weren't binding. A critic accuses them of believing

> . . . nothing is sin, if a man himself do not count it sin
> and so make it sin unto himself. . . ([J03],472).

And since they were "God" religion had nothing to teach them.

> They admitted that Paul had the spirit, when he wrote,
> but they said: "Have not *I* the Spirit, and why may not I
> write Scriptures as well as Paul, and what I write be as
> binding and infallible as that which Paul writ?"
> ([J03],474).

The Ranters held an uncommon view of scripture, diametrically opposed to the usual orthodox view. Both views are extreme. Science has a more balanced view of its writings. It doesn't believe Einstein's theories are divine, flawless, and absolutely true. Yet it recognizes genius and relative worth, and doesn't equally value the theories of all physicists and would-be physicists.

The Ranters were a short-lived sect. Yet even ancient, established mystical traditions sometime fall into wild errors. For example, Sufism has produced many sincere and saintly mystics. It also produced an early extremist sect, the Malamatiya, who believed

> . . . the true worship of God is best proved by the contempt in which the devotee is held by his fellow-men . . . ([A08],70).

The Malamatiya committed scandalous sins to generate condemnation so they could show their contempt of public opinion. During the 15th and 16th centuries C.E. attitudes similar to the Malamatiya's became prevalent throughout Egypt.

> To live scandalously, to act impudently, to speak unintelligibly - this was the easy highroad to fame, wealth and power. ([A08],119).

For example, one Sufi, 'Ali Wahish,

> . . . made a special point of displaying his bestialism on the common highway whenever opportunity presented itself. ([A08],120).

Today, the misuse and abuse of mystical ideas and principles continues. The last illustration we'll consider is Bhagwan Shree Rajneesh. His followers believe he lived in an elevated mystical state we'll discuss later, a state above the duality of liking and disliking, a state of pure is-ness.

> Bhagwan does not either *need* or *like* anything! . . . *He is*. In is-ness there is no like or dislike. ([M15],157).

Yet, although he didn't need or like anything, at one point Rajneesh owned 80 Rolls Royces. His method of raising funds for things he didn't like or dislike was inventive. At a fund-raising interview a follower was given a brandy and told Rajneesh

> . . . could see you were ready for a great jump, but . . . [y]our money is a barrier . . . The time has come to . . . give everything! . . . [Y]ou are going to feel so *good*, so *clean*, with all this . . . off your hands! ([M15],222).

Later, the method was refined: the brandy was laced ([M15],290) with "Ecstasy," a euphoric drug.

Once obtained, the funds might have been better spent for food and medicine rather than Rolls Royces.

There were to be many deaths . . . from hepatitis and
other diseases which could have been cured with
proper medical attention. Rajneesh never gave
enough money for food . . . and was not concerned
when we worked too hard or slept too little. ([M15],97-
8).

Nonetheless, Rajneesh was popular, partly because early in his
career he had

. . . a reputation as the 'sex guru'. This description
seemed to refer both to his personal tastes and the
content of many of his lectures. ([M15],55).

And he had ([M15],115-8) sexual intercourse with numerous female
disciples.

Unrestricted sex was common among his followers themselves,
also. To avoid pregnancy, sterilization ([M15],159) and vasectomy
([M15],160) were "recommended."

It was only possible to avoid the operations by being
adamant . . . and such refusal sometimes meant
having to leave the . . . workforce. ([M15],160).

Some women were as young ([M15],321) as fourteen when they were
sterilized. Years later, the doctor received ([M15],320) hundreds of
letters asking if the operation was reversible.

Since unrestricted sex was so common, it's perhaps not surprising
that when money was needed

. . . many of the girls turned to prostitution.
([M15],154).

Other followers raised money by dealing illegal drugs.

Whenever a disciple was about to make a drug run,
they would ask Bhagwan whether it was a good time
to go . . . ([M15],155).

About 1980, Rajneesh and a few select followers fled India,
eventually settling on a ranch in Oregon. His other followers found
themselves abandoned. Some who had little or no money turned to
parents or friends. Not surprisingly, others turned to

. . . the now traditional . . . ways of making a quick
buck - drug running and prostitution. ([M15],191).

The Oregon ranch soon had thousands of followers, including
over a hundred trained ([M15],288) with automatic weapons. Eleven
armed watch towers ([M15],291) were built. Eventually, bugging and
wiretapping equipment ([M15],288) was installed.

For followers who had thoughts of leaving the ranch mind-altering drugs ([M15],290) were prescribed. Such drugs were also mixed in the food ([M15],292-3) of thousands of homeless people who were bussed to the ranch to help win a local election. As legal troubles increased, some followers ([M15],295) plotted the murder of a local political official.

Traditional Measures of Mystical Truth

Mysticism is powerful. Historically, its ideals and ideas, its practice and theory, have shown themselves capable of deeply influencing the lives of individuals and entire civilizations. Unfortunately, what is powerful is often dangerous, too. For example, nuclear energy and biogenetic engineering are very powerful and very dangerous. Mysticism's power demands safeguards. If it is to remain healthy and sane, a mystical culture must always guard itself against ills such as superstition, charlatanry, mystification, degeneracy, and anti-intellectualism.

How?

One safeguard is evaluating the alleged mystics themselves. Are they interested in God or their followers' bodies and money? Are they above liking and disliking or very much drawn to people and things? A later chapter discusses how mystics themselves might be examined. Another safeguard is testing and evaluating not the alleged mystic but rather their observations and statements.

But how can alleged mystical statements be tested? How can we judge the private visions of an individual? And how are we to distinguish the assertions of healthy mystics from deranged mystics?

It's the function of a way of knowing to test and verify statements. As we've seen, a way of knowing is a way of answering questions such as: "How can I acquire knowledge? How can I be sure my knowledge is true?" Applied to mysticism these questions become: "How can I acquire knowledge of God? How can I be sure my knowledge is true? How can the mystical be differentiated from the magical, psychic, occult, and demonic? How can healthy mysticism be distinguished from unhealthy or perverted mysticism?"

But which way of knowing should we use to judge alleged mystical statements?

Traditionally, religion's way of knowing has been used to evaluate such statements. Typically, religious systems accept as true only mystical visions and experiences that agree with their divine, complete, and final scripture. That is, mystical declarations are subjected to the test of religious orthodoxy; experiences and statements that disagree with scripture are declared wrong, and the alleged mystics are subjected to varying degrees of rejection - from disbelief to condemnation to torture to death.

Measuring statements against scripture has a few shortcomings. First, it inherits all the shortcomings of the revelational way of knowing. Second, it's sometimes sadistic: condemning writings should be sufficient, is it also necessary to torture and murder the writer? Third, it sometimes condemns seers whose only "crime" is seeing clearly, and honestly telling what they see. That is, mystics are sometime condemned for their truthfulness, vision, and forthrightness.

For first-hand knowledge often disagrees with second-hand knowledge. Suppose some official keepers of the "Truth" know New York City only by books they've read. Suppose they choose a list of canonical books, "true" books, about the city. Suppose they "harmonize" any disagreements among the books with appropriate hermeneutic principles. Finally, suppose they solemnly declare the result the "One and Only Truth." Anyone who actually travels to New York is liable to see things that don't agree with the "Truth." And anyone who has the forthrightness to tell what they saw may suffer at the hands of the official "Truth" keepers. Mystics have often expressed truths which did not fit into their religion's established dogma or world view - and suffered the consequences.

The Scientific Way of Knowing as the Standard
Today, there's another way of knowing, the scientific way. Physics, chemistry, sociology, psychology, anthropology, and many other fields accept it as their way of knowing. Some previously followed a way of knowing much like the way used by religion: they relied on authority to decide truth. Let's look at some examples.

As we've seen, physics once accepted a way of knowing similar to the revelational way. Aristotle's teachings were once believed *because* they were Aristotle's teachings. Eventually, physics abandoned that way of knowing and adopted the scientific way. How much more advanced would we be today if physics had changed its way of knowing earlier? We can only guess, but it's suggestive that

> Aristarchus of Samos, about 270 B.C., proposed a system identical with the Copernican . . . [I]t attracted few, if any, followers, however, and there was talk of a charge of impiety being brought against him. ([T02],30).

Today, only a few fields still use the revelational way of knowing. One is astrology, which doesn't accept the scientific way of knowing and therefore is not a science. In contrast, geometry long ago abandoned the revelational way and accepted a method of knowing that eventually evolved into science's way. Because it did, even Euclid, perhaps the greatest geometer of all time, couldn't merely declare something true - he had to prove it. And his geometric theorems remained open to question, criticism, revision and refinement.

For example, for over two thousand years geometry was based on the work of Euclid: geometry was Euclidian geometry. Yet, when mathematicians discovered Non-Euclidian geometries in the nineteenth century, they weren't declared heretics or burned at the stake. No doubt some people initially questioned the usefulness of the new geometries. Eventually, however, Einstein based his theory of Relativity on one of them.

For fields such as physics that once accepted the revelational way of knowing, adopting the scientific way has proven a great step forward. In field after field the revelational way of knowing has been abandoned for a superior way of knowing, the scientific way. And in field after field, this change has led to great progress.

So, even though alleged mystical statements have traditionally been evaluated with religion's way of knowing we can ask if science can evaluate such statements. That is, can the scientific way of knowing be used to decide religious questions and questions of ultimate value? If it can, then science may someday be able to discuss topics that presently lie outside of its domain, inside the domain of religion. It may someday be able to create a "scientific religion."

But is it reasonable - and is it in keeping with the spirit of science - to ask science to discuss "supernatural" questions? Arthur Compton, who won the Nobel Prize in physics, was one scientist who believed it was. Compton believed the scientific method could be applied to part of the religious domain, to what he calls the "supernatural" realm of "visions and hope and faith."

> Those whose thinking is disciplined by science, like all others, need a basis for the good life . . . They need a faith to live by. . . . [V]isions and hope and faith are not a part of science. . . . They are beyond the nature that science knows. Of such is the true "supernatural" that gives meaning to life. This supernatural is as real as the natural world of science and is consistent with the most rigorous application of the scientific method. ([C15],369).

Of course, Compton's "supernatural" realm isn't the same as religion's domain, because most religions include more than visions, hope and faith: they include dogmas and a God who is a Person.

Religion Without Dogma or a God Who Is a Person
The scientific way of knowing rejects blind faith and insists on understanding and proof. Therefore, a religion that wanted to employ it couldn't teach its truth or revelation is beyond the power of the human mind to discover, understand, criticize, test, modify, or reject. Instead, all of its beliefs would be subject to testing by the scientific way of knowing. And any belief that couldn't be proven would have to be abandoned or, at best, accepted as a theoretical construct.

Perhaps existing religions could justify abandoning dogma as intellectual humility. After all, religion has sometimes insisted on some dogma (that the sun rotates around the earth, for example) which it later admitted was wrong. So perhaps existing dogmas could be relabeled as the official expression of religious truth according to admittedly fallible human thinkers. Dogma, then, might be downgraded from divine, unerring truth to ideas that are open to adaptation and change, able to conform to new insights and truths.

But could existing religions ever abandon their Gods who are Persons? Would they have to? Would a religion necessarily have to abandon the idea of a God who is a Person if it sought to employ the scientific way of knowing?

Gods who are Persons may be divided into two types: Gods like Jehovah and Allah who have not assumed human form, and Gods like Jesus and Krishna who have. The scientific way of knowing does not recognize superhuman god-men. Mozart, his almost supernatural

156

musical gift notwithstanding, was still an ordinary fallible mortal. Gauss, his huge mathematical gift notwithstanding, was one, too. The ideas and theories of Mozart and Gauss are subject to disagreement, correction, and revision. Therefore, because the scientific way of knowing insists on understanding and rejects blind faith, the dogma that Jesus or Krishna was God in human form is incompatible, because the dogma is based on faith and not open to scientific testing or proof.

What about Gods like Jehovah and Allah, for whom no human incarnation is claimed? As we've seen, there's natural monism and religious monism. The two types of monism have much in common. In fact, a scientific religion can be built upon the common area they share. But neither type of monism accepts a God who is a Person as the ultimate entity. Natural monism ignores such a God entirely; religious monism stipulates a Godhead, an Ultimate Ground of Existence, upon which all Gods who are Persons depend for their existence.

Therefore, a scientific religion built on monism probably wouldn't include any actual Gods who are Persons. As we'll see later, however, a scientific religion might emphasize an aspect of the Uncreated that is very similar to Gods who are Persons. In fact, the idea of Gods who are Persons may have originally derived from this aspect of the Real: they may be this aspect of the Uncreated, misperceived or misunderstood.

But isn't the idea of a God who is a Person a necessary part of religion? Albert Einstein, for one, didn't think it was. In fact, he believed the highest type of religion was free of this idea. Einstein described three types of religion: a primitive "religion of fear" ([E03],37), a more advanced "moral religion" ([E03],37), and a third type, "cosmic religious feeling" ([E03],38), which has

> . . . no anthropomorphic conception of God . . . no dogma and no God conceived in man's image . . . ([E03],38).

Einstein believed that

> . . . teachers of religion must have the stature to give up the doctrine of a personal God . . . ([E03],48)

and that

[t]he main source of the present-day conflicts between
the spheres of religion and of science lies in this
concept of a personal God. ([E03],47).

And he believed science could help "purify" religion of the idea of a
God who is a Person, as well as give our life spiritual meaning.

. . . [S]cience not only purifies the religious impulse of
the dross of its anthropomorphism but also contributes
to a religious spiritualization of our understanding of
life. ([E03],49).

Would a religion without dogma and a God who is a Person still
be a religion? It would. In fact, a few existing religions already meet
these requirements. Some Buddhist groups make no claims to
revealed truths or to founding by a god-man or God who is a Person.
To these sects, Buddha was a man who discovered certain important
truths in a natural, human way, just as Euclid, Einstein, or Gauss
made their discoveries. Some ultra-liberal Christian groups, and
perhaps analogous groups in other religions, have similar beliefs.
Such groups might easily apply the scientific way of knowing to the
religious domain.

However, most major religions, which believe in Gods who are
Persons and in revealed, unchangeable scripture, would have to
change what they've taught for hundreds, even thousands, of years to
meet the requirements of the scientific way of knowing. They
probably never will.

But there's another possibility: science itself could extend its
domain by applying its way of knowing to the raw data that the
mystics provide, the descriptions of their experiences. If it did, the
extension would include some of what has traditionally been in the
domain of religion. The extension would be a scientific religion.

What would a scientific religion be like? Like any other scientific
discipline, it would have the following elements: 1) its domain of
knowing, 2) its raw facts, the outcomes of experiment and
observation, 3) its generalizations of fact, that is, hypotheses and
laws, and 4) its explanations of fact, theories. Let's discuss each
element in turn.

The Domain of Knowing

Biology's domain of knowing is living creatures. Sociology's domain of knowing is groups of people. What would be a scientific religion's domain?

We've seen that science's domain already includes the Ultimate Ground of Existence, because the eternal Basis of the universe is already studied objectively, "from the outside," by nuclear physics and, in a sense, by all the sciences. Now, through the acquisition of a scientific religion, science's domain would gain the study of the Uncaused Cause "from within." Direct experience - mystical experience - of the Eternal Root would be incorporated into science's domain. A scientific religion would use the instrument of mystical awareness to study and explore the Eternal Ground of Existence.

But mystical experience isn't limited to the Eternal. Some mystics describe the Root as It relates to the external universe. And others describe the Center as It relates to our deepest selves. Therefore, a scientific religion might have something to say about the external universe and our deepest selves, too. Questions such as Who are we? Why are we here? What is our place in this world? and What happens when we die? could be addressed and perhaps answered.

Moreover, mystics have recommended certain values, attitudes, and actions, and have censured others. Either explicitly or implicitly, they've addressed the questions What is the best way to live one's life? What is life's greatest good? and How can life's greatest good be obtained? Therefore, these questions, as well as the morals and ethics that derive from them, might also fall within the scope of a scientific religion.

As we've seen, Gods who are Persons would not be in the domain of a scientific religion. Therefore, no claims would be made about Gods such as Jesus and Jehovah, Krishna and Allah. Statements such as Jesus is the only begotten Son of God, Jews are God's chosen people, Muhammad is the Seal of the Prophets, etc., would remain in the domain of religion.

The Data

A scientific religion would take as its raw data the observations of those who claim mystical experience. It would be based on the statements and experiences of healthy mystics, no matter what their world view, religion, nationality, or era. By applying the scientific way of knowing to such raw data, a scientific religion would eventually distill its hypotheses, laws and theories.

Pharmacology developed in a similar way. For centuries, various cultures used a variety of folk remedies. Some remedies worked, some didn't, but none had a scientific basis. Pharmacologists investigated the remedies, discarding the useless, extracting or synthesizing the active principle of the useful, then scientifically proving effectiveness. For example, Quinine, used to treat malaria, was extracted ([S17],9) from cinchona bark; ephedrine, helpful for asthma, was isolated ([N05],v11,825) from the mahuang herb; and reserpine, derived from the rauwolfia plant, has recently been used ([N05],v11,825) to treat high blood pressure.

A scientific religion would perform an analogous process. For centuries, various religions have believed supposed mystical insights and principles, some healthy and true, some nonsense, but none having a scientific basis. A scientific religion would examine these ideas, incorporating the true ones into its own framework of hypothesis, law, and theory. And it would reject statements it found untrue, no matter who said them. So, while it might accept some of the ideas of Jesus, Moses, Mohammed, Ramakrishna, and Buddha, it might reject others. A scientific religion would not be Christian, Jewish, Islamic, Hindu, or Buddhist but, no less than pharmacology, would constitute its own integral discipline.

Returning to the analogy, after the active ingredient is isolated, there's no need to take the herb or plant in its natural state for the remedy to be effective. And there's certainly no need to adopt the folk beliefs of the culture that originally discovered the herb or plant. Though it's been taken out of context twice - from the plant in which it occurs, and from the folk culture that discovered it - the active ingredient works.

Similarly, a scientific religion would use mystical insights and ideas removed from their original context, the religion or culture that discovered them. But rather than forming an incompatible mishmash the insights and ideas would fit like pieces of a puzzle, because they're different aspects of a single Truth.

The Data: Agreement and Objectivity

But don't mystics of different religions disagree? The Christian mystic talks to Jesus while the Hindu mystic has a vision of Rama or Kali, who is a woman. The Islamic mystic knows nothing of Buddha, the Buddhist mystic knows nothing of Allah. Mystics of different religions generally experience different Gods who are Persons. Such contradictory experiences are what we'd expect of dreams, not perceptions of an objective reality. How can they fit together like pieces of a puzzle?

They can't - yet another reason why a scientific religion would have little to say about Gods who are Persons, why it would leave such Gods in the domain of religion. Nonetheless, such Gods, though not ultimately real, may be as real as you or I. A later chapter returns to this issue.

Experiences of Gods who are Persons often contradict each other, but experiences of the God which is not a Person generally agree. This explains why

> . . . many works on mysticism tend to either of two . . . positions . . . One school supposes that mystics are basically all alike, all representatives of the *philosophia perennis* which transcends religions and cultures; the other that all non-christian mystics are not mystics at all. ([R03],213).

The second school bases its judgement on the experience of mystics who identify God with some particular God who is a Person, Jesus Christ for example. To such mystics, the experience of Kali or Allah isn't experience of (their) God. Therefore people who experience such Gods aren't (in their view) mystics at all. This is one reason why religion has often denied the validity of other religions. Indeed, it was not too long ago Christianity considered all non-Christian groups, be they Hindu, Buddhist, Islamic, Jewish, or other, to be "of the devil." Other religions have shown similar prejudices. (Fortunately, however, ecumenicalism has recently fostered a mutual recognition of different religion's merits, so that some religions grant the validity of other religions.)

In contrast to the second school, the first bases its judgement on the experience of mystics who experience the Ultimate Ground of Existence, an Entity that is objective and universally the same. Therefore, it finds - rightly - that the experiences are "basically all alike," that they transcend religion and culture, and are all expressions of an common experience, described by the *philosophia perennis,* the "perennial philosophy."

What is the perennial philosophy? The phrase was coined by Leibniz ([H11],vii) but popularized by Aldous Huxley, who recognized common elements in the mystical writings of various eras, places, cultures, and religions, in

> . . . Vedanta and Hebrew prophecy, in the Tao The King and the Platonic dialogues, in the Gospel according to St. John and Mahayana theology, in Plotinus and the Areopagite, among the Persian Sufis and the Christian mystics of the Middle Ages and the Renaissance . . . ([S18],11-12).

Huxley collected these common elements in a book which he called *The Perennial Philosophy*. About 1955, Schrodinger wrote:

> Ten years ago Aldous Huxley published a precious volume which he called *The Perennial Philosophy* and which is an anthology from the mystics of the most various periods and the most various peoples. Open it where you will and you find many beautiful utterances of a similar kind. You are struck by the miraculous agreement between humans of different race, different religion, knowing nothing about each other's existence, separated by centuries and millennia, and by the greatest distances that there are on the globe. ([S07],139)

Huxley's *The Perennial Philosophy* [H11] is one attempt to capture in writing this world view, based on the insights of the mystics. *A Treasury of Traditional Wisdom* ([P12]) is another. It's an extensive compilation of quotations from the world's religious and philosophical traditions, grouped by topic, that records many of the mystical utterances upon which the perennial philosophy is based. These and other compilations could provide much of the raw data necessary for a scientific religion. The mystical statements they contain exhibit such a degree of agreement and consistency - exactly what we'd expect if mystical experiences are experiences of an objective, universal reality - that science could take them as its raw data and fuse them into a single system, a scientific religion.

The Data: Observability

Yet, isn't science based on observations that are open to the average person? Anyone can witness an eclipse or the outcome of a chemical experiment. But how can they see what the mystic sees?

They probably can't. Yet, this does not make the observations any less scientific.

After all, many scientific observations are not open to the average person but are only accessible to someone who has the proper equipment. As Huxley ([H11],x-xi) observes, astronomy isn't based on the observations of uneducated, naked-eye observers. A person can prove to themselves a certain galaxy exists by looking through a telescope, but without the telescope the galaxy is unobservable. Astronomy's objective, testable, and replicable - to the person with proper equipment.

Moreover, some scientific specialties demand above average intelligence, that is, part of the "equipment" they require is an above average intellect. Such sciences, therefore, aren't testable by the average person. Certainly, many a person with even an above average intellect could attend course after course and still remain unable to understand, and therefore verify, some of science's more esoteric areas - quantum mechanics or algebraic topology, for example.

Indeed, some scientific journals regularly publish papers which are understood by, perhaps, less than fifty living specialists. Yet, the fields aren't considered less than scientific, even though the papers remain unverified and unverifiable to most people. And the handful of specialists aren't considered less than scientists. For their work is verifiable - to anyone with the necessary talent, time, motivation, intelligence, and education.

Mystical experience is also open to those with the proper aptitude who are willing to do the necessary preparation. Not that after a fixed number of prayers, fasts, or meditations, a direct experience of the Eternal is guaranteed, of course. The Unconditioned certainly isn't subject to human compulsion. But if the past is any indication, the Source does appear to those who truly want to experience, and prepare for it.

Moreover, even if the Eternal Substance could not be seen by any and all who wished, this would not necessarily invalidate its objective reality. To those who are blind, the existence of certain galaxies will forever remain an article of faith. Yet, -the galaxies exist.

The Data: Replication

Even if the mystics observe an objective Reality, science can't blindly accept their declarations. It needs a way to test them. Religion uses fixed and unalterable scripture to test supposed mystical declarations. How can science test mystical observations for itself? Science tests its data by replication - others repeat the experiment, trying to obtain the same result. Will science's method work with mystical observations, even if it's granted they are observations of an objective Reality? After all, even if mystics have more or less the same vision, and experience more or less the same Entity, they view that Entity "from within." Are such experiences replicable?

Partially. In an exact science like mechanics, the behavior of a falling metal ball can be replicated at will. The ball falls the same way each time. If mystical experiences were as replicable they could be repeated, exactly, by any other mystic. They aren't. Any particular mystical experience may fail to match, even remotely, any other. How, then, is mystical experience replicable?

It's replicable in a statistical sense. Even some branches of physics, thermodynamics and quantum mechanics, for example, have phenomena that defy exact replication. In such cases, exact laws don't hold, but statistical laws do. For example, quantum mechanics can't predict when a particular uranium atom will fissure into two barium atoms. It only knows the atom's average behavior. Similarly, actuarial science can't predict exactly how much longer a particular 30 year old man will live, but it can predict how much longer the average 30 year old man will live. Quantum mechanics and actuarial science can only describe statistical behavior. Yet both are sciences.

In the same way, it would be quite sufficient if many or most mystics have similar experiences. As we've seen (and shall see again), many do. Their differences don't necessarily disprove the reality of their experiences. Indeed, would we expect someone visiting New York City to have precisely the same experience as someone else? Or have the same things to say about it, or describe it in the same way?

Hypothesis, Law, and Theory

The type of mystical observation that's objective, testable and statistically replicable, could serve as raw data for a scientific religion. Though religions disagree with each other, a scientific religion could begin with mystical observations, by mystics of any or no religion, and proceed from there - just as, if each folk culture declared all other remedies useless, pharmacology would test and verify each remedy for itself, and accept those it found effective.

What would the hypotheses of a scientific religion be like? Some might be similar or identical to the raw data, because what is a fact to a mystic is a hypothesis to the rest of us. For example, that the Eternal Light underlies all that exists is an observation, a datum, to the mystic who can "see" It. To the rest of us, it's an hypothesis, or a law if we've somehow proved it to ourselves, using modern physics perhaps. Another hypothesis is that it's possible to have direct experience of the Eternal Substance. Another, that the most intimate form of such experience transcends the triad of knower, knowing, and knowing, uniting the knower with the known; while in less intimate forms, the experiencer sees the Eternal as something different from themselves. Yet another hypothesis is that a person can so intimately unite with the Eternal that their individuality is lost.

Such hypotheses may be true or false, but they are derived directly from the statements of recognized mystics.

What would the laws of a scientific religion be like? Some scientific laws express a static relation, like Einstein's $E=mc^2$, which expresses an unchanging equivalence of energy and matter. The law that all things are a manifestation of an eternal substance expresses a similar static relationship. Other scientific laws express a cause and effect relationship, such as Newton's law that for every action there is an equal and opposite reaction. Buddha expressed a law of this type in his famous claim that desire leads to suffering.

Coming chapters contain many other hypotheses and laws: among them that 1) pleasure and pain are co-dependent and inseparable, but perception of the Eternal is beyond them both; 2) that such perception offers escape from pain and pleasure; 3) that dying to self is immortality; 4) that we are not our body, emotions or thoughts - rather, we perceive them; 5) that the Ultimate Ground of Existence is our true and enduring self; 6) that we have no true and enduring self;

7) that perception of the God which is not a Person is truer and purer than perception of any God who is a Person; 8) that relating to the Absolute as if It has a personality makes sense and, furthermore, may be how the idea of Gods who are Persons originated; 9) that component entities, entities with relative existence, and actions, all lack an absolute identity; 10) that decay is inherent in all component objects; 11) that a person may make mystical experience a goal; lastly 12) that meditation and contemplation aid the mystical quest.

Once we've established facts and laws, we can attempt to explain them; that is, we can construct theories. How are pleasure and pain co-dependent? Why does desire lead to suffering? The next chapter uses the theoretical construct of yang and yin to explore those questions. In what sense is perception of the God which is not a Person truer and purer than perception of any God who is a Person? We return to that question in the ninth chapter. Why is decay inherent in all component objects? The tenth chapter examines that question on an abstract, theoretical level. How do meditation and contemplation aid direct perception of the Real? Part III discusses that question.

Of course, a scientific religion's hypotheses, laws, and theories would always remain subject to improvement and revision. A scientific religion would have its beliefs, but it would have no dogma.

Uniqueness?

Coming chapters explore some of the hypotheses and laws we've discussed. In doing so, they demonstrate what a scientific religion might be like. This brings up two questions: 1) is what those chapters present actually a scientific religion? and 2) if so, is it the only such religion possible?

The answer to both questions is no.

The answer to the first question is "No" because the material of Part II and III isn't fully scientific. Why? Because it's the view of just one person, the author. Science is a group enterprise. Fact, hypothesis, law and theory must be tested, replicated and judged by other scientists, to be fully scientific. Though coming chapters try as far as possible to construct a scientific religion, the most they can do is serve as a starting point, a seed, a beginning. If other people accept, extend, prune and nurture that beginning, however, it may someday evolve into a scientific religion.

Or, perhaps, multiple scientific religions. This brings us to the second question, whose answer is also "No".

It's probable that two or more scientific religions could exist, at least initially. That is, two scientific religions might share the same domain, the same raw data, but nonetheless derive different "mental models" - different laws and theories. In the classic *The Structure of Scientific Revolutions*, Thomas Kuhn describes this situation in other sciences, past and present.

For example, Kuhn writes that in the early study of electricity,

> . . . a number of theories, all derived from relatively accessible phenomena, were in competition. ([K09],51).

One school of scientists tried to understand electricity ([K09],51) as a kind of fluid. Their efforts to bottle it led to the discovery of the Leyden jar. In their struggle to understand electricity, scientists developed other mental models. Eventually, Benjamin Franklin introduced a single body of law and theory - a single mental model - that explained known electrical phenomena so satisfactorily, other researchers accepted it as true. Where there had been different competing theories of electricity, now there was one.

Scientists studying electricity were fortunate to reach agreement after a relatively short time. Sometimes it takes much longer: for example, witness the long road from alchemy to chemistry. Indeed, even today consensus eludes some sciences. Kuhn notes that "it remains an open question what parts of social science" ([K09],12) have achieved the unity of a single body of theory and laws. Moreover, Kuhn offers no guarantee that the social sciences will ever accept a single theoretical framework, a unified body of hypothesis, law and theory. He observes:

> Philosophers of science have repeatedly demonstrated that more than one theoretical construction can always be placed upon a given collection of data. ([K09],63).

What does the future hold for scientific religion? If we suppose a future similar to other sciences's past, then a few obvious possibilities come to mind. First, perhaps a single, unified scientific religion will develop relatively quickly. Second, perhaps a unified scientific religion will develop only after a long time, after several competing religions find a body of law and theory they can all accept. Third, perhaps competing scientific religions will exist indefinitely, each with different laws and theories about mutually accepted data. Of course, another possibility is that no fully scientific religion will ever exist, for one reason or another.

Of these possibilities, the third is the most interesting. The first and fourth are simple and conclusive - either it happens or it doesn't. The second is a bit more complex, but just as conclusive - after, perhaps, a very lengthy time a single scientific religion emerges.

The third possibility, however, which parallels the present-day situation of the social sciences, is indeterminate. With this scenario, applying the scientific way of knowing to the religious domain doesn't yield one, single world view, one mental model. Rather a family of different world views are derived. Differing hypotheses and theories are constructed depending on which mystic's declarations are accepted, which declarations are rejected, and how accepted declarations are interpreted.

The third possibility has both positive and negative consequences. On the negative side, the existence of more than one scientific religion again seems to imply no objective Reality exists, that the mystics "observations" are, in fact, dreams or personal delusions. Yet, many people, I suppose, agree that human nature exists, that all human beings, by the very fact of being human, share certain traits. But the social sciences have yet to discover a single mental model, a single theory, that explains all these traits. Why? Because a single human nature doesn't really exist? Or because human nature is very complex?

Perhaps, a single, comprehensive theory of human nature is possible, but centuries away. After all, it took centuries for alchemy to turn into chemistry, and for chemistry to understand the nature of chemical interactions. Kuhn points out:

> History suggests that the road to a firm research
> consensus is extraordinarily arduous. ([K09],12).

But not hopeless, because

> [i]n the free community of scientific discourse,
> untrammeled by doctrinal bounds, convergence of
> opinion yet takes place. ([S03],10-11).

Moreover, if multiple scientific religions fail to reach agreement in the short term, the consequences need not be entirely negative. A positive consequence of multiple mental models follows from a surprising fact: our mental model can affect our perceptions. We'll discuss two illustrations.

In medieval Europe, the stars were thought fixed and unchangeable; this theoretical model seems to have influenced perception!

> Can it conceivably be an accident . . . that Western astronomers first saw change in the previously immutable heavens during the half-century after Copernicus' new paradigm was first proposed? The Chinese, whose cosmological beliefs did not preclude celestial change, had recorded the appearance of many new stars in the heavens at a much earlier date. ([K09],95).

Another, and more dramatic, example of how mental models can effect perception - and more - occurs in religion. Western Christianity emphasizes the suffering Christ; Eastern Christianity, the risen, transformed Christ and the Uncreated Light. Vladimir Lossky, who claims "spirituality and dogma, mysticism and theology, are inseparably linked" ([L08],14), makes the following observation.

> No saint of the Eastern Church has ever borne the stigmata, those outward marks which have made certain great Western saints and mystics as it were living patterns of the suffering Christ. But, by contrast, Eastern saints have very frequently been transfigured by the inward light of uncreated grace, and have appeared resplendent, like Christ on the mount of Transfiguration. ([L08],243).

Multiple models, it seems, foster a wider range of observation and even experience. Although these many insights probably make the victory of one, single model much more difficult, any theoretical model that does unite such a diverse body of observations will probably be much closer to the truth than if it had a lesser number of insights to explain. The final result might well be a more comprehensive picture of truth.

To conclude, science, which investigates so many other objective, statistically replicable phenomena, could investigate mystical experience, too. It could develop a religious world view by subjecting the declarations of the mystics to the test of the scientific way of knowing. Such a scientific religion would be descriptive and explanatory. It would describe and explain the visions and statements of the healthy mystics, past and present. It would develop theories to explain their experiences. (One such theory, a world view derived from mystical visions, is presented in subsequent chapters.)

A scientific religion would be descriptive and explanatory. Could it be experimental, too? For non-mystics, as for those who study continental drift or supernova, there would be no possibility of direct experimentation. But for those who decide to undertake the mystic quest, a scientific religion would also be an experimental science. The effect of practices and beliefs on experience would constitute experiments.

Could a scientific religion ever, in any sense, be predictive or exact? We'll return to this question much later.

Fundamental Premises
Now that we've now completed the first and most fundamental part of this book, let's review our basic premises. They are:

- the scientific way of knowing is superior to the
revelational way:
- the God which is not a Person is the basis of all that
exists;
- both science and religion seek to know the God
which is not a Person; both scientist and mystic seek to
know the single unity behind seeming diversity, the
difference being the scientist seeks to know mentally
while the mystic seeks to know by experience;
- some mystics have had direct, first-hand knowledge
of the God which is not a Person;
- the experience and statements of such mystics exhibit
a significant degree of agreement, and may be
considered varying expressions of a single, underlying
world view, called the "perennial philosophy;"
- the perennial philosophy could supply the initial laws
and theories for a scientific religion;
- a scientific religion would use the scientific way of
knowing to evaluate the declarations of past and
present mystics;

- a scientific religion would not necessarily be unique,
more than one scientific religion might be possible;
- the laws and theories of scientific religions would
always remain subject to revision and improvement; a
scientific religion would have no dogma, no truths
above human understanding.

Testing the declarations of the mystics with the scientific way of knowing would yield comprehensive, integrated world views. Extending science's domain to include part of what is now the domain of religion would yield truths satisfying the demands of both science and religion. The resultant scientific-mystical discipline would truly be both science and religion, a scientific religion that satisfies the opening quotation of this book. Moreover, the incorporation of a scientific religion into science would transform our present-day agnostic science into a science which includes religion. We would no longer have separate scientific and religious world views, but integrated world views, fully scientific and fully religious. Science would offer an explanation of not only the behavior of electric current in a circuit, but of our place in the universe as well.

Looking Back, Looking Ahead

We've completed the most basic material, but an important question remains: what would a scientific religion look like? The remaining chapters attempt to give one answer to this question. Based on the declarations of acknowledged mystics, a particular world view is presented. A necessarily personal view of one possible scientific religion is described.

To someone who is not a mystic much of the previous material is speculative. The Ultimate Ground of Existence and Gods who are Persons aren't actual, vivid experiences, but beliefs, ideas, theoretical constructs. What we actually experience may be divided into two separate realms, an exterior world of people and things, and an interior world of thoughts, feelings, and bodily sensations.

We look out upon these two worlds but do not see down to their deepest level. The Ultimate Ground of Existence - if It truly is ultimate - must underlie the universe of people and objects, as well as the deepest level of our psyche. The mystics have had much to say

171

about the relationship of the self-existent, eternal Basis to our two worlds.

Therefore, the following chapter describes the relationship mystics see between the Eternal Substance and the outer world, the universe. The subsequent chapter describes the relationship they see between the Real and our inner world, ourselves. Then, the relationship between the Unformed and Gods who are Persons is explored. Lastly, some ideas which apply to all three - the universe, ourselves, and Gods who are Persons - are examined.

Also, up to now we've been content mostly with description. Now, in addition to describing mystical observations and experiences, we'll construct a theoretical framework within which these observations and experiences are understood.

The Tale of the Scientific Alchemists and Religious Newtonians

We'll close Part I with a short allegory. About three hundred years ago, when alchemists were still vainly trying to turn lead into gold, Sir Isaac Newton discovered some fundamental equations that accurately describe the physical world. In our tale, let's give Newton's discoveries to people who regard them religiously. We'll call these people "religious Newtonians." The religious Newtonians are religious because they follow the revelational method of knowing truth. They're Newtonians because they accept Newton's theories.

And let's give Alchemy to people who regard it scientifically. We'll call these people "scientific Alchemists." The scientific Alchemists are scientific because they follow the scientific way of knowing truth; they are Alchemists because they accept the theories of Alchemy.

So "religious" or "scientific" indicates the way of knowing, the way of finding new knowledge. And "Newtonian" or "Alchemist" indicates the theories currently accepted as true. Our tale will illustrate that the method used to find and test beliefs may be more important than the initial beliefs themselves.

Our tale opens in the seventeenth century. The religious Newtonians believe in calculus and the basic laws of Newtonian physics. They worship Newton as a god and venerate his writings as divinely inspired and perfectly true. Following the ideas and theories in his writings, in "holy scriptures," the religious Newtonians are beginning to understand the natural world. New discoveries in mathematics, mechanics, astronomy, and navigation are being made almost daily.

The beliefs of the religious Newtonians are substantially correct and many centuries of progress await them.

Our other group, the scientific Alchemists, follow not Newton but Aristotle, particularly his theory of the four basic elements: earth, water, fire, and air. According to Aristotle's ideas, it's possible to turn lead into gold. And that's what the scientific Alchemists are trying to do. Into their crucibles, flasks, mortars, and pots, they put eggs, toads, snakes, herbs, urine, entrails, lead, mercury, sulfur, and saltpeter. They grind, mix, filter, hammer, and heat them. They describe their experiments with bizarre symbols such as toads, dragons, birds, stars, crowns, keys, and planets.

The beliefs of the scientific Alchemists are wrong and their quest is doomed to failure.

Notice that we've given the religious Newtonians a lot of correct physical knowledge. We've given them an kind of head start in the race toward more and more truth about the physical world. But we've given them a poor way of knowing, a way that binds them to a "divine and unchanging" truth.

In contrast, we've given the scientific Alchemists a serious handicap in the form of erroneous physical theories. But we've given them a better way of knowing, a way that allows revision and progress. Which will prove more important in the long run, the knowledge currently accepted as true, or the method of testing current knowledge and discovering more knowledge? Let's return to our tale.

As time passes, the scientific Alchemists slowly and independently discover some laws of nature that the religious Newtonians believe to be divine and unchangeable truth.

"You've found," say the religious Newtonians, "but a tiny portion of our divine Dogma. Surely, your mortal, imperfect minds will never uncover all of our complete and perfect truth. God gave us our revelation. It's far beyond what we fallible humans can find, alone and unaided. Why then do you not give up your slow, painful search for truth and embrace our Truth?"

"Never," reply the scientific Alchemists. "Truth is to be earned, to be understood. You are satisfied to follow blindly, without understanding. We are not. Even though some of our truths now match your faith, one day we may find other truths of which you are ignorant.

As the decades pass, the scientific Alchemists independently uncover, test, and accept more and more of the truths held by the religious Newtonians.

"For many decades now," say the religious Newtonians, "our sacred scriptures have held the full and complete truth. Ignoring these writings, you have been winning, bit by bit, through much labor and suffering, what was already fully given to the fathers of the fathers of our fathers. Our way to truth, the way of divine revelation, the way of our fathers, is ancient and sure. Why then do you not cease your needless searching and accept out divine revelation?"

"Never," reply the scientific Alchemists. "No book can hold the full and perfect truth. Our way of knowing is a never-ending process of observation, hypothesis, theory, and experiment. Even as knowledge is limitless, the search for knowledge must be unending. This is our way of knowing. One day our knowledge shall surpass yours."

By the end of the nineteenth century, the scientific Alchemists have independently found and verified all the beliefs of the religious Newtonians.

"For centuries now," say the religious Newtonians, "you have groped in the dark while we, following the divine knowledge given in our holy scriptures by our god, have lived in the light. Now, after much error and effort, you have finally reached the Truth. Will you not now admit the inspired nature of our religion and join us in our worship?"

"Never," respond the scientific Alchemists. "Your way of blind acceptance is not our way. We are pledged to follow the truth; you to follow your holy books and god. We are free to go wherever the truth leads; you are bound to a fixed, limited knowledge now hundreds of years old. One day we shall go beyond your knowledge."

So for centuries, the *religious* Newtonians have gone nowhere, they've stayed bound to their "holy and eternal" truth. But the *scientific* Alchemists have outgrown their initial "knowledge" and have acquired - earned - a truer, more accurate knowledge. One way of knowing has led nowhere, the other has discovered more and more knowledge.

In the early twentieth century, a thinker named Einstein claims the theories now accepted by both religious Newtonians and scientific Alchemists are not actually true, but only a near approximation of the truth. He proposes radically different theories, superior only in that they explain the orbit of the planet Mercury a bit better. The new theories demand, however, a drastic, new view of space and time.

"Blasphemy!" shout the religious Newtonians. "Heretical, perverse, mind-twisting ideas of an iconoclastic rebel. Surely our Holy Faith, the faith of our fathers, will prevail against such diseased drivel!"

"It seems to be the truth!" reply the scientific Alchemists. "We shall test it and, if true, we shall accept it. We are long accustomed to molding ourselves to the truth, not molding the truth to ourselves."

Twenty years later, the two camps welcome the theory of Quantum Mechanics in much the same manner. The religious Newtonians reject Quantum Mechanics as heretical nonsense; the scientific Alchemists test and then accept it. Using the Theory of Relativity and, more significantly, Quantum Mechanics, the scientific Alchemists begin to surpass the religious Newtonians in their understanding and control of the physical world. Using Quantum Mechanics they discover atomic energy, semiconductors, lasers, and computers. The religious Newtonians, bound as they are to a way of knowing that limits what they can know, refuse to accept or use the new discoveries. The world beyond their holy scriptures, the world of computers, lasers, nuclear energy, and space-time, is a world which they, as believers, can never enter.

Our tale attempts to dramatize that a *way* of knowing can be more important than initial beliefs. The scientific Alchemists were given a lot of erroneous beliefs based on Alchemy. But they were given the scientific way of knowing. Since their method of acquiring and testing knowledge was sound, they eventually corrected their initial misconceptions. The religious Newtonians, on the other hand, were given a lot of accurate physical knowledge based on Newtonian physics. But they were given a religious way of knowing. Since their method of acquiring and testing knowledge was faulty, eventually their beliefs became outmoded, a hindrance to finding more truth.

So even if scripture is eternal and inerrant truth (and this is debatable), the religious way of knowing hinders the search for more truth. And even if science's ideas are all wrong (this, too, is debatable), its way of knowing leads to more and more truth.

Our tale compared the scientific way of knowing, the way of knowing accepted by science, with the revelational way of knowing, the way of knowing often accepted by religion. It showed the scientific way of knowing the superior method - at least, for understanding the natural world. Is it a better way for understanding the "supernatural world" too?

Let's see.

Part II: World View

7

- The Universe -

Chapter Summary: This chapter discusses the relation of the Ultimate Ground to the objects and events we experience, and explores various dualistic descriptions of the universe. The problems of evil, suffering and pain are also discussed.

Christianity recognizes three types of hindrances to salvation - the world, the flesh, and the devil. We'll discuss salvation and similar concepts later. For now our concern is the three domains the hindrances imply: the external natural world, the internal natural world, and the supernatural world. Two ideas dominate Part I: ways of knowing and the Ultimate Ground of Existence. In Part II - in this and the next three chapters - we'll investigate the relation of the Real with the external natural world, the internal natural world, and the supernatural world.

How is the Perfect related to the world we see around us? The Ultimate Substance is unchanging, eternal, and perfect. Yet the world we experience is changing, transitory, and often imperfect. If entities we experience really have their existence grounded in an eternal, perfect, and unconditioned Reality, how can they themselves be impermanent, imperfect, and conditioned? How is impermanence grounded in Permanence? How is imperfection grounded in Perfection? How is the conditioned grounded in the Unconditioned? And how does this world have its root and source in the Divine Ground? What relation exists between the temporal and the Eternal, the imperfect and the Perfect, the universe and the Ultimate Ground of Existence? This chapter addresses those questions.

Since there are people for whom the universe's Self-Existent and Eternal Basis is not a doctrine but an observation, a self-evident fact, we'll begin with their testimony.

Modes of Light

As we've seen, some mystics transcend the triad of knower, knowing and known, to achieve unitive knowledge of the Real. For them, the vision of pure Isness is so absorbing they lose consciousness of the universe. Some never return from that vision; after a few days their body dies. For those that do return, the unitive vision begins to fade. Perception of thoughts, emotions, objects and events slowly returns. During this transition period - especially if the Eternal has been recognized as the universe's Ground of Existence - the mystic sees the Real shining through the apparent, seeing

> . . . the Perfect One self-revealed in the Many.
> ([U01],254).

Simultaneously beholding the Eternal and the perishable, the One and the many, the Unconditioned and the conditioned, the mystic sees the second as a manifestation, an emanation, of the First.

Not all mystics reach such heights, of course. Many remain conscious of the universe, but a universe transformed, suffused with God, the Uncreated Light. Each and every entity, from the worm to the cathedral, shines with a transcendental splendor. As Evelyn Underhill writes,

> [I]lluminated vision in which "all things are made new" can afford to embrace the homeliest and well as the sublimest things; and, as a matter of experience, it does do this, seeing all objects . . . as "modes of light."
> ([U01],262-3).

All things are seen to be identical in that they are all modes of the same Light, all bundles of a single Energy.

Probably because of such visions, many mystics declare that the universe's entities are all grounded in a single Uncreated Light. Let's examine a few of their statements.

The Hindu sage Ramana Maharshi declares

> . . . the world is a phenomenon upon the substratum of the single Reality . . . ([T03],17).

And Ramakrishna expresses Vedanta philosophy, as well as his own direct insights, when he says:

> He who is realized as God has also become the universe and its living beings. One who knows the Truth knows that it is He alone who has become father and mother, child and neighbour, man and animal,

good and bad, holy and unholy, and so forth.
([G03],328).

Ramakrishna echoes the ancient sage who found

God alone . . . has become the universe and all its
living beings. ([G03],648).

Turning to Buddhism we find that "Mind" signifies ([A15],13)
"the Absolute as it is expressed in the temporal order" and that

. . . "the essential nature of the Mind" is unborn and is
imperishable. . . . [T]herefore all things . . . are, in the
final analysis, undifferentiated, free from alteration,
and indestructible. ([A15],32-3).

In Islam, Shaikh Abd al-Razzaq Jhanjhana believes:

Everything is manifest because of the Light of God. It
is not possible for this world to exist without the
Presence of God. All is He. ([R04],305).

And other Sufis, says Nicholson,

. . . conceive the universe as a projected and reflected
image of God. ([N11],96).

He writes:

The unique Substance, viewed as absolute and void
of all phenomena, all limitations and all multiplicity, is
the Real . . . On the other hand, viewed in His aspect
of multiplicity and plurality, under which He displays
Himself when clothed with phenomena, He is the
whole created universe. Therefore the universe is the
outward visible expression of the Real, and the Real is
the inner unseen reality of the universe. ([N11],81-2).

Jewish Hasidism also sees the universe as the clothing, veil, and
external form of the Eternal. Alan Unterman, in the introduction of his
The Wisdom of the Jewish Mystics, writes:

Reality is the clothing of the Godhead. Behind the
solidity of the inorganic world and the living, breathing,
organic world is the Godhead itself towards whom the
mystic breaks through by penetrating the everyday
thought forms and perceptions which only tell us about
the clothing, not about that which is clothed.
([W08],24-5).

He continues:

This identification of the world with God, albeit in the
sense that it is a masking of the divine, developed a
sense of the holiness of the profane amongst

> Hasidism and greatly disturbed some of their
> opponents. ([W08],25).

Perhaps because of this belief (that the world is a manifestation of
God), the Jewish Hasidic mystic

> . . . never doubted that his separation from God was
> illusory, nor that his role in life consisted of stripping
> away the illusion. ([E04],121).

In Christianity Meister Eckhart, in the words of Rufus Jones,
declares that

> . . . the temporal world is a show or reflection, but a
> reflection of an eternal reality. ([J03],229).

And "Dionysius" also believes the world is an appearance of the Real.
He teaches, according to C. E. Rolt, that the Godhead

> . . . in Its ultimate Nature . . . is beyond all
> differentiations and relationships, and dwells in a
> region where there is nothing outside of Itself, yet on
> another side of Its Nature (so to speak) touches and
> embraces a region of differentiations and relationships
> . . . ([D08],6).

The region of differentiations and relationships is, of course, the
universe. Rolt continues that the Godhead

> . . . is therefore Itself related to that region, and so in a
> sense belongs to it. Ultimately the Godhead is
> undifferentiated and unrelated, but in its eternal
> created activity It is manifested under the form of
> Differentiation and Relationship. It belongs
> concurrently to two worlds: that of Ultimate Reality and
> that of Manifested Appearance. ([D08],6-7).

He goes on to say that because our minds are part of this realm of
"Manifested Appearance" there is

> . . . the possibility not only of Creation but also
> Revelation . . . . Just as the Godhead
> creates all things by virtue of that Aspect of Its Nature
> which is (as it were) turned towards them, so It is
> revealed to us by virtue of the same Aspect turned
> towards our minds which form part of the creation.
> ([D08],7).

Therefore, knowledge of the Real is knowledge of our own inner Self.
And vice versa. The next chapter explores our true and enduring self.

The Kingdom of Heaven
When vision of the One is absorbing, only the Real is perceived.
When it's less absorbing, but still very strong, other things (thoughts,
feelings, people, objects) are seen but seem shadowy, unsubstantial,
unreal, because their Isness is foreground while their particular
qualities are background. When vision is less absorbing still, things
seem solid and real but are still quite plainly manifestations of the
Real.

To see things in this way, as manifestations of the Real, grounded
in the Eternal, is to see things as they are in the Kingdom of Heaven,
according to the English mystic William Law.

> Everything in temporal nature is descended out of that
> which is eternal, and stands as a palpable visible
> outbirth of it . . . In Eternal Nature, or the Kingdom of
> Heaven, materiality stands in life and light; it is the
> light's glorious Body, or that garment wherewith light is
> clothed . . . (quoted in [U01],263).

If Law is correct, then (as Jesus claims) the Kingdom of Heaven is
indeed among us - if we have the eyes to see.

The Kingdom of Heaven - the world transfigured, where all things
are made new and sparkle with the Uncreated Light - is also the world
of Eden. From a Christian hymn:

> Mine is the sunlight! Mine is the morning
> Born of the one light, Eden saw play! ([H12],389).

It's the world Adam saw before the fall, and anyone who sees the
world in this way sees the world that Adam saw. For example, George
Fox, Quakerism's founder, had an experience where he entered the
Kingdom of Heaven, the

> . . . paradise of God. All things were new, and all the
> creation gave another smell unto me than before,
> beyond what words can utter. I knew nothing but
> pureness, and innocency, and righteousness, being
> renewed up into the image of God by Christ Jesus, so
> that I say I was come up to the state of Adam which
> he was in before he fell. (ch.2,[J05],27).

The vision of the Kingdom of Heaven, of Eden, of the Eternal
shining through the transitory, is perhaps the most common mystic

vision. It occurs, at least in its weaker forms, to many people at one
time or another.

> To "see God in nature," to attain a radiant
> consciousness of the "otherness" of natural things, is
> the simplest and commonest form of illumination. Most
> people, under the spell of emotion or of beauty, have
> known flashes of rudimentary vision of this kind.
> ([U01],234).

Indeed, even supposedly non-mystical experiences of beauty, truth, or
love may be dim, unrecognized visions of Eternal Light. As the
mystic Plotinus claims:

> Beauty is the translucence, through the material
> phenomenon, of the eternal splendor of the 'one'.
> ([Q01],67).

Look deeply into beauty, truth, or love, and you may see the Absolute
behind them; it may be literally true that beauty is God, truth is God,
and love is God. Writes Rabbi Abraham Isaac Kook:

> Love in its most luminous aspect has its being beyond
> the world, in the divine realm, where there are no
> contradictions, limits and opposition; only bliss and
> good . . . ([K05],227).

Dualiam: The Fall from Eden

We are tracing a path down, from absorbed vision of Isness to the
universe we know. (It is the same path that consciousness must
traverse - in the opposite direction, of course - to reach its final end,
union with the Eternal.) When someone falls from absorbed
perception of the Eternal, and falls from mixed vision of a universe
suffused with Uncreated Light shining everywhere, they fall into
perception of the universe we know.

Some religions describe this fall in the language of myth. For
example, the biblical story of Eden may be interpreted as an allegory
of such a fall. In Eden,

> [i]n the middle of the garden stood the tree that gives
> life and the tree that gives knowledge of what is good
> and what is bad. (Gn2:9, [G02],2).

What is the "tree of life"? In our interpretation it's the Kingdom of
Heaven, a way of seeing the world where the Isness of things is their
most obvious quality. And what is the "tree of knowledge of good and

bad"? It's seeing the world in the usual way, seeing things in terms of good and bad. Such a way of seeing is call "dualism".

Good and bad, however, isn't the only dualistic system. Others are mind and matter, form and substance, and yang and yin. But in any dualistic system, vision of the One has been mostly lost and dualistic perception - vision of the "two", of "pairs of opposites" - has arisen.

Adam and Eve were forbidden to eat of the tree of knowledge of good and bad.

> You may eat the fruit of any tree in the garden, except
> the tree that gives knowledge of what is good and
> what is bad. (Gn2:16, [G02],2).

That is, they were forbidden to see dualistically, to view the world in terms of good and bad. Why?

When we see things in a dualistic way, in terms of good and evil, for example, by that very fact, we lose perception of the One. Adam and Eve's descent into dualistic vision was itself sufficient to rob them of vision of the One, of the Kingdom of Heaven, of Eden

By the way, religions don't invariably use myth to describe the fall into dualism. For example, a Buddhist scripture is quite direct.

> In due course of evolution sentiency appeared and
> sense-perception arose . . . And the world split in
> twain: there were pleasures and pains, self and
> notself, friends and foes, hatred and love. ([C04],255).

A Christian spiritual treatise also has a direct description of the fall into duality, which it calls the fall into "flesh."

> [I]n our downfall, we fell away from God . . . and fell
> into the flesh; thereby we went outside ourselves and
> began to seek for joys and comforts there. Our senses
> became our guides and intermediaries in this. Through
> them the soul goes outside and tastes the things
> experienced by each sense. ([U02],128).

But why call it a descent into "flesh"? Probably because the body, the "flesh," is the instrument of sense perception. When consciousness is aware of the body, it's usually aware of pairs of opposites - pleasant/unpleasant, hot/cold, etc.

Whether stated directly or in myth, it's true that turning from seeing the One as the One, to seeing It as two - and therefore to seeing the created order as real - is the first and original separation from God. Therefore, it's the first and "Original Sin." Our very existence as separate, individual persons is based on separation from the Source and Root. We are separated from God by a "veil of createdness" ([S04],143). Of this, a modern follower of Vedanta writes:

> The belief, then, in existence apart from God is the major sin; ultimately, it is the only sin, error or misconception. ([V01],351),

while a Sufi writes:

> When I said: What have I sinned? she answered: Your existence is a sin with which no sin can be compared. ([S04],142).

Most of the time, we perceive duality, the many, the pairs of opposites. Single, unitary awareness, mystical awareness, of pure Isness, of the Ultimate Ground of Existence, if it exists at all, lies deep in our mental background. We live in a universe of objects, people, feelings and thoughts - and fail to see their Ultimate Ground of Existence.

Illusory Dualism

Not all dualisms are identical; there are various types. We've already seen one way dualistic systems can differ: one dualism is based on good and evil while another, on mind and matter. Dualistic systems can differ, however, even when they accept the same pair of opposites. In the next few sections, we'll examine a few different dualistic systems, all based on good and evil. We'll see that dualisms based on good and evil have differing "strengths" depending on how much reality is granted to evil.

In the weakest kind of dualism, illusory dualism, good and evil are both considered in some sense illusory and ultimately unreal. Vedanta expresses illusory dualism when it teaches good and evil are appearances, not Reality. Shankara, for example, writes:

> The world appears as if real only so long as *Brahman* which is the non-dual substrate of all has not been known . . . ([S09],13).

So, from the viewpoint of vision of Brahman,

> . . . there is neither good nor evil, neither pleasure nor pain. ([V01],110)

because they exist in the world, not beyond it. Similarly, the Sufi poet Attar writes:

> So long as you are separate, good and evil will arise in
> you, but when you lose yourself in the sun of the
> divine essence they will be transcended by love.
> ([A12],116).

It might seem someone who believes good and evil are illusions
would have no basis for morals and ethics, no way to judge right and
wrong. Not true, as we'll see later.

An idea that can follow from illusory dualism is that the world is,
in some sense, a show. As we saw above, when the world is
transfigured, suffused with the God which is not a Person that is its
Ground, it may appear to be a "show world," a world created, literally,
as an image of Eternal Light. Illusory dualism says the images are
ultimately unreal, a play of Light. It says only the Light really exists.
"Maya" is a Hindu term which refers to the magical illusion, the
show, the universe created by the dance of Energy. Ramana Maharshi
said maya

> . . . makes one take what is ever present and all
> pervasive, full to perfection and self-luminous and is
> indeed the Self and the core of one's Being, for non-
> existent and unreal. ([T03],18).

Conversely, it makes one mistake

> . . . for real and self-existent what is non-existent and
> unreal, namely the trilogy of world, ego and God.
> ([T03],18).

Of course, the universe, the show, and Maya all feel quite real.
Therefore even though good and evil exist only within Maya, they
feel real too.

> So long . . . as we are experiencing pleasure and pain,
> so long do both good and evil exist as empirically real.
> . . . Vedanta thus recognizes both good and evil, and
> pleasure and pain, as . . . facts of . . . our empirical
> lives . . . ([V01],110-1).

Apparent Dualism

Vedanta sees God (the God which is not a Person) as beyond good
and evil, which exist only in the show world. Other religions identify
God (usually a God who is a Person) as Goodness in its highest
degree. For these religions good really exists since God really exists.
They differ, however, in the amount of reality they grant evil.

Some religions teach apparent dualism, which says good really exists but evil only seems to exists. For example, Mary Baker Eddy, the founder of Christian Science, writes that

> . . . evil is but an illusion, and it has no real basis. Evil
> is a false belief. . . . If sin, sickness, and death were
> understood as nothingness, they would disappear.
> ([E02],480).

According to Christian Science, then, evil is a mere appearance, an illusion having no reality - an idea similar to illusory dualism in that both agree evil is an illusion, but differing about whether good really exists. Christian Science teaches that the universe is actually good, and any evil we think we see doesn't actually exist.

Augustine has a similar teaching; he says evil has no actual existence but is merely an absence of good, just as darkness has no actual existence itself but is merely an absence of light.

> . . . [E]vil has no positive nature; but the loss of good
> has received the name "evil." ([A13],354).

Therefore it follows, as William Law observes,

> [e]vil can no more be charged upon God than
> darkness can be charged upon the sun . . .
> ([P12],476).

This assumes, however, that God, like the sun, is not omnipresent, that there are places where God is not.

"Dionysius" holds good to be omnipresent, and explains evil as a deficiency of good rather than a total lack. By analogy, heat, which is known to be molecular activity, is always present in some measure or other. Cold is a deficiency of heat, rather than a complete absence. This answer grants God omnipresence but assumes God is sometimes present only weakly.

Dependent Dualism and the Problem of Evil

Vedanta and Christian Science see evil as ultimately an illusion. Augustine and others grant evil existence only as an absence or deficiency of good. Other systems grant evil full and real existence, differing on whether evil is less powerful than good or equally powerful.

The kind of dualism commonly accepted by religion is a dependent dualism of good and evil. In such a system, good is fully real and exists independently (because God is fully real and self-existent). Evil is real, too, but it's weaker than good, and depends upon good for its existence. Typically, an all-good God who is a Person is also all-powerful, and therefore could destroy evil - but chooses not to.

Why? One explanation is destroying evil would violate the laws of the universe, or destroy our free will. Harold Kushner argues this in his popular book *When Bad Things Happen to Good People*.

> . . . God has set Himself the limit that He will not
> intervene to take away our freedom, including our
> freedom to hurt ourselves and other around us. He
> has already let Man evolve morally free, and there is
> no turning back the evolutionary clock. ([K11],81).

Whatever the explanation, if God permits evil to exist, then evil depends upon God for its continued existence, because if God didn't allow it to exist, it could not.

To many people, dependent dualism seems to suggest God really isn't all-good. For can a God who permits evil - starvation, disease, rape, murder, incest, war, etc. - really be called "all-good"? Of course, it might be argued that God is all-good, but not all-knowing (and so doesn't know evil exists) or not all-powerful (and therefore can't stop it). Many religions, however, say God is all three, which brings us to the problem of evil.

Suppose that God is in fact all-knowing, all-powerful, and all-good. Suppose, also, that evil exists. An all-knowing God is aware of any evil about to occur. An all-powerful God is capable of preventing that evil, destroying the evil-doer if necessary. An all-good God has no love of evil, no wish it should exist. So if evil does exist, then it would seem God is not all-knowing, all-powerful, or all-good. It seems God can't be all three - a logical dilemma for any religion that teaches the contrary.

Such religions usually "solve" the dilemma by labeling it "the mystery of evil" or "the problem of evil." To illustrate, a Catholic theologian first acknowledges Augustine's solution to the problem of evil: he writes that evil

> . . . is to be found in what God created. Yet, because it
> is essentially a lack of goodness, it cannot claim God
> for its direct author; for He made what is good, not the
> lack thereof. ([P15],77).

But he adds:

> But evil remains a problem for us, even after we have
> given assent to this piece of Christian philosophy. God
> is almighty, and therefore we ask: since He has full
> control of the whole created universe, why doesn't He
> eliminate evil? The question is easy to ask; it is
> complicated to try to answer. . . . We are left, then,
> with the mystery of evil. It is here in the world, and
> somehow God is willing that it be here. He is not its
> direct cause, but He does not exclude it from His
> world, even though He is almighty. ([P15],77).

The Problem of the Origin of Evil

The continued existence of evil is not the entire mystery and problem;
its origin is also a difficulty. How could evil originate in the creation
of an all-good God? A common explanation is creatures exercise their
free will and make the wrong choices. From a Seventh-day Adventist
publication:

> Mystery of mysteries, the conflict between good and
> evil began in heaven. How could sin possibly originate
> in a perfect environment? . . . Although sin's rise is
> inexplicable and unjustifiable, its roots can be traced
> to Lucifer's pride . . . ([S10],99).

Christianity attributes evil to the misuse of free will. Lucifer was
the first to misuse his free will by a prideful rebelling against God.
Therefore, Christianity associates Lucifer's rebellion and his
successful tempting of Adam and Eve with the origin of evil. (A
question: Is another rebellion possible in heaven even today?
Presumably, the saints and angels still enjoy free will. Will some of
them try another revolt? If free will still exists in heaven, doesn't the
possibility of evil still exist there, too?)

But explaining the existence of evil as the result of free will puts
the cart before the horse. If only good existed, then free will would
mean the freedom to choose between one good or another. Doesn't
evil, or at least the potential for evil, have to already exist before it
can be chosen? There's an anecdote about Henry Ford, whose early
cars were always painted black. One day someone asked if he'd
consider giving the buying public a choice of car color. Henry replied,
"They can have any color they want as long as it's black." Obviously
cars must come in at least two colors before real choice is possible.

Similarly, the nature of the universe must allow evil before evil can be chosen. Sadly, as daily newscasts constantly illustrate, our universe allows the possibility and actual existence of all sorts of evil. The choice to do evil is ours, but who first created evil so choice would be possible?

> Oh Thou who didst with Pitfall and with Gin
> Beset the Road I was to wander in . . .
> Oh Thou, who Man of baser Earth didst make,
> And who with Eden didst devise the Snake . . .
> ([R05],123-4).

And why was the world furnished with so *many* opportunities? How could an all-good God, personal or not, create creatures such as ourselves, a world such as ours, and even an angel, Lucifer, all having such capacity for evil?

Independent Dualism

The axioms of an all-knowing, all-powerful, and all-good God seems to prohibit the real and full existence of evil. Therefore, thinkers such as Mary Baker Eddy, Augustine, and Dionysius teach evil doesn't really and fully exist. If they're wrong and evil does exist, and if God really exists, then God isn't all-good, or doesn't know evil exists, or isn't powerful enough to stop it. Independent dualism chooses the last alternative. Evil exists because God isn't strong enough to eliminate it - at least, at present. Some independent dualisms teach the ultimate triumph of good over evil.

Independent dualism is the strongest kind of dualism. In it, evil really exists, independently of good. Independent dualist systems were influential in the past. If fact, good and evil, were originally part of the ancient vision of Zoroaster. Later Zoroastrianism, if not Zoroaster himself, taught good and evil exist independently of each other, though it did predict good would eventually vanquish evil. Other independent dualistic systems include Marcionism, which flourished in the second century C.E.; Manichaeism, which survived for over 1,000 years; and that of the Cathari, who arose in Western Europe in the 12th and 13th centuries C.E. and were probably influenced by Manichaeism. To better understand independent dualist systems, we'll discuss one of these systems in more detail.

Mani, for whom Manichaeism is named, lived in the 3rd century C.E. Like Mohammed over two hundred years later, he accepted earlier prophets such as Zoroaster, Buddha and Jesus. But he believed written records of their teachings were distorted since they themselves hadn't written them. In contrast, Mani himself recorded his teachings, thus assuring their integrity over time. Like Mohammed, he was regarded ([N04],v14,783) as the "seal of the prophets". That is, since the pure and perfect teaching had finally been captured in writing, no other prophet would be needed.

In Manichaeism belief, the kingdom of Light and Spirit and the kingdom of Darkness and Matter had originally been separate and independent. But at the border, Darkness and Light mix, creating the world we live in. Overpowered by Matter, Light forgets its own nature. Consciousness becomes aware of matter and forgets Itself. For salvation - that is, to free the soul, a particle of Light, from the domination of Matter - a person must practice strict asceticism. Only then can the soul finally return to its original home, the kingdom of Light.

Manichaeism disappeared centuries ago. Yet, similar ideas exist even today. Kushner seems to embrace an independent dualism of God and

> . . . chaos, in those corners of the universe where God's creative light has not yet penetrated. And chaos is evil; not wrong, not malevolent, but evil nonetheless . . . ([K11],53).

He writes:

> . . . [T]he earthquake and the accident . . . are not the will of God, but represent that aspect of reality which stands independent of His will, and which angers and saddens God even as it angers and saddens us. ([K11],55),

and:

> God does not want you to be sick or crippled. He didn't make you have this problem, and He doesn't want you to go on having it, but He can't make it go away. That is something which is too hard even for God. ([K11],129).

Though independent dualism still exists, it's rarely accepted, especially by established religions. It's easy to see why: in an independent dualistic system the following three statements are true. (1) The bubonic plague, which ravaged Europe centuries ago, was too hard for God to stop then, but could be stopped today with modern drugs. (2) God cannot prevent a man from abusing a woman or child, but a policeman with a gun can. (3) A man, a woman and their three children are asleep in their home. Somehow a small spark has ignited the curtains in the living room; the fire is beginning to spread. The family is all sound asleep and will die in the fire if not awakened soon. God cannot awaken them to their danger. But a pet dog, if they had one, could.

Even though they aren't widely believed, independent dualistic systems do have one advantage over other systems: the continued existence of evil is explained (even though the origin of evil may remain unexplained). Evil exists simply because good is not powerful enough to destroy it. In theistic terms, good and evil exist simply because good and evil exist in God. Or if God is all-good, then good and evil exist because God is not always powerful enough to destroy or hinder Evil.

Recognizing Evil

Perhaps only philosophers and theologians worry about the origin and continued existence of evil. After all, evil exists - either independently, dependently, apparently, or as an illusion. A practical person intent on avoiding evil doesn't need to know how evil originated or why it continues to exist. They cannot, however, ignore a much more vital question: "How can evil be recognized as evil?"

The question is not easily answered because not everything pleasant is good, and not everything unpleasant is evil. A child might think a bad-tasting but life-saving medicine is evil, and a sweet-tasting poison, good. Evil is usually defined as something contrary to the will of God, but how are we to know the will of God? Innumerable wars have been fought where both sides' religious leaders decided their country was doing the will of God, and the other wasn't.

For example, in the Second World War, Italian Catholics killed and were killed by Italian-American Catholics. And German Lutherans killed and were killed by German-American Lutherans. Christianity allows participation in a war only if the war is "just." But could the war have been just for both sides? Or did some churches misjudge evil as good? If religious leaders can't always accurately identify evil, what chance has the common person? (These questions are, perhaps, naive. Over the past fifteen hundred years Christian churches have routinely given *carte blanche* to their nation's leaders for war. I can remember reading of only one war that a major Christian church declared unjust. The Catholics forbidden to fight were attacking the Vatican.)

Islam has a tale that illustrates how difficult it can be telling good from evil. The Koran (Sura 18:66,83) tells a story of Moses and an angel. Moses wants to accompany the angel. The angel agrees if Moses promises not question his actions. First, the angel bores a hole in the bottom of an unattended ship. Then he slays a youth. And finally he rebuilds the crumbling wall of those who refused Moses and himself food. Troubled by the destruction of property, the murder, and the kindness to those who turned them away, Moses breaks his promise and questions the angel.

The angel explains. A king is commandeering all available boats; disabling the ship saved it for its owners. Killing the youth saved his parents, true believers, from his wickedness. A man had hid a fortune under the wall. His sons will dig it out when they've grown. Repairing the wall kept those who had refused food to Moses and the angel from discovering the fortune. What seemed evil to Moses was actually good. Distinguishing good and evil, it seems, is not always easy.

The System of "Good" and "Bad"

For over three thousands years, writers have used good and evil to explain religious observations, principles, and laws. There's probably isn't much that hasn't already been thought, said, and written, a hundred times over. Yet, dualisms based on good and evil still haven't explained how evil originated, how it continues to exist, and how to recognize evil as evil. Sometimes a question defies solution because it's the wrong question, because it's asked in the wrong way, or because the ideas it's based on are faulty. That the "problem" and "mystery" of evil has remained unsolved for so long suggests another approach might work better. But a much more compelling motive is that good and evil don't describe very much of the world we live in. Let's consider an illustration.

Imagine Joe, driving down a street, late for work, feeling tired and tense. A cheerful song comes on the radio and Joe sings along, more relaxed now. Parking his car, Joe notices the day is cold. Shivering, he walks quickly to the building. The lobby is quite warm; the receptionist smiles pleasantly. Joe returns the smile, but then remembers someone he forgot to call yesterday and feels regret. Joe enters his office, takes off his coat, and sits behind the desk. Does he have any appointments today? He's uncertain. He sighs. Another day has begun.

In our illustration, Joe experiences - not good and evil - but "pairs of opposites" such as relaxed/tense, cheerful/sad, warm/cold, happy/regretful, pleasant/unpleasant, energetic/tired, and certain/uncertain. We, too, continually experience the universe as pairs of opposites. Our minds are almost always filled with them. The light and dark, the positive and the negative, the desirable and undesirable, the yang and the yin - the two faces of drama in endless variation. From an introduction to the *I Ching*:

> The list of contraries is inexhaustible. ([I02],22).

But while pairs of opposites may be inexhaustible they aren't necessarily good and evil. Relaxed need not be morally good, and tense need not be morally evil. Warm isn't good, and cold, evil. And cheerful isn't good, and sad isn't evil.

Or are they? Many people would say that cheerful is good and sad is, not evil, but bad. And they're right, because they aren't using "good" and "bad" in the moral sense but in another sense entirely. Most people use "good" for the agreeable, desirable side of a pair of opposites, rather than any moral good. When someone says a back rub feels good, they don't mean it feels morally right. (In fact, some religions frown on sensual pleasure.) They mean the back rub is pleasurable. Similarly, if someone learns their favorite sports team won a game, they say "good" meaning "I'm happy to hear that" not "the win is a morally good event." Similarly, when a person's back hurts they say they're feeling bad, not evil. When their team loses, they say "that's too bad." Any number of examples could be given.

It seems we stumbled on yet another dualistic system: the system of "good" and "bad", a system used throughout the world, very different from any system based on moral good and evil. "Good" and "bad" is how most people, most of the time, view their universe. It's a system we'll investigate in detail.

Yang and Yin

Our aim in this and the next few sections is to explore the dualistic system of "good" and "bad." Our first task is to find better labels, to avoid confusing "good and bad" with "good and evil."

The terms I'm going to use - "yang" and "yin" - are drawn from the Taoist tradition. Ancient Chinese Taoist texts, such as *Tao Te Ching* and the *I Ching,* suggest "yang" and "yin" are equivalent to "good" and "bad." It's difficult to be certain, however, because translations vary. Nonetheless, I'll use "yang" and "yin" in place of "good" and "bad," offering my apologies if the usage differs from what a Taoist would consider correct.

So in this book, "yang" refers to what most people call "good" - the pleasant, the agreeable, the desirable. And "yin" is what most people call "bad" - the unpleasant, the disagreeable, the undesirable. If we imagine the two faces of drama, then "yang" is the smiling face and "yin" is the sad face.

Yang and yin differ from good and evil in fundamental ways. For one, good and evil are usually thought of as moral absolutes that inhere in the entity (kindness *is* good; killing *is* evil). But yang is anything agreeable - and agreeable must be agreeable to some person. Like beauty, yang and yin are in the eye of the beholder. In other words, yang and yin don't exist independent of the observer. Rather, a particular person creates yang and yin qualities. How? We'll discuss two answers.

The first answer is interactive projection, which says that an observer interacts with an entity, experiences a private (yang or yin) sensation, and projects that sensation onto the entity. Interactive projection says that ice cream's taste exists only in the observer, not the ice cream. Therefore, "This ice cream tastes good." isn't accurate. It's more accurate (but much more wordy) to say: "Interacting with this ice cream causes me to experience a pleasant taste. Of course, someone else might experience an unpleasant taste. Therefore, the pleasant taste must be my own private sensation, something I project onto the ice cream."

After all, if the pleasant taste was entirely a property of the ice cream, then the ice cream would always taste the same. It doesn't. The first spoonful tastes good, the thirtieth spoonful tastes neither good or bad, the hundredth spoonful may make you ill. Each spoonful of ice cream is identical, but its qualities change. This shows that the observer does more than observe; the "observer" and entity interact to bring a quality into existence.

The second answer is interactive invocation, which says that an entity possesses innumerable yang and yin qualities in a potential state. A particular quality is brought into actual existence when it's "invoked" by the entity and an observer interacting. Interactive invocation says that the ice cream has the potential of tasting good, bad, rich, thin, flavorful, bland, healthy, sickening, etc., but only when a particular person and the ice cream interact are one or more potential qualities "invoked", that is, brought into actual existence.

Projection and invocation are probably equally valid ways of thinking about yang and yin. In what follows, however, I'll usually choose the invocation viewpoint. With either view, however, the quality is created when an observer and an entity interact. Quantum mechanics, by the way, has a similar idea.

> The crucial feature of atomic physics is that the human observer is not only necessary to observe the properties of an object, but is necessary even to define these properties. In atomic physics, we cannot talk about the properties of an object as such. They are only meaningful in the context of the object's interaction with the observer. ([C03],140).

Of course, an observer can interact with an entity and invoke no qualities at all. As an illustration (which we'll return to), suppose two men compete for a job. The first man is chosen, the second rejected. That the first man was hired and the second rejected is a fact, a datum, an entity. From this single entity, an indifferent person (who, perhaps, doesn't know either of the men) invokes no qualities. The first man, however, invokes yang qualities - satisfaction, a sense of success and achievement. The second man invokes yin qualities - disappointment, a sense of failure. All three people interact with the same entity but invoke different (or no) qualities.

There's another way that yang and yin differ from good and evil. As moral absolutes, good and evil are the same for everyone but yang and yin aren't, because they depend on the observer. Something that's yang to one observer may be yin to another. Some kids say chocolate ice cream tastes good, others say it tastes rotten. The pleasant and unpleasant taste is invoked by the particular kid. Observers interact with the same entity but invoke different qualities. In fact, an identical quality can be yang for one observer and yin for another. That a particular car is expensive, is yang to someone who desires the prestige of such a car, and yin to someone who would like to purchase the car but can't afford to.

Because yang and yin depend on the observer, basing moral values on them might seem difficult. It isn't. In fact, much of Part III concerns goals, values and morals. We'll see how good and evil seem to require a God who is a Person to define them, but how yang and yin are much more compatible with the idea of the God which is not a Person.

Inseparable Interactive Invocation, an Inseparable Dualism
Once we understand what yang and yin are, we can look for a connection between them. Is there a connection between the yang and yin we experience?

The question, in one form or another, is ancient and has often been asked of good and evil: Why do good people suffer? Why do evil people prosper? Why did this good (or evil) thing happen to me? The Biblical axiom "As you sow, so shall you reap" is one answer. So is the Hindu law of Karma. Both say that the good we experience is connected to the good we've done; that the evil we experience is a result of the evil we've done.

Many people find such answers unsatisfying, however. They ask: What evil could a child have done to deserve some painful disease? They see a person suffer some crippling accident or contract some horrible illness, and ask the same question. The law of Karma answers that the evil was done in some past life. Many religions offer no answer, and advise faith in the justice of God's inscrutable will.

Are yang and yin connected? The *Tao The King* says they are, in a statement reminiscent of Newton's law that "for every action there is an equal and opposite reaction.

> It is because we single out something and treat it as distinct from other things that we get the idea of its opposite. Beauty, for example, once distinguished, suggests its opposite, ugliness. . . . [A]ll distinctions naturally appear as opposites. (II,[T01],12).

Let's examine this idea. It says that by making a distinction, by invoking one side of a pair of opposites, we somehow invoke both -

that yang and yin are two sides of a single coin we create. We'll call this idea "inseparable interactive invocation", or sometimes just "inseparable invocation" for short.

We can distinguish two senses of inseparable invocation. In the strict sense, it says that when we invoke a quality, say beauty, that we somehow also invoke its opposite, ugliness. In a weaker sense, it says that when we invoke a yang quality, we also invoke some yin quality that may not be the strict opposite. Inseparable invocation in the weaker sense is easier to understand.

For example, continuing our illustration, looking back years later, the first man sees some yang and yin aspects of the job he accepted. The salary was adequate, but had he rejected the job he might have started his own business and eventually enjoyed greater success. If the company had been bigger it would have offered more opportunity for advancement. Yet he would have felt lost in a large impersonal company. He enjoyed his company's family atmosphere. But he didn't enjoy working with the owner's inept and abrasive son.

When he won the job, the man saw yang qualities. But looking back years later, he sees yin qualities, too. Notice how the qualities are inseparably invoked. The job gave financial security, which has a yang side (the man knew he could pay the mortgage and provide for his family) and a yin side (he was less likely to start his own company). The company was small and had a family-like atmosphere, which means the man enjoyed (yang) the personal, friendly environment, and had less opportunity (yin) for advancement.

The strict form of inseparable invocation is harder to see (and perhaps isn't always true). By deciding that chocolate ice cream tastes good, how do we invoke the opposite? Perhaps because things that once tasted good become bland in comparison. The kid who once liked potatoes now thinks they taste rotten and would rather have chocolate ice cream.

In any event, inseparable invocation (in either sense) says that when we interact with an entity, we can remain indifferent, or we can split it into yang *and* yin qualities, but not solely yang or solely yin. This splitting is reminiscent of an idea we had earlier, that the decent from unitary vision involves a fall into duality, that the shift from unitary to dualistic vision invariably endows the one substantial thing with two sides. Alternately, it could be said that yang and yin are dual expressions of the Eternal. One treatise on Taoism says exactly that:

> The dialectics of Yin and Yang are the double
> manifestation of the one and only eternal, undividable,
> and transcendent principle: Tao ("the Way").
> ([N05],v15,1068).

Inseparable invocation, then, says yang and yin properties, when they are arise, arise together. As the *Tao The King* claims:

> Every positive factor involves its negative or opposing
> factor. . . (XI,[T01],18),

Inseparable interactive invocation is in some respects similar to other dualisms we've discussed. it's similar to illusory and apparent dualism because the dual pairs don't exist in and of themselves but depend upon the observer. It's similar to dependent dualism because yang and yin, in a sense, depend on each other for existence because they don't exist alone. They are brought into existence together.

Yet, inseparable interactive invocation is different from the other dualisms we've seen. It's an "inseparable dualism" because yang and yin arise together, inseparably. In other dualisms, good is separate from evil, mind is separate from matter, etc. But yang is not separable from yin, nor yin from yang. We'll use inseparable dualism, in the form of inseparable interactive invocation, as our theoretical model for understanding the universe.

A Symbol of Yang and Yin

There are a few more ways that yang and yin differ from good and evil.

First, because yang and yin are inseparable they are complimentary rather than opposed and antagonistic. Good and evil war with each other, but yang and yin co-exist, complimenting each other like the positive and negative terminals of an electric battery. A battery must have both terminals, and neither terminal is superior to the other. As a translation of the *I Ching* has:

> Yang is not superior to yin, nor is yin superior to yang.
> ([I02],22).

Because yang and yin are equally important, inseparable dualism is similar to independent dualism, yet it differs, too, because good and evil are antagonistic while yang and yin are complimentary.

Second, something can be entirely good or entirely evil but it can't be entirely yang or entirely yin. As Parrinder writes:

> There is no Yang without Yin in it, and no Yin without
> Yang. ([P05],173).

Therefore, an entity is like a coin: seen non-dualistically it has no sides, but seen dualistically it has two.

Because yang and yin are inseparable, it follows that whenever we "grasp" an entity - that is, whenever we interact and split an entity into yang and yin - we grasp both sides. Like the moon, entities sometimes show their light yang side, sometimes their dark yin side, sometimes both. Interact with the entity indifferently and neither side is invoked, so neither side affects us. But grasp the entity and you take both sides, though only one side or the other may be apparent at the moment.

There's an ancient symbol that expresses much of what we've discussed: the "yang-yin" symbol, a circle divided into two parts by an S-shaped line. To understand this symbol, we'll begin with what an undivided circle symbolizes.

An undivided circle symbolizes an entity seen non-dualistically, either because the observer is above all liking and disliking, and sees the entity as it is in itself, as a "mode of Light" (a rare situation); or because the observer is indifferent (a much more common situation). Let's pause to examine the second situation.

Suppose your car develops serious engine problems. This situation is an entity, a fact, a datum. Of itself, it's neither yang or yin, and it certainly isn't morally good or evil. Now, suppose you're tired of the car and welcome an excuse to replace it. Then you see the situation as "good," as yang. Suppose, on the other hand, you're rather not spend the money, either to fix the engine or for another car. Then you see the situation as "bad", that is, yin. Now, suppose it's not your engine that has the problem but someone else's. If you don't know the person, you may feel neither yang or yin. You may feel nothing about this datum. You're indifferent and therefore see neither yang or yin. You have, not unitary vision, but indifferent vision.

Indifferent vision, then, is similar to unitary vision in that the pairs of opposites are not in the forefront. It differs, of course, in that pure Isness isn't perceived, although a lower form of isness is perceived. The entity just "is", with no yang or yin qualities. (We return to the subject of indifference later.)

So, an entity, seen with unitary vision or seen indifferent vision, is symbolized by an undivided circle. But when it's grasped - when it's seen dualistically so that yang or yin qualities are invoked - the entity is symbolized by a circle divided by the "S" shaped line. In fact, the observer, by the act of invoking yang and yin qualities, draws the "S" shaped line on what was previously an undivided circle.

The circle/S-line symbol expresses that yang and yin are inseparable and complimentary because the S-shaped line invariably divides the circle into inseparable, complimentary parts. And it expresses that what's yang to one observer is sometimes yin to another, because it's the observer who draws the S-shaped line and different observers draw different lines. If you like the chocolate flavor of ice cream you draw the line so that the flavor is on the yang side of the line. If you don't, then your line puts chocolate flavor on the yin side.

The symbolism reminds us of ideas we've already seen, but it also contain an idea we haven't yet discussed. The S-shaped line divides the circle into two equal parts. That is, the symbol says yang and yin components are more than inseparable, they're equal.

> . . . [W]e profit equally by the positive and the negative
> ingredients in each situation. (XII,[T01],19).

This idea is hardly intuitive. In fact, it sharply contradicts common sense. The yang aspects of losing a dearly loved, only child, of being tortured, of losing a limb or sight, are, to put it mildly, not very obvious. The yin aspects of fame and success, wealth, beauty, and prestige are perhaps a bit more apparent in the lives of famous people who had them but were miserable. Yet, there's no shortage of volunteers who would like to experience this particular brand of misery.

But if the yang and yin aspects of any entity do indeed ultimately balance, then in the final analysis our pleasures and pains balance. Even if life occurs at random, it nonetheless offers us entities which have equal yang and yin. When we interact with an entity, we absorb both its yang and yin. In the final analysis, we receive equal measure of yang and yin. This idea contradicts the common sense belief that life can grant us pleasant and unpleasant things in any amount.

It's certainly not obvious that the yang and yin aspects of any entity are equal. Perhaps they aren't. For now, let's just tentatively accept the idea of yang and yin as theoretical constructs, as inferred entities. There'll be no scientific proof for their existence. We'll simply use them to build a theory for understanding things we see later. For theoretical constructs, the critical question is not if they're true, but if they "work," if they successfully explain various observations, principles, and laws.

Inseparable Dualism and the Problem of Suffering and Pain

While some people are interested in questions of good and evil, or questions of yang and yin, others find pain and suffering a more immediate concern. For even we knew the answers to such questions, we would still experience life's pain and suffering. Even if evil is actually an illusion, it's not completely non-existent since illusions possess some reality. A mirage is real in that observers see it; although water doesn't exist, the *appearance* of water does. And evil's appearances in the form of pain, suffering, terror, and starvation - whether illusion or not - hurt. This hurt, by the way, has a yang aspect: it makes distinguishing pain and suffering from the pleasant quite easy. Even a baby - who certainly can't distinguish good and evil - can recognize pleasure and pain.

But recognizing pain and suffering doesn't eliminate them. The questions still remain: Why do we suffer? and Is there a way to avoid suffering? Buddha considered these to be life's central questions. His answer is logical: understand the cause of suffering and eliminate it; as a consequence, suffering itself will be eliminated. Buddha identified desire as the root cause of suffering. Let's see why.

Paul sees another child eating a chocolate ice cream cone and begs his mother for one. She thinks he's already had enough sweets for one day, and refuses. Paul feels unloved, frustrated, and angry - feelings which cause him pain and suffering. He cries and has a tantrum.

Frustrated desire causes Paul's suffering. Were he not ruled by his desire for ice cream, his mother's refusal wouldn't cause him to suffer. His desire is the root cause of his suffering. As Buddha observed.

> And what, brethren, is the root of pain?
> It is this craving . . . ([B08],31).

Similarly, St. John Climacus writes that

> . . . he who has an attachment to anything visible is
> not yet delivered from grief. ([C09],12).

For even if achieved, the object of our desire is liable to change, turning perhaps into something we don't like. Or it's liable to be lost, stolen, or die. Therefore, the logical way to eliminate suffering, Buddha reasoned, is to eliminate desire.

But where does desire come from?

Before Paul acquired a liking for chocolate ice cream, being deprived of an ice cream cone wouldn't have caused him pain. What changed him?

For Paul, chocolate ice cream has a yang aspect - its pleasant taste - and a yin aspect - his increased desire for it. As Buddha observed:

> . . . whatever in the world seems lovely and pleasant,
> here when it arises doth craving arise . . . ([B07],77).

Each time Paul eats chocolate ice cream, he absorbs both aspects, even though only the yang aspect - the pleasing taste - is apparent at the moment. Eventually, Paul's desire for chocolate ice cream is frustrated. The resultant pain and suffering are the flip side of the pleasure he's already derived from eating chocolate ice cream.

Thus, our desires are simply the debt we've incurred for the pleasures we have enjoyed. And since desires lead to suffering, we may add that our sufferings result from the pleasures we've enjoyed.

Conversely, suffering often builds up a reserve of pleasure, not in any sadistic sense, but in the sense of the French proverb "Hunger is the best spice." For example, a person standing in the cold for a short time may feel some relief when entering a warm house and sitting by a fire. As they stood in the cold they only saw the unpleasant side of the cold, its stinging coldness. At the same time, however, they were being primed to enjoy the warmth of the house. If they had stood in the cold until they were chilled "to the bone," they'd probably enjoy a proportionally greater relief when they enter the warm house. On the other hand, someone coming into a warm house from a warm car might take little or no pleasure in the house's warmth. In fact, they might not notice it at all.

Often, the longer we wait for something and the harder we work for it, the more enjoyment we derive when we finally have it. The potential for enjoyment is built up while we're waiting and working. So, discipline and hard work, in a word suffering, are prerequisite for many pleasures. Conversely, those who don't work and suffer for a thing often take little pleasure in having it. The cliche of jaded rich kids comes to mind.

This explains why we can't listen repeatedly to a recorded song and derive the same satisfaction each time. Suppose you enjoy a song. Play it over and over. You eventually tire of it. The song itself certainly doesn't change. You change. Your potential for pleasure runs out, and you'll have to wait until it's "recharged" before you enjoy the song again. Endure a boring day at work or a frustrating experience at the supermarket and perhaps you'll enjoy listening to the song again. This, by the way, also shows pleasure, a yang quality, doesn't inhere in the entity itself - the song - but is invoked by the interaction of entity and experiencer.

A Solution?

Buddha believed all entities contain an element of imperfection because they all contain the seeds of suffering. Within the framework of yang and yin, the imperfection and suffering inherent in entities are easy to understand - they are simply manifestations of the entity's yin side. Thus, no entity offers pure pleasure. There is always a measure of pain.

Since our pains are the result of past pleasures we've enjoyed, there is, as Buddha observed, a straightforward way for a person to eliminate pain. If a person looks

> . . . upon whatever in the world seems lovely and pleasing as impermanent, as suffering, as not good, as disease, as danger, they put off craving. They who put off craving . . . put off suffering. They who put off suffering are liberated from birth, old age, death, from grief, lamentings, sufferings, sorrows, despairs, yea, I declare they are liberated from ill. ([B07],78).

Buddha presents a logical cure for suffering, but it's easy to feel the cure is worse than the disease. True, life contains much pain and suffering. But life offers many delights and enjoyable moments, as well. Must they really be given up before suffering is eliminated? Suppose we abandon taking pleasure in anything. Suppose, further, this eliminates pain and suffering. What remains? An unceasingly gray, joyless, featureless life? Is this what the Buddhist ideal life, Nirvana, is? Many people wouldn't consider such a life worth living.

To fully answer this question we must first understand what it means to achieve the unitive life, a subject of a later chapter. To anticipate a bit, according to the mystics the ideal life is the life of unitive knowledge of God (either Person or not).

A mystic who enjoys the vision of, or even union with, the God which is not a Person no longer regards the universe's light show, with its pleasure and pain, as fully real. Acquaintance with unitary vision reveals the illusory nature of dualistic vision. Enjoying the Real, the enjoyment of pleasant dualistic visions is readily, even eagerly, forsaken. Such a unitive life is quite the opposite of an unceasingly gray, joyless, featureless life.

Ignorance of the Real

We began this chapter with the aim of investigating the relationship between the Source and the universe.

We saw that some mystics see no relationship between the two because they are completely united with the Source; for them the universe does not exist. Other mystics, however, see each and every entity as a mode of Light, an expression of the Eternal. In the vision of such mystics, the One is foremost; the entity's particular qualities are secondary. Theirs is a mystical vision of the universe, because the universe is seen as an expression of Ultimate Reality.

Other people, mystics or not, see the universe's objects and people as expressions of not one, but two fundamental entities, like good and evil, mind and matter, yang and yin. We may call such insights a metaphysical vision of the universe.

Scientists see the universe as a group of entities related by scientific laws. They discuss, not modes of Light or yang and yin, but space, time, matter, energy, and other empirically-verifiable entities. The scientific world view is the most elaborate and detailed vision of the universe.

As we progress from mystical to metaphysical to scientific visions of the universe, we progress from simplicity to complexity, from the One to the Many. Yet, all three visions seek to reverse the progression by finding unity amid diversity; they seek to reduce the Many to some simpler system.

Ultimately, we progress to the lowest level where any relationship to the Source is forgotten, where the observer perceives a multitude of transitory objects, all apparently unrelated and unconnected. It's the world of ignorance in that the One is ignored and the Many are taken to be Real. In the Yoga Sutras of Patanjali, we find:

> Ignorance consists in mistaking the transient for the permanent, the impure for the pure, evil for good, and the apparent self for the real self. ([YO1],83).

In the world of ignorance, there are pleasurable and unpleasurable things which apparently can be had in any quantity. In this world, some individuals ruthlessly pursue the things they desire. Others allow scruples to moderate their pursuit. Some scrupulous people are motivated by a natural fairness and regard for other people, others by a fear of retribution in the afterlife. Yet, all these individuals live in a world which contains much suffering, a world which is ultimately unsatisfying. Rufus Jones quotes Meister Eckhart's view that our "lower" consciousness

> . . . is able to deal only with the particular and finite - its sphere is the show world . . . The life in this lower stage is always restless and unsatisfied, for it is endeavouring to anchor upon fleeting, vanishing things. ([J03],229-230).

Only when a person acquires some measure of wisdom and discrimination in the religious, philosophical, or metaphysical sense, do they seriously begin to look beyond the vanishing things of this world and seek higher consciousness and the Eternal. Until then, they see the play of the Eternal Light, but do not see the Light Itself that underlies that show. They see the particular and finite, but do not see the Universal and the Infinite that underlies all particular and finite entities. They experience the change and restlessness which are inherent in the show, and never reach That which is changeless, stable, enduring. The entities which they experience always have a measure of imperfection, of suffering. And even if those entities bring satisfaction, they are liable to vanish, to be lost, stolen, or die.

Not only do we not see down to the Basis of this world, we do not see our own Basis. It is to the basis of our own personal identity to which we now turn.

8

- Personal Identity -

Chapter Summary: This chapter explores the question of personal identity: what exactly do we mean by "I"? Various answers are discussed. The possible relation of personal identity to the Uncreated is explored.

Landmarks: Ancient and Contemporary
The last chapter discussed the external world we live in, the world of people, places, events, and things. This chapter discusses the other world we live in, the inner world of feelings, thoughts, and awareness - mostly with the aim of investigating personal identity. To discuss personal identity, we'll need some landmarks. The landmarks we'll use are body, emotion, intellect, and awareness.

Of course, "intellect" means not any particular thought but our cognitive faculty, often called the "mind." Similarly, "emotion" means not any particular emotion but our affective faculty, figuratively called the "heart", the part of us that feels. Obviously, there is a connection between mind and brain, and perhaps between the emotional heart and the physical organ of the same name. Nonetheless, we'll think of the brain and heart as parts of the body, different from the mind and the emotional "heart."

So, we'll mentally divide a human person into four parts - body, emotion, intellect and awareness - to discuss personal identity. But our four-part division doesn't mean a person isn't actually a holistic entity. It doesn't mean a person can actually be divided into parts. A person is a unity that transcends the division of body, emotion, intellect, and soul. (A later chapter returns to the unity of the human person and the intimate relation between the four "parts.")

But if a person is actually a holistic unity, why the division, even if it's only mental? For the same reason we divide the earth in parts such as continents, oceans, countries and states: to better discuss and describe it. The earth transcends various political and geographical boundaries; it's a unity, a whole. Yet, to describe it we must divide it into parts such as land and sea, hill and valley. These divisions are ancient and natural.

Our "landmarks" - body, emotions, and mind - are ancient and perhaps natural, too. They've been used for millennia, in many cultures. We'll consider a few examples, beginning with some symbolism.

In *Man and his Symbols* ([J04]) psychiatrist Carl Jung writes:
> Animals, and groups of four, are universal religious symbols. ([J04],21),

and gives an example from the ancient world: the four "sons" (three animal and one human) of the Egyptian sun god Horus. Christianity uses a similar symbolism: the ox, lion, eagle, and man or angel. (Refer Ezekiel 1:10 and Revelations 4:7.) These symbols are often used to represent the four evangelists, so that the ox symbolizes Matthew, the lion, Mark, the eagle, Luke, and the man or angel, John. These symbols may also be taken as symbols of the four parts of a human being. How? The massive ox may be thought of as a symbol of body; the ferocious lion, heart; the soaring eagle, the bird's eye view of intellect; the angel, the spark of God in us, that is, awareness or soul.

Not only does the Roman Catholic tradition accept the ox-lion-eagle-man/angel symbolism, it also accepts the four-part division of a human being into body, emotion, intellect and soul: for it teaches we have a soul, and recommends we direct our mind, heart, and body to God. From an elementary school catechism:
> What must we do to gain the happiness of heaven?
> To gain the happiness of heaven we must know, love, and serve God in this world. ([N08],12).

Knowing, of course, is a function of the mind, loving, of the heart, and service, of the body.

India also accepts our four-part division. It divides the yogas, or ways to God, into four broad types. Karma yoga is the way of works, of using the body to draw nearer to God, performing duty to family, neighbors, and country, feeding the hungry, helping the poor. (By the way, to some people "yoga" means only Hatha yoga, that is, bodily postures and purification. Hatha yoga is part of Karma yoga.) Bhakti yoga is the way of devotion, of using feelings of love and adoration to draw nearer to God, feelings that may include love of a divine incarnation such as Krishna. Jnana yoga is the way of knowledge, of using the intellect to draw nearer to God via the study of scriptures, philosophy, and even science. Finally, Raja yoga is the way of awareness, of using consciousness to draw nearer to God. It's the way of meditation.

Our four-part division of a human being occurs in the modern world, as well. Often, people are often classified as extravert or introvert. The extravert is outgoing, lively, and friendly, while the introvert is thoughtful, quiet, and perhaps shy. Not surprisingly, the crude extravert-introvert scheme has been subdivided into finer classes. Jung, for example, notes:

> If one studies extraverted individuals . . . one soon
> discovers that they differ in many ways from one
> another, and that being extraverted is therefore a
> superficial and too general criterion to be really
> characteristic. ([J04],60).

Jung himself believed human beings have four faculties; sensation, feeling, thought, and intuition. There's an obvious, loose correspondence to our four components of body, heart, mind, and awareness.

In another classification scheme, Dr. William Sheldon, a trait theorist ([H11],150 and [A01],153), divides the extravert into two subtypes, the physical "somatotonia" who loves action, adventure, and competition, and the emotional "viscerotonia" who loves people, comfort, and food. Sheldon's third type, the "cerebrotonia," describes the introverted intellectual. Sheldon believes a person has some characteristics of each type, but one predominates. If we add to Sheldon's system a fourth element, the awareness that's conscious of body, feelings, and thoughts, we again arrive at body, heart, intellect, and awareness. And if we understand "soul" as meaning awareness then the four elements also may be called body, heart, mind, and soul.

But should "soul" be used as a synonym for awareness? "Soul" has a bewildering array of meanings, for example,

> 1. The rational, emotional, and volitional faculties in
> man, conceived of as forming an entity distinct from
> the body. 2. *Theol.* **a** The divine principle of life in man
> . . . The moral or spiritual part . . . The emotional
> faculty . . . ([F08],1280).

But, although "soul" can mean any sort of non-physical faculty,

> [t]raditional definitions of the soul have usually
> emphasized man's consciousness of his psychological
> and mental processes . . . ([N04],v20,924A).

Therefore, using "soul" to mean awareness or consciousness is justifiable.

In general, however, I'll use the more definite terms "awareness" or "consciousness" to indicate the fourth element, the awareness that at various times is conscious of body, feelings, or thoughts. And when I use the word "soul," I'll use it as a synonym for "awareness" and "consciousness."

Of course, someone might argue that the soul isn't awareness or consciousness, that it's something entirely different. But if the soul isn't our body, heart, mind, or consciousness - if it's entirely distinct and separate - then why should we be concerned where it goes after we die? We would never know where it was or what it was experiencing, so why should we care? Certainly, people who believe in heaven and hell expect to experience one or the other - that is, to be conscious of one or the other - in the after-death state. But if our awareness moves on to heaven or hell and stays there eternally, how can it differ from our soul?

Relative Personal Identity

To discuss personal identity we'll think of a person as consisting of body, emotion, intellect and awareness. But what do we mean by "personal identity"? We can mean two things: either a relative personal identity or an absolute personal identity.

What is a relative personal identity? It's an identity I have that's relative, i.e., that depends on someone or something else. Let's consider an illustration.

Al and Sue decide to be married. At that moment, Sue gains a new relative identity: Al's fiancee, an identity that depends on someone other than herself - Al. Similarly, Al gains the relative identity of Sue's fiancé. When they marry, Sue will lose the relative identity of fiancee and gain a new relative identity, Al's wife. When their infant son John is born, Al and Sue will gain the new relative identity of parents, an identity that depends on someone other than themselves - John. Later, if Al and Sue have more children John will gain the relative identity of older brother. Throughout his life, John may gain and lose relative identities such as student, taxi driver, apprentice, engineer, farmer, or lawyer.

The idea of inseparable interactive invocation, by the way, applies to relative personal identity. A man has many potentials, for example, to be brother, uncle, father, or son. However, a particular potential, for example to be someone's uncle, isn't invoked - brought into actuality - until a niece or nephew is born. The birth of niece or nephew "interactively" invokes the relative identity of uncle in the man. To the niece or nephew, the man is uncle, but to the daughter or son, the man is father. Different observers invoke different qualities and the invocation is inseparable: there can be no uncle without a niece or nephew, and no niece or nephew without an uncle or aunt.

Because a relative identity exists relative to someone or something else, it depends on that someone or something else for its own existence. It comes into existence when it's invoked, and may go out of existence when the interaction ends. When I'm hired I become an employee of a certain company. My identity as employee depends on the company; if the company goes out of business then I cease to be that company's employee, I lose my relative identity of employee.

Because it is dependent, a relative identity may be temporary. People become a teacher or student for a while, and then cease to be so. However, a relative identity may be more permanent. For example, someone will always be the son or daughter of their parents, even after their parents have died. Such an identity is unchanging and permanent, but it may not be unique - I'm my parent's child but so is my sister. Do I have an identity that's unique to me? that's unchanging? that's independent of anyone or anything else? In other words, do I have an absolute personal identity?

Absolute Personal Identity

Suppose Ann, Sue's college friend, last saw John ten years ago, when he was five years old. Suppose she meets him again. John now has different clothes, a different personality, different emotions and thoughts, a different set of relative identities, and a different body - even a different set of teeth. Yet, Ann believes she's meeting the same person she met ten years ago. What do we mean by "same person"? What about John is the same? Of what does his "I-ness" consist? When John says "I" exactly what does he mean?

211

He seems to mean four things. First, he means something that's unique. Why? Because if two people have it, then "I" means both of them. When John says "I" he means something that's unique, because in all the world no one but John is John. Second, he means something that has endured, unchanged, at least from the moment he was born. Why? Because if everything about John has changed then he has no right to say *he* was born fifteen years ago. Rather, someone else was born fifteen years ago who eventually changed into the "I" that is John today. John is a descendant of that person. Third, John means something that's absolute, not relative to anyone and anything else, because John is John independent of anyone and anything else. When he says "I" he means his deepest self, independent of anything and anyone else.

So when John says "I" he refers to something that's unique about himself, that's unchanging, and that's independent of anyone or anything else. He also means (and this is the fourth quality) something that is him, as opposed to something he possesses. Suppose at birth he inherited his father's custom-made watch. The watch is unique to John and has endured, unchanged, since he was born. It exists as an object independent of any other object (but not independent of the Eternal). But it's certainly not what John means when he says "I". When John says "I" he doesn't mean his watch or any other possession. "I" is a possessor, not a possession.

Therefore, the question we asked at the end of the previous section - "Do I *have* an absolute personal identity?" - was poorly put; it doesn't really make sense Why? Because it treats absolute personal identity as a possession. But I can't *have* an absolute personal identity because, as we've defined it, it *is* me. "I" and it are identical. It's what the word "I" means. It's what I was when I was born and shall be when I die. Therefore, an absolute personal identity can't be something I have; rather, it must be something I am.

Nonetheless, we'll sometimes find it useful to speak of absolute identity as a possession when the alternative is awkward. For example, it's easier to ask "Do I have an absolute personal identity?" than "Is there something about me that constitutes an absolute personal identity?" Yet, it should be remembered that absolute personal identity, if it really exists, isn't a possession.

For if it was a possession we could ask: "*Whose* identity? *Whose* possession?* Who is it that possesses a watch, name, body, feelings, thoughts, and even an absolute personal identity?" And if we couldn't find anything that possessed all those things, we'd conclude that no possessor really exists, that the word "I" refers to no single thing. Rather, it refers to body, emotions, thoughts, and identities, an assembly of things all existing relative to someone or something else, all not necessarily unique, and all subject to change.

No permanence is ours; we are a wave
That flows to fit whatever form it finds:
Through day or night, cathedral or the cave
We pass forever, craving form that binds. ([H04],397).

Some people - Buddha, for example - see "I" in this way, as a conventional term that refers to a changing assembly of things rather than to any single thing that's unique, unchanging, and absolute. (We'll see more of this viewpoint in this and a coming chapter.) Is Buddha correct? Or does absolute personal identity really exist? It seems we should answer this question before trying to find one - just as Ponce de Leon, DeSoto and others should have made sure a Fountain of Youth really existed before they spent the time and trouble searching for one in Florida and the Bahamas. Unfortunately, discussing the existence of an absolute personal identity would require an abstract theoretical discussion that might leave us no wiser. A better course of action will be to assume for the moment that an absolute personal identity exists, and try to determine exactly what it might be. If our search for it is successful then our assumption will be proven. Hopefully, we'll be more fortunate than Ponce de Leon and DeSoto.

Identity as Body, Heart or Mind
One way to search for absolute personal identity is to mentally eliminate everything that's a possession. If anything is left when we're done, then perhaps it's our searched-for absolute personal identity. John's watch is a possession, as is John's clothes, his books, etc. What about John's hand? Is it a possession or a part of John, part of what he means by "I"? John refers to his hand as a possession if he says "my hand hurts" but he might just as well say "I hurt." Which is it? Is John's hand a possession, or is it John?

Some people, particularly people who aren't religious, would say that John's hand is a part of John. They would say that their body is their self, what they mean by "I". They would answer the questions "Does absolute personal identity really exist? Does the word 'I' refer to something that's unique, unchanging and absolute? and, if so, exactly what is it?" by saying "Yes, an absolute personal identity exists. 'I' refers to something unique, unchanging and absolute. In fact, here it is. It's my body." They might point to their body and say "Here I am, this is me. So what's the problem?"

The problem is that the matter that composes our body is constantly changing, constantly flowing through it.

> . . . [T]he human body is changing constantly, and in fact goes through a complete change every seven years. ([B16],121).

Fuel passes through an engine but leaves the individual parts - the pistons, the valves, the block, etc. - unchanged.

Once, an educated person might have believed food and drink flow through the body like fuel through an engine. Had they been right, then our body might possibly remain the same over the years, and therefore might be our unchanging identity. True, it grows bigger and smaller over time, but a balloon can grow bigger and smaller all the while remaining the same balloon.

But our body does not remain the same. It changes. Not merely in appearance (a few wrinkles or grey hairs) but in substance. Its individual atoms are constantly being replaced. In seven years, every one has changed. True, our cells may remain, but the individual atoms that compose them do not.

Imagine replacing each and every part of a car - the engine, the doors, the windows, the tires. When you're finished, is it the same car? Imagine replacing every part of a building - the foundation, the bricks, the walls. When you're done, is it the same building? It may look the same if each part was replaced with a similar part. But is it the same building? In a conversational pragmatic sense perhaps yes, but not in any strict sense.

Our body is like the car and the building. Over the years each and every atom is replaced. In fact, our body resembles a whirlpool or flame more than any solid thing. Imagine a whirlpool. The water is constantly flowing through it. In a short time, all the water has been replaced. Or imagine a candle flame. New wax is constantly entering the flame and being burned; no part of the flame remains the same. Our body is like the whirlpool or flame, and whirlpools and flames can have no absolute identity because no part of them remains unchanged over time.

Besides, even if the atoms of our body weren't all eventually replaced, how could our identity be based on them? on a mere arrangement of atoms, atoms that existed before we were born and will exist after we've died? (By the way, the atoms of our body were created in supernovas, ages ago. Our body is quite literally stardust.) How can the atoms of a sandwich, atoms that were created millions of years ago, become "me" for a few years and then become "not me" when they leave my body?

They can't. Whatever the absolute "me" may be, by definition it's not something that changes. Buddha makes this point in a dialogue reminiscent of Socrates.

> Is body permanent or impermanent?
> Impermanent. . . .
> Now what is impermanent . . . what is of a nature to
> change, - is it proper to regard that thus: "This is mine,
> This am I. This is my self"?
> Surely not . . . ([B09],270-1).

What's true of the body is true of the emotions and thoughts, too. Our emotions and thoughts also undergo change, much more often than once in seven years. Someone who is fifteen is usually quite a different person emotionally than they were at five. They're usually very different intellectually, too. Old feelings, thoughts, and memories are lost, and new ones take their place. This process of constant change, though it often slows in older people, continues throughout our life.

Therefore, our changing emotional and intellectual natures can't be our absolute identity. Buddha made this point, too.

> Is feeling permanent or impermanent?
> Impermanent . . .
> Is perception . . . are the activities. . . permanent or
> impermanent?
> Impermanent . . .
> Now what is impermanent . . . is it proper to regard
> that thus: "This is mine. This am I. this is my self"?

Surely not . . . ([B09],271).

Nothing that changes can constitute an absolute personal identity because such an identity, if it exists, must be unchanging. Our body, emotions and intellect change constantly. What then constitutes our unchanging self, our real identity? Do we have one? Is there anything about us that is unique, unchanging, and absolute, and therefore capable of being our absolute personal identity?

Hierarchy of Body, Heart, Mind and Awareness

An illustration involving a table previously brought us to the unchanging and eternal Basis of the physical universe. We placed the table at the top of an inverted hierarchy and by going deeper into the table, towards its center, we found its Root and Source. Using a procedure that's somewhat analogous we may find that which is unique, unchanging and absolute in us.

In this procedure, the body, like the table, is placed on the highest, most physical level. This is appropriate because only the body, certainly not emotions, intellect, or awareness, may be seen and weighted. In the past, by the way, some religious people attempted to weigh the soul, trying to find objective proof of its existence. They noticed the body lost some fraction of an ounce at the moment of death and claimed this was due to the loss of the soul. These claims were shown to be false, however, and I don't know of anyone who currently takes them seriously.

We've placed the body on the highest level. Now, let's take the next step: which of the three remaining components - heart, mind, or soul - is closest to the physical level?

The link between body and emotions is well known. Anger, for example, has obvious physiological effects, such as rapid breathing and redden face. Chronic stress can cause ulcers. Joy and sadness manifest in facial expression or voice tone. So let's place emotions on the next level, a level roughly analogous, perhaps, to the table's wood molecules.

While not as obvious as emotions, thoughts also have some physical manifestations. The furrowed brow, fixed gaze, and abstracted inward attitude of someone in deep thought are subtler than signs of anger or joy, but they are nonetheless bodily effects. So let's place the intellect on a plane below emotions, a level we may think of as analogous to the atomic level.

Now, only awareness or soul remains. It exists at the deepest, least physical, level. Therefore, it's nearest to the ultimate level. We've seen that our absolute personal identity, if it exists, cannot be our body, emotions or thoughts. Therefore, awareness or soul is the only remaining part of us that might be our unchanging self, our unique and absolute personal identity. Whether it actually is such an identity is a question we'll return to shortly.

A Unitary Psychophysical Language?

The hierarchy of body, emotion, thought, and consciousness is similar to the hierarchy of table, wood, molecules, atoms, and Ultimate Ground of Existence, but there may be an important difference. True, emotion exists on a less physical, more subtle level than body, and intellect exists on a less physical, more subtle level than emotion. But in the table hierarchy, a level is the ground of existence of the next higher level. Atoms are the ground of existence of molecules, wood is the ground of existence of the table. Is emotion, in any sense, the ground of existence of body? Is intellect, in any sense, the ground of existence of emotion?

For our purposes, it doesn't matter if they are, or if body, emotion, and thought are separate, independent manifestations. There is, however, some interesting material that argues for something like the first alternative. Let's discuss it.

Thought is often considered to be the ground of the physical world. As Ken Wilber writes,

> . . . the idea of the physical realm being a "materialization of thought" has extremely wide support from the perennial philosophy. As Huston Smith points out in *Forgotten Truth*, the perennial philosophy has always maintained that matter is a crystallization or a precipitation of mind . . . ([Q01],145).

In this view, an intermediary - thought or mind - exists between physical objects and the Self-Existent. If it does, that would explain, as Wilber goes on to point out, a question that troubles some philosophers and scientists. We'll describe the question by beginning with some observations of James Jeans.

The physicist James Jeans observes:
> Our remote ancestors tried to interpret nature in terms
> of anthropomorphic concepts . . . and failed.
> ([J01],158).

For example, the explanation of the sun as a fiery chariot driven by a god across the sky doesn't explain the known facts very well. (What's an eclipse?) Picturing the earth in orbit around the sun explains the facts better. (An eclipse occurs when the moon moves between the sun and the earth). In general, Newtonian mechanics explains nature better than anthropomorphic ideas do.

Yet, Newtonian explanations aren't perfect. Mechanical ideas don't describe the behavior of subatomic particles like the electron very well. Jeans perhaps refers to this when he writes:
> The efforts of our nearer ancestors to interpret nature
> on engineering lines proved equally inadequate.
> ([J01],158).

(I certainly don't agree they are *equally* inadequate.) Jeans continues:
> On the other hand, our efforts to interpret nature in
> terms of . . . pure mathematics have, so far, proved
> brilliantly successful. ([J01],158),

and concludes
> . . . the universe appears to have been designed by a
> pure mathematician. ([J01],156).

This brings us to the question that troubles certain philosophers and scientists: why does the physical universe so often obey mathematical laws? Following the famous mathematician Rene Descartes, many people believe the outer and inner worlds are distinct and separate. Mathematical patterns and ideas "live" in the inner world, in the mental realm, a world of pure thought. Yet they often govern the outer world's physical phenomena; that is, natural processes often follow mathematical rules. Why?

If the outer physical world and the inner conceptual world are truly separate and distinct, then why do physical phenomena obey mathematical rules? Why does gravity diminish as the square of the distance? Why does energy equal mass times the speed of light squared? It certainly seems strange, puzzling, and wholly unexpected.

On the other hand, if the outer world is, in some sense, a crystallization of the inner world, then the problem vanishes and everything seems quite natural. Indeed, if the two worlds are intimately related, one being a more concrete manifestation of the other, then laws which reveal themselves to the intellect would naturally inhere in the physical, too. The intellect sees the more subtle aspect. Physical sight senses a grosser aspect. But both sense one and the same thing.

What of emotion? Can we fit it into this scheme? Perhaps. Both emotion and thought are functions of the psyche, so what applies to the psyche may apply to both. Carl Jung believed that the psyche's core or ground - what he called the "dynamic *nuclei* of the psyche" ([J04],304) - consists of the unconscious and the archetypes. He suspected there might be

> . . . a possible ultimate *one-ness* of all life phenomena
> . . . [T]he unconscious somehow links up with the
> structure of inorganic matter . . . [A]n archetype shows
> a "psychoid" (i.e. not purely psychic but almost
> material) aspect . . . ([J04],309).

Unlike Descartes, Jung believed the physical and psyche might somehow be two aspects of the same thing. Parallel phenomena in psychology and nuclear physics led him to suspect an intimate relation between the deepest regions of psyche and matter.

Physicist Wolfgang Pauli had a similar idea. Pauli (as his friend, Werner Heisenberg, writes)

> [i]n the alchemistic philosophy . . . had been
> captivated by the attempt to speak of material and
> psychical processes in the same language. Pauli
> came to think that in the abstract territory traversed by
> modern atomic physics and modern psychology such
> a language could once more be attempted. ([H03],35).

Pauli believed such a "unitary psychophysical language" would be

> . . . a mode of expression for the unity of all being . . .
> a unity of which the psychophysical interrelation, and
> the coincidence of a priori instinctive forms of ideation
> with external perceptions, are special cases.
> ([H03],35-6).

It might be worth dissecting that last sentence. The "psychophysical interrelation," of course, is what Jung suspected, that the deeper reaches of psyche are the same as the deeper reaches of matter. "a priori instinctive forms of ideation" refers to instinctive ideas and images, such as archetypes (for example, the hero) and mathematical ideas. "External perceptions," of course, refers to the physical world. Why the external world should obey "a priori instinctive forms of ideation" is the problem we've been discussing. It should, if the physical is the gross aspect of thought, and thought is the finer aspect of the physical, that is, if the deepest level of matter and the deepest level of the psyche is one and the same, if there is in fact a "unity of all being."

All of the above, of course, doesn't quite yield an neat hierarchy where consciousness is the ground of thought which is the ground of emotion which is the ground of body. But is does offer some food for thought.

Awareness as Absolute Personal Identity

As we've seen, awareness exists on the deepest, most subtle level. This allows two possibilities. Suppose that this level is the ultimate level. Then, because only one entity exists on the ultimate level, awareness would have to be identical with the Ultimate Ground of Existence, which being unique, unchanging and absolute, might be our absolute personal identity. On the other hand, suppose awareness is on some plane above the deepest, ultimate plane. Then it's changeable, like body, heart, and mind, and can't be our absolute identity. This would mean none of the four components constitute an absolute identity, that we have no absolute personal identity.

This section discusses the first possibility. A later section discusses the second. Surprisingly, the two possibilities don't differ as much as might be expected.

In many religious systems, the soul is considered the part of a human being that's closest to the ultimate level, to God. Just how close is the point of difference. Some religious and philosophic systems identify God with the self of only one individual (Jesus in Christianity, for example), or a few select individuals (Rama and Krishna in Hinduism, for instance). Other religions, Judaism and Islam for instance, insist on the absolute transcendence of God and deny the existence of any human Divine Incarnation.

Yet, even religions that teach the human soul is different from God acknowledge it's nonetheless quite close to God, that knowing our Self is akin to knowing God.

> "Look in your own heart," says the Sufi, "for the
> kingdom of God is within you." He who truly knows
> himself knows God, for the heart is a mirror in which
> every divine quality is reflected. But just as a steel
> mirror when coated with rust loses its power of
> reflexion, so the inward spiritual sense . . . is blind to
> the celestial glory until the dark obstruction of the
> phenomenal self, with all its sensual contaminations,
> has been wholly cleared away. ([N11],70).

Such religious and philosophic systems teach that the soul is not God, but can reflect God, like a mirror. Just as a mirror seems to contain the sun, the soul seems to contain God.

Other systems explicitly teach the inherent equivalence of each and every human awareness or soul with the Ultimate Ground of Existence. For instance, the early Greek philosophers believed

> . . . man had a soul which was, in some way, the real
> part of him . . . ([F06],155),

and

> . . . thought of this soul as the least material form of
> the particular substance out of which everything in the
> universe was made. ([F06],155).

The equivalence of absolute personal identity and the Eternal also occurs in the Hindu tradition, where for example Ramana Maharshi teaches

> [t]he body and its functions are not 'I'. Going deeper,
> the mind and its functions are not 'I'. . . . 'I' must
> therefore be the unqualified substratum . . .
> ([T03],116),

and also teaches that ([T03],25) Self is pure Light and pure Consciousness. Shankara, too, equated the self with consciousness or "witness."

> Know the self as different from the body, sense-
> organs, mind, intellect and primal nature, and as the
> witness . . . ([S09],30).

Similarly, the contemporary philosopher Alan Watts describes our identity with the Source thus:

> At this level of existence "I" am immeasurably old; my
> forms are infinite and their comings and goings are
> simply the pulses or vibrations of a single and eternal
> flow of energy. ([W02],12).

One objection to the identification of awareness with the Unchanging and Unconditioned is awareness does apparently change and is apparently conditioned. Buddha, for example, taught that

[a]part from condition there is no origination of consciousness. ([C13],312).

However, he doesn't seem to attach the same meaning to the word "consciousness" as we.

It is because . . . an appropriate condition arises that consciousness is known by this or that name: if consciousness arises because of eye and material shapes, it is know as visual consciousness. ([C13],314).

He goes on to describe auditory, olfactory, gustatory, tactile, and mental consciousness.

. . . if consciousness arises because of mind and mental objects, it is known as mental consciousness. ([C13],315).

In our model, awareness is unchanging. It remains constant over different physical, emotional, and mental states, just as a mirror reflects different scenes, but remains itself unchanged. Even though our awareness reflects physical, emotional, and mental phenomena that are changing and conditioned, it itself remains unchanging and unconditioned.

Yet even if it's granted that awareness is constant and unchanging, it does seem to disappear in deep, dreamless sleep. In this state, awareness seems lacking; our real self seems to have vanished. How then can awareness be our unique, unchanging, absolute self? An answer to this objection, in the words of Swami Rama Tirtha, is as follows.

When you get up, you say, "I slept so profoundly that I saw nothing in my dreams." Vedanta says that this statement is just like a statement made by a man who said that at the dead of night, at such and such a place, there was not a single being present. . . . "Is this statement made on hearsay, or is it founded on your own evidence? Are you an eyewitness?" asked the judge. He said, "Yes, I am." "All right. Then, if you were an eyewitness, and if you wish us to understand that your statement is correct, that there was nobody present, then in order that your statement may be

> right, you, at least, must have been present on the
> scene." ([P14],185).

Saying our sleep is dreamless says that no dreams occurred, which we can only know if we are aware, at some level.

Because we are in some sense aware, we know that time has passed. When I wake from a dreamless sleep, I have the feeling that I went to sleep some time ago. If I fall into a dreamless sleep in a dark room, when I awake the next morning I don't feel the room has suddenly changed from dark to light. Rather, I feel some time has passed. Had I been, in fact, totally unconscious when I slept, I wouldn't know time had passed. I would experience a dark room, close my eyes, and open them (apparently) a moment later to find the room suddenly changed from dark to light. But this is not my experience, which again says that in dreamless sleep I'm dimly aware of the passage of time.

If awareness is unchanging and always present, then it may be our true absolute Self, our absolute personal identity. By the way, the belief our real self is not the body but awareness or soul underlies a Hindu phrase for death - "giving up the body." In this view, the soul, which we are, gives up a possession, the body. In contrast, a Western phrase for death is "giving up the ghost." These phrases express differing views of what we really are.

Consequences of Awareness as Absolute Personal Identity

A few consequences follow if we accept the Eternal as our absolute personal identity, our true Self. We'll discuss four.

First, if our consciousness is, in fact, our true self and is identical to the God which is not a Person then, like any other reference to deity, "self," "consciousness," and "I" deserve to be capitalized. There is, perhaps, an obscure intuition of this identity embodied in the practice of capitalizing the words "God" and "I".

The idea that consciousness and God are identical is also suggested by the "All Seeing Eye" which tops the pyramid in the Great Seal of the United States. The Seal appears on back of the United States one-dollar bill. Since some founders of the U. S. were known to be Masons ([P03],529) and others may have been, the Eye is sometimes considered a Masonic symbol of Deity ([P03],403). However, a publication of the Department of State has:

> Use of the eye in art forms . . . as a symbol for an
> omniscient and ubiquitous Deity was a well
> established artistic convention quite apart from
> Masonic symbolism . . . [I]t seems likely that the
> designers of the Great Seal and the Masons took their
> symbols from parallel sources, and unlikely that the
> seal designers consciously copied Masonic symbols
> with the intention of incorporating Masonic symbolism
> into the national coat of arms. ([P03],531-2).

Whatever its origin, the all-seeing eye, an acknowledged symbol of
God, may be taken as a symbol of consciousness as well.

Second, we have two kinds of selves: a single, eternal, real,
absolute Self and a set of changing, temporal, phenomenal, relative
selves, often called the ego. Like the water underlying the ocean's
foam, our Self is the Substance that stands under our self, or selves,
the ego. As Ramakrishna declares:

> The water and its bubble are one. The bubble has its
> birth in the water, floats on it, and ultimately is
> resolved into it. So also the individual ego and the
> supreme Spirit are one and the same. The difference
> is in degree; the one is dependent, the other
> independent. ([T04],11).

Like a relative self, a bubble is born, changes, and dies. Like our
eternal, absolute personal self, the water is the basis, the substance,
that remains. So while the bubbles that comprise our ego may
temporarily disappear in deep sleep, the underlying awareness, like
water, remains.

Following many other writers, we'll use capitalization to
differentiate awareness, our true *Self*, from the ego or personality,
which is commonly considered a person's *self*. A secular person may
see capitalization merely as a device for distinguishing our true
essence from the ego. Some religious people, however, may also see
it as distinguishing deity, the part of us that's God, from our lesser
selves.

Third, truly knowing our own deepest Self is equivalent to
knowing God, to enlightenment. Thus the Tao Te Ching's teaches

> Knowing others is wisdom;
> Knowing the self is enlightenment. ([L01],ch.33).

Similarly,

> [i]t is an axiom of the Sufis that what is not *in* a man he
> cannot know. The gnostic . . . could not know God and
> all the mysteries of the universe, unless he found them
> in himself. . . . In knowing himself as he really is, he
> knows God . . . ([N11],84-5).

Lastly, since our absolute Self is the world's Substance, the
enlighten person sees

> [t]his entire world, verily, is the self; other than the self,
> there is nothing. He sees all as the self, even as (one
> sees) pots, etc., as (but) clay. ([S09],69).

In fact, the entire mystical journey may be described as a process of
Self-realization, a process where the ego awakens to, and eventually
comes to know - comes to directly experience - its own basis, its
higher, absolute Self.

No Absolute Personal Identity

Earlier, we decided that an absolute personal identity must be
something that's unique, unchanging and absolute, something that I
am rather than something I possess. The Eternal certainly is unique,
unchanging and absolute. Moreover, It's something I am (or, better, I
am something that It is) rather than something I possess. In fact, It's
my deepest, realest self. But we also decided that an absolute personal
identity should be unique to me because if two people have it then "I"
would mean both of us.

Is the Eternal in any sense unique to me? Can it differentiate me
from everyone else? No. The spark of the Eternal that creates me in
no way differs from the spark that creates you. The Eternal is one,
undifferentiated, the same. So even if each and every one of us has a
consciousness, a soul, an Ultimate Ground of Existence that's
identical, It cannot be our absolute personal identity. So where is my
unique, unchanging, absolute and *distinct* identity? It's nowhere to be
found! Writes Nicholson:

> There is no real existence apart from God. Man is an
> emanation or a reflexion or a mode of Absolute Being.
> What he thinks of as individuality is in truth not-being;
> ([N11],154).

Thus, "I" as an enduring and distinct personal identity doesn't really
exist! Thus, as Ramakrishna observed:

> [J]ust as when one goes on removing the coats of an
> onion nothing is left over; so, in order to ascertain the
> self, when one goes on eliminating the body, the mind,
> the intellect, etc. and makes sure that none of these is
> the self, one finds that there is nothing separate called
> "I" but everything is He (God) and nothing but He . . .
> ([S01],394).

A different line of reasoning seems to have led Buddha to a similar insight. As we've seen, Buddha places consciousness in the same category as body, heart, and mind; that is, as caused and having a dependent type of existence.

> Have I not said, with many examples, that
> consciousness is not independent but comes about
> through the Chain of Causation and cannot arise
> without a cause? ([W07],64).

and

> [I]s consciousness permanent or impermanent?
> Impermanent . . .
> Now what is impermanent . . . is it proper to regard
> that thus: "This is mine. This am I. this is my self"?
> Surely not . . . ([B09],271).

He taught that a human being consists of five elements: body, feelings, perceptions, tendencies, and consciousness. Because ([B16],122) none of these, individually or in combination, is a self, there is no such thing as self. Thus

> . . . early Buddhists did not believe in a permanent self
> or ego . . . ([B16],122).

Rather, Buddhists believe the five elements

> . . . come together at birth, are dispersed at death, and
> therefore can be regarded only as convenient names
> for those basic elements of a human being, all of
> which are impermanent, involve suffering, and have
> no ego. ([B16],122).

So Buddha denies the existence of any absolute identity, distinct or otherwise, to a human being.

If, like Buddha, we consider consciousness changeable and conditioned then we have no absolute identity. On the other hand, if we consider consciousness identical to the Eternal then we have no *distinct* absolute personal identity. In either case, these word of Buddha apply:

> Just as the word 'chariot' is but a mode of expression
> for axle, wheels, the chariot-body and other
> constituents in their proper combination, so a living
> being is the appearance of the groups with the four
> elements as they are joined in a unit. There is no self
> in the carriage and there is no self in man . . . The
> thought of self is an error and all existences are as
> hollow as the plantain tree and as empty as twirling
> water bubbles. ([C04],115),

and

> The foolish man conceives the idea of 'self', the wise
> man sees there is no ground on which to build the
> idea of 'self' . . . ([C04],242),

although I would substitute "unenlightened" for "foolish." The
enlightened person, on the other hand, has realized their true Self. As
Nicholson writes:

> Gnosis, then, is unification, realisation of the fact that
> the appearance of 'otherness' beside Oneness is a
> false and deluding dream. . . . Gnosis proclaims that 'I'
> is a figure of speech . . . ([N11],85).

Thus, someone who has looked deeply into the basis of their own
personal existence realizes they possess no unique, unchanging,
absolute and distinct self.

In Buddhism, the insight that absolute personal selfhood has, in
fact, no real basis (i.e., doesn't actually exist) is labeled "nonself."
Buddha identified three traits that inhere in all entities:
impermanence, suffering, and nonself. We've already seen entities
lack permanence and always include yin traits that may considered as
imperfections conducive to suffering. Now we've seen how the third
trait, nonself, applies to us. Later, we'll see how it applies in general.

Two Views of Absolute Personal Identity

We have many relative personal identities, many relative selves,
selves that are created, change, and eventually vanish. Relative
personal identities exist (just as waves exists), but they do not reach
all the way down to the ultimate level. On the ultimate level, God
exists. And on that ultimate level, there's only one Entity. So, we can
follow Buddha and say we have no absolute personal identity, since
our identity is a created, transitory, created thing. Or we can follow

Hindu Vedanta and say we have a common absolute personal identity, and it's identical with God. But to say we have a distinct absolute personal identity - different from God and from other people's personal identity - and to say this personal identity is eternal, seems to say that more than one entity exists at the ultimate level. Moreover, it says that our union with God can only be partial, that we shall forever be in some measure separate from God.

There are religions that say this - Christianity is one, the Yoga philosophy that underlies the Hindu *Yoga Sutras* (refer, for example [Y01]) by Patanjali is another - but I do not believe it, and it's not part of the monist world view being presented. I realize that saying either we're really God or we really don't exist is a non-intuitive idea that goes against common sense but those are the only two alternatives monism allows. Moreover, that's exactly what many mystics see.

So, we may take two attitudes toward our absolute personal identity: the first, that it exists and is identical with the Real; the second, that we have no absolute personal identity. Mystics have employed both viewpoints.

The second attitude - that we have no absolute personal identity - is the safer attitude, but it's more discontinuous with our normal way of thinking. The first attitude - that our true Self is the Absolute - has historically been the more dangerous choice. Such identification was often interpreted by followers of a God who is a Person as the blasphemous claim "I am God." Their response was not always charitable. Therefore, the viewpoint that we have no true Self, that only God exists, is safer. Followers of a God who is a Person might interpret this claim as displaying intense - even excessive - humility. But I know of no case where a mystic suffered torture or death for excessive meekness.

Yet the second attitude demands we give up the idea of the thing nearest and dearest to us, the idea of our own self enduring over time. For if the relative selves are the only selves we possess, then we have no enduring unchanging self. The person we were at five years of age is dead and gone. True, the person we are now has descended from that person in a more intimate sense than we've descended from our ancestors, but we've descended nonetheless. We are not that person, but someone different. Just as the candle flame or whirlpool is descended from, but something different than, the candle flame or whirlpool of a moment ago.

This viewpoint, by the way, in a sense answers the afterlife problem - we need not be concerned with the afterlife since we do not even endure over one lifetime. Only a descendent of the person we are now endures. But it's not a complete answer since the existence of an afterlife is still an open question. Possibly, a descendent endures, in some sense or another, beyond death. Possibly, it does not. (A later chapter returns to these questions.) But in either case, strictly speaking we are dead and gone as soon as we lose or acquire one or another relative self. Just as adding or subtracting a single letter changes a word into another word or just a series of nonsense letters, adding or subtracting a relative self changes us into a different person.

Although the first viewpoint is more dangerous, it accords more with our normal way of thinking. We are accustomed to referring to our self, to assuming we have a self. This viewpoint says, indeed, we have a self and moreover that self is quite close to, and even identical with, God's Self. What prevents this viewpoint from leading to megalomania is the essential qualifications that, first, our Self is no different than the Self of anyone else and, second, we don't, in any real sense, "possess" this Self since the Self can be the possessor but never the possession.

The Ego as Possession

If all of the above seems shocking or silly, recall our empirical, every-day identity is not being denied. Certainly, our relative selves exist. Certainly, an "I" exists in the conversational and practical sense. What is being denied, however, is an identity that is eternally separate and distinct from God. On the ultimate level, there is nothing unchanging that distinguishes me from you or anybody or anything else. On the ultimate level, nothing but the Eternal exists.

Of course, in everyday life it's quite useful, even necessary, to use the concept of personal self. The word "I" is useful, even if what it refers to doesn't exist in the deepest sense.

Sometimes an inferred entity, a theoretical construct, is known not to exist but is nonetheless used. As an example, imagine a checker board where only one square is empty. We could keep track of which piece is moved where, but if we are only interested in which square is empty, we could track the location of the "hole" instead. As pieces are shifted, the hole moves from one square to another. Something similar is done in semiconductor electronics, where a missing electron is tracked. What actually happens is an electron hops from a full atom to the atom with the "hole" (missing electron), thereby transferring the hole in the reverse direction, to the full atom. But instead of concentrating on what's actually happening, it's easier and more useful to follow the motion of the hole. Thus, something lacking positive existence is considered as having positive existence, an interesting parallel to Augustine's view of evil.

Thus, a distinct absolute personal identity is an inferred entity, a theoretical construct, that, in fact, doesn't actually exist. "I" as it's used in the everyday sense is a theoretical construct that doesn't really exist. This was implied by and easily follows from two quotes we saw in the fourth chapter:

> God . . . is not one Being among others, but . . . dwells on a plane where there is nothing whatever besides Himself. ([D08],4),

and

> For all other things, ourselves included, compared to that pure and perfect Substance, are not even shadows. ([A06],101).

So as we approach the level of the Uncaused Cause, the level where nothing exists but the Ultimate Ground of Existence, then of necessity we approach a level where our separate existence and identity vanish. Perhaps this is why Ramana Maharshi says:

> One cannot see God and yet retain individuality. ([P12],213).

Another point: whether our non-distinct absolute identity is our awareness or soul, or whether it fails to exist at all, it follows our body, feelings, and thoughts are our possessions rather then our selves. Therefore, referring to the body, feelings, and thoughts as possessions, as something other than ourselves, would be appropriate. Some mystics have done so.

For example, the Christian mystic Henry Suso (1295-1365) typically writes of himself - or rather his body, emotions, and thoughts - in the third person ([U01],218). He calls them the "Servitor." Thus, Suso is speaking of himself when he writes:

230

> One night after matins, the Servitor being seated in his
> chair and plunged in deep thought, he was rapt from
> his senses. ([U01],404).

Swami Rama Tirtha is another mystic who habitually refers to himself
in the third person - as "Rama" or "he" rather than "I". Of the many
examples that can be given, one is particularly interesting. (Note: "he"
in the first sentence refers to Rama, not the scientists.)

> The scientists may or may not agree today with Rama,
> but he is fully convinced that even the smallest particle
> of the dust of this universe is a storehouse of energy
> which may be possible to be released, under suitable
> conditions, more or less like fire from fuelwood or like
> heat from coal. In other words, energy is condensed
> into matter which can be reconverted into energy
> which is stored in it. ([P14],109).

These words were spoken in 1905, about the time Einstein was
coming to the same conclusion.

A Living, Conscious Absolute

If our consciousness isn't the Everlasting then we have no enduring
identity, no absolute personal identity. But what if consciousness is
the Absolute? Then the Absolute and consciousness are one and the
same. If our consciousness is the Absolute, then the Absolute is in
some sense conscious. In a sense, It's alive. Therefore, It might have
some of the attributes we usually associate with persons. In other
words, It might have a personal side as well as an impersonal side.
Therefore, thinking of and relating to the Eternal in a personal way, *as
if* It is in some sense an actual, distinct Person might make sense.

Someone who regards the Root *as if* It's a distinct Person is
obviously similar to someone who regards their God *as* an actual
Person. That is, believing the impersonal Ultimate Ground of
Existence has a conscious, personal aspect is very close to believing
in a God who is a Person. There is a difference, but is it important?
The next chapter, which investigates the relationship of the Eternal to
the "supernatural," explores this question.

9

- The God Who is a Person -

Chapter Summary: This chapter explores the idea of relating to God as if He/She is a Person and relating to the Uncreated *as if* It is a Person. The power of such ideas of God is discussed, as well as their advantages and disadvantages.

We've discussed the relationship of the Real to our external and internal worlds. What of Its relationship to the supernatural? Does a supernatural realm even exist? In particular, do Gods who are Persons exist? Let's begin our discussion of these questions by examining some more ideas about the perennial philosophy.

As distilled by Aldous Huxley, the perennial philosophy's "highest common factor" is ([S18],13) a nucleus of four fundamental principles and an fifth optional principle. The first principle says the world of people and things is a manifestation of a single Eternal Ground without Which it could not exist; the second, that we can directly experience this Ground and even consciously unite with it; the third, that we possess two selves, an ego and a Self. We've already discussed these ideas. The fourth principle, which we'll see more about later, is that life's ultimate end and aim is unitive knowledge of the Eternal. Huxley's fifth principle is of concern to us in this chapter. We'll begin discussing it by briefly re-examining the first principle.

Monism and Gods who are Persons

The first principle says that everything is a manifestation of one Eternal Substance, a single Reality. It's been called "pantheism" and "monism." Religious monism is the idea that God is the One and only, the sole Reality, the One without a second. We've seen a monist description of the universe and ourselves.

In contrast to monism, monotheism is the more familiar idea that some God who is a Person, a Person supreme among all persons, has created the universe but remains distinct from it. In monotheism, God is one entity among many. God, people, animals, and inanimate objects all exist and are distinct.

So, in monotheism there is only one God. In monism there is only One. Period.

Religious monism says that everything - us, a lamp, a worm - has the same Ultimate Ground of Existence, God. It may seem absurd when it's first encountered, perhaps because the idea of God as some Person is so ingrained. In his youth, Swami Vivekananda met the monist viewpoint in the teachings of Ramakrishna. He was less than impressed.

> What's the difference . . . between this and atheism?
> How can a created soul think of itself as the Creator?
> What could be a greater sin? What's this nonsense
> about *I am God, you are God, everything that is born
> and dies is God?* The authors of these books must
> have been mad - how else could they have written
> such stuff? ([I04],205).

Vivekananda's statements were based on a misunderstanding of monism. Saying the God which is not a Person is the Ultimate Substance of both any person and any God who is a Person doesn't say the two are equal or identical. Just as saying both ice and steam are water doesn't say they're equal or identical. Ice and steam are different, but share a common ground. People and Gods who are Persons are different but share a common Ground.

So, monism doesn't equate the creature with any God who is a Person. It doesn't equate any human soul with the Creator. And it doesn't confuse a God who is a Person, the Creator of the universe who is distinct from it, with the God which is not a Person, the creator and upholder of the universe at this very moment in the sense of being its Eternal Substance. Rather monism maintains a distinction between these pairs of very different ideas.

For example, Shankara, one of religious monism's foremost spokesman, carefully maintained the distinction between the creature and "Iswara," his term for the God who is a Person. In an introduction to his *Crest-Jewel of Discrimination*, the translators write:

> We can become Brahman, since Brahman is present
> in us always. But we can never become Iswara,
> because Iswara is above and distinct from our human

233

> personality. . . . [W]e can never become rulers of the
> universe - for that is Iswara's function. ([S11],23-4).

They even label ([S11],24) the desire to become Iswara as madness
and Lucifer's sin. However, they seem to slightly misstate our relation
to Brahman, perhaps for the sake of the parallelism "we can become
Brahman . . . But we can never become Iswara." In fact, we already
are Brahman, the Source, the Ultimate Ground of Existence. Our
conscious realization of this fact is all that's lacking. Our ego has not
yet realized its Basis.

So monism doesn't, as the young Vivekananda feared, teach that a
person can become some God who is a Person, just as it doesn't teach
that a person can become a rock. However, it does say that people,
Gods who are Persons, and rocks all lack the ultimate reality
possessed by the Real. It denies the ultimate reality of people, the
universe, and Gods who are separate, distinct Persons, calling them
"Maya," illusions, projections of Uncreated Light. As Ramana
Maharshi taught:

> The Self alone exists and is real. The world, the
> individual and God are . . . imaginary creations in the
> Self. They appear and disappear simultaneously.
> Actually, the Self alone is the world, the "I" and God.
> All that exists is only a manifestation of the Supreme.
> ([T03],16).

The idea that the God who is a Person isn't fully real also occurred
to certain early, heretical Christians who

> . . . insisted on discriminating between the popular
> image of God - as master, king, lord, creator, and
> judge - and what that image represented - God
> understood as the ultimate source of all being . . . .
> "the depth" . . . an invisible, incomprehensible primal
> principle. ([P01],38).

Similarly, "Dionysius" taught, in the words of Rolt, that

> . . . God is but the highest Appearance or
> Manifestation of the Absolute. ([D08],40).

So from the monist point of view, the God who is a Person and
mundane entities have this in common - their existence is grounded,
as is all entities, in the Ultimate Ground of Existence. Later in his life,
Vivekananda came to understand monism. He shocked Christians by
claiming there was no essential difference between Jesus and the

lowliest of God's creatures, since they are both manifestations of the same Godhead, the same Eternal Substance.

The Religious Monist

Since Gods who are Persons aren't absolutely and ultimately real, why not just dispense with them entirely? A certain kind of religious seeker often does. This kind of person is usually capable of deeply loving entities which are just abstractions to other people, such as Truth, Love, or the Ultimate Ground of Existence. For someone with this temperament such things are quite real.

For example, the German philosopher Friedrich Nietzsche

> . . . lived with his intellectual problems as with realities, he experienced a similar emotional commitment to them as other men experience to their wife and children. ([N12],11).

Imagine someone with Nietzsche's temperament who is also religious. Isn't it likely they'd naturally relate to God intellectually and philosophically? For example, might they not tend to think of God as the Root and Source? And might they not prefer this relationship over an emotional one where, for example, they are the child and God is the Divine Parent?

Certain mystics of Nietzsche's temperament have found the intellectual and philosophical approach congenial. They've been as devoted to the God which is not a Person as Nietzsche was to his intellectual pursuits, and as other people are to their spouses and children. For such mystics, Huxley's first four principles - his "simple working hypothesis" - are sufficient. Their religion doesn't require a God who is a Person.

> A man who can practice what the Indians call Jnana yoga (the metaphysical discipline of discrimination between the Real and the apparent) asks for nothing more. This simple working hypothesis is enough for his purposes. ([S18],17).

Someone who discriminates between the Real and the unreal practices what's called "Jnana yoga" in India. So, they're a "Jnana yogi." Typically, the jnana yogi is a religious person with an introverted, cerebrotonia personality. They're strong in intellect, and often wary of emotion. I could use the term "jnana yogi" to refer to such an individual. Although it's a Hindu term, such individuals appear in any culture. Instead, I'll use the term "religious monist" or simply "monist."

Monists naturally tend to philosophical and metaphysical approaches to Reality. They often disdain highly emotional presentations of religion based on the life of some religious Personality or Incarnation. Sometimes they even condemn such presentations as mere histrionics. For instance, Totapuri, one of Ramakrishna's teachers, criticized him for worshiping Krishna with dancing, chanting, and clapping hands. Totapuri sarcastically asked ([L07],166-7) if Ramakrishna was clapping bread dough between his hands. Totapuri was a strict monist. He regarded Gods who are Persons as mythological, that is, as having less real existence than you or I.

Are Gods who are Persons mythological? Or are They real? We've seen that our own existence and identity vanish as we approach the level of the One and the All. The existence of God considered as an individual Person, as one entity among many, necessarily suffers the same fate.

> . . . Brahman only appears as Iswara when viewed by
> the relative ignorance of Maya. Iswara has the same
> degree of reality as Maya has. ([S11],23).

All manifestations exist in the Ultimate Ground of Existence and lose their separate existence and identity as we approach the Absolute.

Yet we may say the same thing in a more positive way: Gods who are Persons may be as real as you, or I, or the world we see around us. Gods who are Persons may have a more than mythological existence, they may have an existence as real as anything else. Of course, they might just as well fail to exist at all. Without proof, a scientific religion could neither deny or affirm the existence of separate Gods who are Persons.

The Optional Principle

Huxley believed the religious monist was the exception rather than the rule. He wrote Jnana yoga was

> . . . exceedingly difficult and can hardly be practiced,
> at any rate in the preliminary stages of the spiritual life,
> except by persons endowed with a particular kind of
> mental constitution. . . . ([S18],17).

Many religious believers don't possess the monist temperament. They aren't capable of loving Truth in the "abstract." Many people, however, are capable of loving a God who is a Person. They find it easier loving someone like Jesus, who died for us, or the cute baby Krishna. In other words, many believers are capable of practicing a religion which includes some form of Huxley's fifth principle. This principle affirms

> . . . the existence of one or more human Incarnations
> of the Divine Ground, by whose mediation and grace
> the worshiper is helped to achieve his goal - that
> unitive knowledge of the Godhead, which is man's
> eternal life and beatitude. ([S18],17).

From the fifth principle follows the bulk of popular religious belief and practice. In Christianity and Hinduism, the story of an Incarnation's life is scripture; reverence and devotion for an Incarnation is piety; and pleas to an Incarnation are prayer. And if we take "Incarnation" in a wide sense, if we view any God who is a Person as an Incarnation of the God which is not a Person, then we find a similar situation in Judaism and Islam. That is, Jewish and Islamic scripture consists of records of the actions of an "Incarnation" (Jehovah and Allah), as well as the actions of prophets. Again, reverence and devotion for an Incarnation are piety, and pleas to an Incarnation are prayer.

Even Buddhism has, in a sense, an Incarnation. Buddhism is perhaps the most impersonal and "unreligious" of religions, so much so that some adherents claim it's not religion at all, but a philosophy. Geoffrey Parrinder describes such believers when he writes:

> It is common nowadays for Buddhist apologists, in
> East and West, to claim that the Buddha was only a
> man, or a man like us . . . ([W07],6).

Yet the needs of some Buddhists have forced Buddhism to undergo some measure of "personalization," so much so that Parrinder (perhaps overstating the situation) continues:

> . . . but no Buddhist thought this in the previous two
> thousand years, since the Buddha was for him the
> object of faith and the means of salvation . . .
> Functionally he is the Supreme Being . . . ([W07],5-6).

In any case, belief in some God who is a Person underlies religion as practiced by - and God as worshiped by - the great majority of believers.

So, religion as it's usually practiced is based on the perennial philosophy's fifth principle, a principle Huxley called optional since some temperaments don't need it. But for most individuals the fifth principle isn't optional at all. It's required. If they're to have any religion at all, it will be a religion based on the fifth principle. For many, monotheism in its most real and actual form is essential. Emotional attachment to some God who is a Person may be their primary, or only, motivation for living a religious life, and for pursuing a spiritual quest.

The Power of the Personal Aspect

Belief in Gods who are Persons is quite powerful. Its influence in civilizations, past and present, is obvious: much of the world's population lives within walking distance of a building dedicated to some God who is a Person. Such Gods play a vital role in the lives of many, if not most, religious people. They function as focus of worship, parent, friend, protector, judge, or teacher. Their most important function, however, may be as a bridge to the Unchangeable, the Real. As Underhill observes:

> The peculiar virtue of . . . Christian philosophy, that which marks its superiority to the more coldly self-consistent systems of Greece, is the fact that it re-states the truths of metaphysics in terms of personality: thus offering a third term, a "living mediator" between the Unknowable God, the unconditioned Absolute, and the conditioned self. ([U01],104).

What's the value of re-stating "the truths of metaphysics in terms of personality"? To some religious monists there's none. They have no need of a mediator, and prefer to approach the Unconditioned directly. But to monists who seek to use emotions as well as intellect in the journey to vision of the Real, the personification of metaphysical truths can be quite valuable.

Transforming one's entire person is a difficult, long-term process. Changing one's daily life to reflect religious and philosophical truths can be an arduous task. Emotions are a powerful aid to this transformation, even to those of the cerebral, jnana temperament. One way the religious monist can engage their emotions is to regard the Uncreated Light *as if* It were a Person. Erwin Goodenough describes some ancient people who seemed to have regarded their Gods in this way. In *By Light, Light,* a book about Hellenistic Judaism, whose most famous exponent was Philo, he writes:

> . . . it is not the mythology itself which matters but the mythology as a symbol of metaphysical truth. The mystery is not a path to Isis or Attis; it is a path to Reality, Existence, Knowledge, Life, of which Isis or Attis is the symbol. The value of Isis, that is, is to make the intellectual concept emotionally realizable, something which can be taken out of the cold words of formulation and made radiantly alive within the longing hearts of mankind . . . ([G01],1).

So, some ancients didn't regard Isis or Attis as actual Gods who are Persons, as real, distinct personalities, as separate Entities. Rather Isis and Attis functioned as personifications of the Ultimate Ground of Existence. Those ancients used the "personal aspect" of the Eternal to help make higher truths "emotionally realizable" and "radiantly alive."

Goodenough observes certain Hellenistic Christians employed the same device. These Christians regarded Christ as a personification of the Eternal Light.

> The same process is illustrated in Christianity. The early Christians seem to have been content with the mythological assertion that Jesus was the Son of God and would return from the clouds to assert his power . . . Such a religion in itself meant nothing to the Hellenistic religious thinkers. Christ almost at once became to them the Logos, the Sophia . . . ([G01],2).

Logos and Sophia, of course, were Greek philosophical concepts. Goodenough continues that after this transformation

> . . . Christianity became another and more adequate means of making emotionally real and accessible the old Hellenistic abstractions . . . ([G01],2).

Now, he observes,

> . . . it was ready to conquer the Graeco-Roman world. ([G01],2).

Actual and Operational Monotheism

We've seen that the separate existence of you, I, the universe, and Gods who are Persons vanishes when we approach the Absolute. For someone temperamentally suited to the worship of a God who is a Person, this leaves two unsatisfactory choices. First, they can choose to worship the ultimately real God which is not a Person. Second, they can choose to worship some ultimately unreal God who is a Person. A very poor choice indeed.

There is, however, an alternative. Someone might choose to relate to, and even to worship, the Self-Existent *as if* It were alive and conscious, *as if* It were a Person.

We've seen that consciousness may be considered identical with the Absolute. If we turn this around, we can say the Self-Existent is Consciousness, is conscious and living. Thus the Root and Source may be regarded *as* a Person, as Friend or constant Companion or Spouse, for example, or as Mother and Father.

After all, the Source gives "birth" to us at this very moment. Our existence depends on the Self-Existent more fully and intimately than it depends on our parents. For we'll continue existing after our parents are gone. But we couldn't continue to exist if the Ultimate Ground of Existence ceased to exist. So, the Real is our Father and Mother in a very actual and literal sense. For It creates us and keeps us at this very moment, and is with us always.

So we can think of the God who is not a Person as a Person, as Mother, Father, Companion, Friend or Spouse, if we desire. If we want, It can be a living, conscious Father and Mother, an eternal Companion and Friend, even a Spouse. So, a monist who has no belief in God as an actual, separate Person, as one entity among others, might nonetheless choose to think of, and even worship, the Absolute and Source *as if* It was a Person.

After all, a person still has some emotions no matter how mind-centered they are. True, some monists choose to ignore their emotional faculties and approach God solely through their intellect. They choose to regard and worship the Real as an entirely non-personal Entity. But others, even if they're predominately mind-centered, choose to also use their emotions in the journey to God. They choose to sometimes regard the Ultimate Ground of Existence as Mother or Father, Friend or Spouse. Such a choice provides them with an Entity able to attract both heart and mind. Such a God (who is *like* a Person) truly is eternal, omnipresent, the Mother and Father of all.

Thus we may distinguish the actual monotheism of the monotheist from the "operational monotheism" of some religious monist. Operational monotheism is a type of *monism*. It acknowledges the separate God who is a Person as ultimately, ontologically false. But it nonetheless worships the Uncreated as if It is a separate entity, a Person among persons.

> . . . [T]o conceive one's self as separate from God is an error: yet *only when one sees oneself as separate from God, can one reach out to God.* (Palmer, *Oriental Mysticism*, in [U01],108).

Operational monotheism is actually a form of monism. It's a pseudo-monotheism. The Eternal is regarded as a Person, even though the believer realizes It is not actually a Person.

The Dual Aspects

A monist who practices operational monotheism regards the Eternal as if It were a Person. Such a monist emphasizes the "personal aspect" of the Ultimate Ground of Existence. Thus we may distinguish two sides or aspects of the One. When we regard It as an non-personal Entity, we emphasize It's impersonal aspect. When we regard It as a Person, we emphasize It's personal aspect. It may seem paradoxical to regard an "It" as a Person, to regard the God which is not a Person as a Person. But the God which is not a Person is also the God which is not a Thing. Thinking of the Uncreated as a Person is as accurate, or inaccurate, as thinking of the Real as a Thing. The Self-Existent is like a person and like a thing. Yet It is neither.

In engineering there are stable and unstable balances. A ball in the bottom of a bowl is in a state of stable balance: shake the bowl and the ball moves but eventually returns to balance. A ball at the top of an inverted bowl, or on the peak of a mountain, is in a state of unstable balance. Disturb the ball and it doesn't eventually return to the top. Instead, it rolls to one side or another and stays there. Only with effort is the ball returned to the mountain peak.

The idea that the Eternal Reality has a living, conscious, Personal aspect and yet is not an actual person, not one entity among many, seems to put the mind into an unstable balance. The mind tends to fall to one side or other of this truth. On one side, the tendency is to think of the Eternal as a non-personal Energy which one *may* regard as a Person (especially if one is a rather fuzzy thinker). On the other side, the tendency is to think of the Eternal as really a Person among other persons, a person who theoretically has non-personal qualities - of no great consequence. A mind trying to hold on to the dual nature of this situation often falls to one side or the other.

It seems the dual aspects of the Ultimate Ground of Existence, the impersonal aspect and the living, consciousness aspect, both represent It imperfectly. To use a familiar analogy from physics, the twin ideas of "particle" and "wave" both represent light imperfectly. Light is a physical entity which has both a particle nature and a wave nature. Yet light is truly neither particle nor wave. In the same way, the Eternal truly has both a personal aspect and an impersonal aspect. Yet the Real is truly neither person nor thing.

Starting now, I'll use the phrase "God *who* is not a Person" to emphasize the Self-Existent's dual aspects, personal and impersonal. So the phrase "God *which* is not a Person" refers exclusively to the impersonal aspect of the Real, as it has throughout this book. And "God *who* is not a Person" refers to both the personal and impersonal aspects of the Ultimate Substance. And as always, "God who is a Person" refers to some God such as Jesus, Krishna, Jehovah, or Allah.

The phrase "God who is not a Person" attempts to remind us of the Eternal's dual aspects. It's easy to forget one aspect or the other. When this happens, confusion and misunderstanding often result. Two instances follow where an author apparently failed to fully grasp the idea of the Self-Existent's dual aspects. One author was writing about Ramakrishna; the other, Guru Nanak.

Confusion of Actual and Operational Monotheism

Sri Ramakrishna worshiped the Eternal as Mother, specifically the Hindu goddess Mother Kali. He deeply yearned for direct experience. He prayed and wept for it and eventually in his frustration almost went mad. He felt ([S01],143) as if his heart and mind were being wrung like a wet towel. Sometimes bystanders assumed he grieved the loss of his human mother, and offered their sympathy. In fact, he grieved that he had not yet had the vision of God.

Eventually, driven by longing and despair Ramakrishna resolved to end his life. As he reached for a sword,

> . . . suddenly I had the wonderful vision of the Mother .
> . . I did not know what happened then in the external
> world . . . But, in my heart of hearts, there was flowing
> a current of intense bliss, never experienced before,
> and I had the immediate knowledge of the light, that
> was Mother. . . . It was as if houses, doors, temples
> and all other things vanished altogether; as if there
> was nothing anywhere! And what I saw, was a
> boundless infinite conscious sea of light! However far
> and in whatever direction I looked, I found a
> continuous succession of effulgent waves coming
> forward, raging and storming . . . ([S01],143).

Ramakrishna declared he had "immediate knowledge of the light, that was Mother." He described his vision as a vision of shining conscious Light. He had prayed for the Mother to reveal herself, and the Mother, Brahman, the Uncreated Light, had revealed Herself - as a shining ocean of Light and Consciousness. The situation seems clear.

Christopher Isherwood's *Ramakrishna and His Disciples* recounts the story of Ramakrishna's life. In it, Isherwood writes:

> Ramakrishna knew that Mother Kali was not other
> than Brahman. ([I04],118).

Presumably then Isherwood knew "Mother Kali" was a personalized label for the God who is not a Person. Yet he wonders if Ramakrishna also saw a woman in his vision, specifically Kali. He writes:

> It is not quite clear from Ramakrishna's narrative
> whether or not he actually saw the form of Mother Kali
> in the midst of this vision of shining consciousness.
> ([I04],65).

Did Ramakrishna see a woman in his vision? Both Isherwood ([I04],65) and no less than a direct disciple of Ramakrishna decide he did. Why? Because afterwards

> . . . as soon as he had the slightest external
> consciousness . . . . he, we are told, uttered
> repeatedly the word 'Mother' in a plaintive voice.
> ([S01],143).

Lack of appreciation of the Eternal Light's dual aspects forces "Mother" to be taken as Mother Kali, a God who is a Person. Lack of appreciation leads to confusion and uncertainty since Ramakrishna talks of the Mother yet describes an experience of Light.

Who was the God Ramakrishna longed for? Was it some God who is a Person, specifically a Hindu goddess, Mother Kali? Or was it the Self-Existent, in his own words

> . . . the universal Mother, consisting of the effulgence of pure consciousness . . . ([S01],255)?

I believe it was the Real. Once, he had declared Brahman

> . . . is Light, but not the light that we perceive, not material light. ([G03],307).

He had also said

> [t]he attainment of the Absolute is called the Knowledge of Brahman . . . ([G03],307),

I believe Ramakrishna wanted - and finally received - first-hand knowledge of Brahman, a direct experience of the Uncreated Light. Ramakrishna was, I believe, a monist practicing operational monotheism.

So if the dual nature is understood there is no cause for confusion. The Mother *is* the Uncreated Light. Thus there is no reason to suppose Ramakrishna also saw some female image. The Mother he referred to was the Mother of the Universe, the Uncreated and Eternal Light.

Like Ramakrishna, Guru Nanak, the founder of the Sikh religion, also seems to have known the dual aspects, the personal and impersonal aspects of God. In *Guru Nanak and the Sikh Religion*, W. McLeod writes ([M09],164) that God for Nanak was "the one" and that Nanak often affirmed "There is no other." McLeod wonders if such statements should be understood in a monotheistic sense -

> Does it refer to the uniqueness of God, to His absolute difference in essence from all other beings . . . ([M09],164).

- or in a monist sense -

> . . . or does it denote the unity which denies ultimate reality to all phenomenal existence? ([M09],164).

McLeod chooses monotheism.

> If we are compelled to choose between these polar conceptions our choice must settle upon the former alternative. Guru Nanak's thought cannot be made to conform to the categories of *advaita* doctrine without

equating his concept of God with the ultimately unreal
Isvara of Sankara's philosophy . . . ([M09],164-5).
Nonetheless McLeod finds elements of monism in Nanak's teaching.

Nanak himself explicitly declares notions of 'duality' . .
. the essence of man's problem, and the overcoming
of such notions to be a vital aspect of man's quest for
salvation. Moreover, we must also acknowledge the
stress which he lays upon divine immanence and
upon the fundamental importance of this immanent
revelation in the quest for salvation. ([M09],165).

Again, there's confusion. Nanak wasn't a monist but often spoke
like one anyway. Is it Nanak's statements which are confused and
inconsistent? Or is it their interpretation?

If the two aspects of the Uncreated are understood, then Nanak's
beliefs are neither inconsistent nor confusing. If Nanak is seen as a
monist who often used an "operational-monotheistic" mode of
speaking, then there is no confusion. Nor is there a compulsion to
choose between the "polar conceptions" of monism and monotheism.
Guru Nanak, I believe, was an operational monist.

Unrecognized Experience of the Personal Aspect?
The dual aspects are easily forgotten, even by the spiritually aware.
The mind easily slips to one side or the other of the unstable balance.
It's natural, therefore, that the personal aspect of the Eternal might
over time in the minds of the average believer become some God who
is a Person.

In Judaism and Christianity Yahweh is the name of a God who is
a Person. How did the word "Yahweh" come to be connected with
God? In Exodus 3:14 we read

And God said unto Moses, I AM THAT I AM: and he
said, Thus shalt thou say unto the children of Israel, I
AM hath sent me unto you. ([H08], Ex3:14).

A footnote explains

I am who am: apparently this utterance is the source
of the word Yahweh, the proper personal name of the
God of Israel. It is commonly explained in reference to
God as the absolute and necessary Being. It may be
understood of God as the Source of all created
beings. ([N02],61).

Thus the phrase "I am who am" may have once referred to the Source, the Root, the Ultimate Ground of Existence. That is, "Yahweh" may have originally indicated the personal aspect of the self-existent and eternal Reality. In time, however, it became the name of a God who is a Person, a God separate from creation, one Entity among others.

In the mind of the average believer, the Eternal's personal aspect is liable to change into a separate, distinct Person. Conversely, in the minds of the mystics a separate, distinct God who is a Person tends to change into a personal aspect of the Eternal. Nicholson describes this phenomena when he writes the Sufi's

> Light, Knowledge, and Love . . . rest upon a
> pantheistic faith which deposed the One transcendent
> God of Islam and worshiped in His stead One Real
> Being who dwells and works everywhere . . . ([N11],8).

So, we may describe experience of a God who is a Person as unrecognized experience of the personal aspect of the Uncreated, as unrecognized experience of the personal side of the God which is not a Person. In other words, as unrecognized experience of the God who is not a Person.

Ramakrishna spoke of such experience when he said:

> For the bhakta He assumes forms. But he is formless
> for the Jnani. ([T04],3).

A Bhakta yogi uses a heart-centered approach to the Eternal, who usually is some God who is a Person. Someone who practices Bhakti yoga uses the emotions to draw nearer to God. Ramakrishna continues:

> . . . Brahman, Existence-Knowledge-Bliss Absolute, is
> like a shoreless ocean. In the ocean visible blocks of
> ice are formed here and there by intense cold.
> Similarly, under the cooling influence, so to speak, of
> the bhakti of Its worshipers, the Infinite transforms
> Itself into the finite and appears before the worshiper
> as God with form. That is to say, God reveals Himself
> to His bhaktas as an embodied Person. Again, as, on
> the rising of the sun, the ice in the ocean melts away,
> so, on the awakening of jnana, the embodied God
> melts back into the infinite and formless Brahman. . . .
> But mark this: form and formlessness belong to one
> and the same Reality. ([T04],3-4).

246

Notice that "*blocks* of ice are formed" and "the Infinite transforms Itself into the *finite*." We've seen the problem with God actually being a Person is that personhood is too limited and finite. So perhaps "transforms" is too strong a word. Perhaps the Infinite *appears* as a finite Person, but remains infinite nonetheless.

Of course, there are different kinds of limited, separate existences. A person is a distinct, separate entity, just as a block of ice is a distinct, separate entity. Yet a person is higher in the "Great Chain of Being" than a rock. Similarly, a God who is a Person is higher than a person. Yet all three are still limited, distinct, separate manifestations of the Uncreated Light. All three are "blocks of ice" in the shoreless ocean of Uncreated Light.

So angels, demons, Gods who are Persons, and other "supernatural" entities - if they exist - are no more supernatural than a rock. Since the Ultimate Ground of Existence underlies their existence as It underlies the existence of the rock, one is no more above nature - i.e, super-natural - than the other. Therefore the entire realm of existence - rocks, angels, and Gods who are Persons - is united, one. It is all natural. Or, if you prefer, it's all supernatural since the God which is not a Person underlies it all. Therefore, the "supernatural" realm is actually a part of the natural universe. However, mentally dividing the natural realm into the interior domain, the exterior domain, and the "supernatural" domain may still be useful on occasion. It should be remember, however, that the "supernatural" realm is actually a part of the natural universe, even if the double quotes are omitted.

The quote also has "He assumes forms." Plural. There are many different Gods who are Persons in the same ocean of Uncreated Light, even as there are many different blocks of ice in the ocean. This brings us again to a question raised in a previous chapter: why mystical experiences of the God which is not a Person generally agree, while experiences of Gods who are Persons often disagree. It's because the Self-Existent assumes different forms for different worshipers.

As a young boy, I heard a T.V. talk show host describe his trip to Japan. He remarked some Christian statues of Jesus and Mary there had oriental features, particularly eyes. At the time, this amused me; I thought such statues ridiculous. Only later did I realize the distinctly

Western, Caucasian appearance of the statues of Jesus, Mary, and the saints I had seen.

Our Gods who are Persons are Gods who are persons *like us*. But suppose creatures existed on another planet and looked like, say, spiders. And suppose these creatures would experience an uncontrollable revulsion at the sight of a human. Then their mystics who experienced a God who is Person would probably experience a God who was a spider, not a human. Therefore, Gods who are Persons are, to some extent, our own creations. I don't mean they are *entirely* our creation, that they are mythological. I mean only that our personalities and limitations condition our experience, that the Eternal seen as some God who is a Person is not seen as It is in Itself. It's seen in a form It assumes for our benefit, as a accommodation to our limitations.

Therefore, there can be as many Gods who are Persons as there are people (or spiders) to see them. As Rufus Jones remarked:

> . . . there is always a profound subjective aspect of interpretation of the Divine . . . in terms of the expectation of the individual, and in terms of the prevailing climate of opinion. ([J02],86-7).

I once knew a man who with perfect sincerity believed Jesus had been tall, light-skinned, blond-haired and blue-eyed. I could, no doubt, find someone else who believes Jesus was short, brown of skin, hair, perhaps eyes, too. Do these two people have the same God who is a Person? Suppose each of them became a mystic, intent on direct experience of their God. Suppose they worshiped and prayed constantly, hoping for an experience of God. Would it be surprising if one experienced a tall, blond-haired Jesus, the other, a short, brown-skinned Jesus? After all, if the Eternal assumes a form for our benefit, might it not assume the form we wish to see?

Advantages of Monism and Operational Monotheism

Relating to the Eternal as if It's a Person is obviously very similar to relating to a God who is an actual Person. There's a difference, but is it important?

The difference may not be important to the average believer. Someone who only knows of a God who is a Person can progress quite far in religious or even mystical life. As Shankara writes:

248

> Devotion to . . . the Personal God, may lead a man
> very far . . . it may make him into a saint. ([S11],23),

More than that, it may make him a mystic who achieves first-hand
knowledge of God, or some restricted kind of "union." Many saints
and mystics have known nothing of the God which is not a Person.

Yet, experience of the Eternal is obviously less conditioned by the
experiencer. It's purer, truer, and more universal than experience of
some God who is a Person. Therefore, Shankara describes such
experience as a higher goal than monotheism offers.

> But this is not the ultimate knowledge. To be
> completely enlightened is to go beyond Iswara, to
> know the Impersonal Reality behind the personal
> divine Appearance. ([S11],23).

To know the Reality behind the appearance of Gods who are Persons
implies either first-hand or unitive knowledge of the Ultimate Ground
of Existence.

Many mystics, Meister Eckhart, for one, valued union far above
first-hand knowledge. In fact, Eckhart placed the Godhead so far
above any God who is a Person he wrote:

> . . . I pray God to rid me of God, for my essential being
> is above God . . . ([M12],219).

This phrase is no doubt shocking as translated. To understand it
properly, we must understand Eckhart's ideas of "God" and
"Godhead."

Eckhart drew a sharp distinction between the God who is a Person
and the Ultimate Ground of Existence. In fact, Rufus Jones claimed
([J03],224) this distinction was at the very core of Eckhart's thought,
which he described as follows:

> He whom we call "God" is the Divine Nature
> manifested and revealed in personal character, but
> behind this Revelation there must be a revealer - One
> who *makes* the revelation and is the Ground of it, just
> as behind our self-as-known there must be a self-as-
> knower - a deeper *ego* which knows the *me* and its
> processes. Now the Ground out of which the
> revelation proceeds is the central mystery - is the
> Godhead. . . . This unrevealable Godhead is the
> Source and Fount of all that is . . . ([J03],225).

Eckhart, it seems, had direct experience of the Godhead. He
wrote:

> When I still stood in my first cause, I had no God, I
> was cause of myself. . . . But when by free will I went
> out and received my created being, then I had a God.
> ([M12],116).

When Eckhart's Consciousness was united with the Real, the First Cause, there was no separate, distinct God who is a Person. When It descended to the plane of duality, then the universe, Eckhart's ego, and a God who is a Person all reappeared.

And so we see why he "prayed to God to rid him of God." Eckhart prayed to stay united with the Real, and not to fall back into duality where a separate God who is a Person exists. Though "for my essential being is above God" is true, I much prefer "for God's essential being is above God" or "for the Godhead is above any God who is a Person." All three versions express the same thought - that the Godhead is above any God who is a Person - but the last two avoid the appearance of blasphemy.

In any version, the meaning is the same: the Godhead, the God which is not a Person, is the Ground of, the Source of, the Basis of - and therefore above - any God Who is a Person. Shankara and Eckhart agree: experience of the God which is not a Person is above experience of any God who is a Person.

So, for mystics who seek the highest goal, union with the Eternal, the distinction between the Eternal Substance and Gods who are Persons may be important. The distinction may also be important to a scientific religion.

We've seen Gods who are Persons are subjective, to some extent. That's why they're experienced differently by different people. The God which is not a Person, on the other hand, is objective. Different mystics experience the same Reality. Therefore, the more universal and objective God which is not a Person is a better foundation for a scientific religion than some particular God who is a Person. Science has found itself more able to study objective phenomena. Physics and chemistry, for example, are often predictive and exact while psychology and sociology often aren't. Therefore, an emerging scientific religion might decide to study the objective Ultimate Ground of Existence, rather than the multitude of different Gods who are Persons.

The distinction could also be important if the human race ever encountered another intelligent species. Any species who can experience some God who is a Person would, presumably, be able to experience Its Ground, too. That is, any species which possesses consciousness would be able to experience God in the same way, as the Eternal Light, Consciousness Itself. The God which is not a Person might be the only God different species could have in common.

10

- Kinds of Existence -

Chapter Summary: This chapter examines some philosophical concepts - component entity, relative existence, action, voidness and emptiness - that apply to the universe, personal identity, and Gods who are Persons. The ideas of absolute existence and identity are also explored. The chapter concludes by summarizing Part II and introducing Part III.

The previous three chapters discussed the relationship of the Real to the universe, to ourselves, and to the supernatural, particularly to Gods who are Persons. We investigated the Eternal's relationship to three domains: the exterior natural universe, the interior natural universe, and the "supernatural."

Although they differ, the three domains all exist in the world of appearance, above the level of the Ultimate Ground of Existence. Therefore, the entire realm of existence is united. The three domains - the external world of rocks and other people, the internal world of emotions, thoughts, and consciousness, and (if it exists) the "supernatural" world of angels, demons, and God who are Persons - are actually sub-domains of a single realm of existence, which is a manifestation of a single Ultimate Ground of Existence.

This chapter explores concepts which apply in general to the world of appearances, and so to perhaps more than one sub-domain. It presents a somewhat theoretical and abstract discussion of the general relationship between relative entities and the Absolute. And it introduces a few new ideas which apply to entities in general. Ideas such as compound entity, component entity, relative existence, and action are discussed. These concepts are from the philosophical field of ontology, a field which discusses theories of existence or being.

Ontology discusses various types of being (existence), such as real being, logical being, ideal being, necessary being, contingent being, etc. One might suppose, therefore, an "ontological argument" is a discussion in the field of ontology. This term, however, has historically been used to refer to a particular argument ([C08],399-401) for the existence of God advanced by Anselm, a Christian saint. Aquinas and Kant considered Anselm's "ontological argument" faulty. The ontological arguments of this chapter are, I trust, sounder.

Component Entities and Relative Existence

The world contains many entities, some apparently simple and having no parts, others obviously compounded of two or more parts. Water for example seems to be a simple entity, an entity which contains no parts. Houses and cars, on the other hand, are entities compounded of smaller parts which are entities in their own right. A house has windows, a distinct sub-entity; a car has a steering wheel.

We'll label any entity which has separate parts a "compound entity" since it's not simply one thing but compounded of different parts. "Component" is a synonym for "part." So a compound entity is also a component entity. I'll use the terms "component entity" and "compound entity" interchangeably.

Let's investigate a particular component entity, a table. A table is a component entity because it's a combination of components or parts. Its components are its top and four legs.

"Relative existence" is another new term. Not only is a table a component entity, it has relative existence. Why? Because more than a top and four legs are needed to make a table. What's needed in addition is for the table's components to have the proper relation relative to each other. Each corner of the top must have a leg, and all legs must be pointing in the same direction. If some legs are fastened pointing down, and others are fastened pointing up, then we don't have a table. Instead, we have a bunch of parts which could make a table if they assumed the correct relation relative to each other.

A car is another example of a component entity with relative existence. Imagine a car has been completely disassembled. The individual components, the nuts and bolts, the engine and transmission parts, the fenders and hood, are all piled in one large heap. The pistons that should be in the engine are lying on top of the windshield, the steering wheel sits on top of the spare tire. The heap is not a car. All the pieces, all the components, of the car exist, but they don't have the proper relation relative to each other for a car to exist.

So tables and cars are component entities which possess relative existence. For them to exist, their components must exist and must have the proper relation relative to each other. Indeed, for any material object to exist, it's relative components, the various atoms, must exist and maintain the proper relation. Change the relation and a different object comes into existence.

> Just as, for instance, the letters *a*, *e*, and *r* make up the words *are*, *era*, *ear*, *area*, and *rear*, so the elements carbon, hydrogen and oxygen appear in a pad of paper, a rubber eraser, a blob of glue, a paste of laundry starch, a lump of sugar and a dry Martini. ([L02],29).

Of course, atoms themselves are component objects. Their components are various subatomic particles.

Words themselves are an excellent illustration of component entities and relative existence. The word "are" has components: the letters "a", "r", and "e". But more than the components are needed for "are" to exist. What's needed is a proper relation between its components, its letters. Confuse the relation and the word "are" vanishes; in its place an entirely different word - "era" or "ear" - appears.

The concept of relative existence is obviously closely related to the concept of component entity. Are the two equivalent? There is a logical principle which says if everything which is A is also B and everything which is B is also A, then A and B are equivalent. For example, if every group of 12 similar items is a dozen, and every dozen has 12 similar items, then the ideas of "dozen" and "twelve" are equivalent. Let's apply this principle.

Anything with relative existence is also a component entity (everything A is B), since if parts have the right relation relative to each other then parts certainly exist. Conversely, considering a component entity as one thing implies a relation (everything B is A), however weak, between the components in question. For example, if a dozen donuts are thought of as a single component entity, then each individual donut is related to the others by being one of the same dozen. It could be questioned whether the relationship between the donuts is a real relationship, but we'll have no need to split hairs that fine; for our purposes, all component entities have relative existence, and anything with relative existence is a component entity. So component entity and relative existence are equivalent concepts, just like twelve and dozen.

In contrast to the Self-Existent which has independent and permanent existence, component entities have an dependent and transitory type of existence. It's easy to see why. A component entity depends for its existence on the continued existence and right relation of its components. As soon as one of its parts ceases to exist or loses its proper relation to the others, the component entity itself ceases to exist. As soon as one letter ceases to exist, "are" ceases to exist.

The same applies to solid material objects. I'll use a somewhat bogus disappearing trick to illustrate.

Find a willing friend and claim you're going to make something disappear before their eyes. Curl your fingers into your palm and fold your thumb over them. Show this to your friend and ask what it is. After they admit it's a fist, slowly open your hand. The fist disappears. Of course, your friend is unimpressed. After all, all you've done is open your hand.

A fist is a component entity (its components are the different parts of a human hand, the palm and fingers) with relative existence (for a fist to exist, the hand's components must have the proper relation to each other, fingers curled into palm, thumb over fingers.) A fist has an unstable, transitory, and dependent type of existence since, as soon as the fingers and palm lose their proper relation to each other (you open your hand), the fist ceases to exist. The fist comes into existence and then goes out of existence - although the underlying substance of the fist, its ground of existence, that is, the hand, exists all the while.

Realizing the transitory nature of component entities, realizing "all things must pass," is basic to Buddhism, by the way. Buddha taught:

> All compound things are transitory: they grow and they decay. ([C04],158),

and:

> . . . [I]t remains a fact and the fixed and necessary constitution of being that all conformations are transitory. ([C04],80).

Indeed, his last words were:

> Decay is inherent in all component things! Work out your salvation with diligence! ([B16],118).

Actions

Component entities with relative existence can be thought of in a more dynamic way: as actions. The action of holding the fingers in a certain way constitutes a fist.

As another example consider Harvard University, which was founded in 1640. A little arithmetic will tell us how long Harvard has existed. Things aren't so easy, however, when we try to decide exactly *what* has existed since 1640. Certainly, none of Harvard's present students or professors were alive in 1640. Harvard today may or may not occupy a building dating back to 1640. Suppose (I don't know if this is true or not) not one of Harvard's presents buildings existed in 1640. Then what has existed since 1640? That is, exactly what constitutes Harvard University?

Harvard University is an action, a process, a flow of students, professors, teaching, research, buildings, money, and academic degrees. Like the flowing water which constitutes a fountain or whirlpool, the fountain and whirlpool we call Harvard has been turning since 1640. If the flow stopped - if the students, faculty, and administration one day decided to stop the educational process and enter the real estate business - then Harvard University would cease to exist on that day. The people and buildings would still exist yet, like the fist and the whirlpool, Harvard University would vanish.

If we generalize "act" to include static states, cars and tables may also be thought of as actions. Just as the dynamic act of folding the fingers together creates the fist, the static "act" of maintaining the fingers together allows the fist to continue existing. Similarly, the dynamic act of assembling the components creates the car or table. The static act of the components maintaining a continuing right relation allows the car or table to continue existing.

Cars and tables are also actions on a deeper level since their sub-atomic components are actions. In the past, matter was thought of as something solid and static. A glass breaks into smaller glass particles, a rock may be ground into gravel. In each case, the "stuff" remains, solid, stable, and unmoving. It seemed matter was the antithesis of action. But the

> . . . discovery that mass is nothing but a form of energy has forced us to modify our concept of a particle in an essential way. In modern physics, mass

is no longer associated with a material substance, and hence particles are not seen as consisting of any basic 'stuff', but as bundles of energy. Since energy, however, is associated with activity, with processes, the implication is that the nature of subatomic particles is intrinsically dynamic. ([C03],202-3).

Today, according to quantum theory,

... particles are also waves ... [P]articles are represented ... by wave packets. ... [M]atter is ... never quiescent, but always in a state of motion. ... Modern physics ... pictures matter not at all as passive and inert, but as being in a continuous dancing and vibrating motion ... ([C03],192-4).

Thus,

Modern physics has ... revealed that every sub-atomic particle not only performs an energy dance, but also *is* an energy dance ... ([C03],224).

Thus, material objects may be thought of as processes, dances of energy, actions.

The Universe as an Action

It's a small step from seeing matter as a dance of Energy to seeing the entire universe as such. Mystics have often taken this step. For instance, in India creation is described as the dance of the god Shiva, a symbol of the Absolute. And the *Ashtavakra Gita* pictures the universe's objects as waves and bubbles of the Eternal.

As waves, foam and bubbles are not different from water, so in the light of true knowledge, the Universe, born of the Self, is not different from the Self. (II,4,[A10],6).

The contemporary philosopher and theologian Alan Watts expressed this idea in the form of a children's story.

God also likes to play hide-and-seek, but because there is nothing outside God, he has no one but himself to play with. But he gets over this difficulty by pretending that he is not himself. This is his way of hiding from himself. He pretends that he is you and I and the people in the world, all the animals, all the plants, all the rocks, and all the stars. ([W02],14).

The universe as a wave on the ocean of God. God playing hide and seek. The images express the universe as an action, a dance, a wave, or a play of the God who is not a Person. Just as waves are a motion of the water, this universe is a motion of the God who is not a Person.

There's another analogy which expresses the relationship between the Eternal and the universe. In a movie, one and only one thing visually exists - light projected on the screen. Although men, women, children, animals, houses, trees, and a thousand other things appear to exist, in reality only light exists. This fact is so obvious we habitually forget it.

Mystics have tried to express a similarly forgotten truth about the universe and the God which is not a Person. Any man, woman, child, animal, house, tree, or other object, like figures on a movie screen, are images of an identical Source and Root. As Attar, a 12th century Sufi poet, wrote:

> Although you seem to see many beings, in reality
> there is only one. . . ([A12],115).

The God which is not a Person "dances" this world into creation. When the mind of the dreamer is quiet there is no dream. Similarly, when the projector has no film, the screen is lit but bare of images. The light is still. When the mind begins to "dance", however, it creates images and a dream results. Similarly, when the film is loaded and running, the light dances and the movie begins.

So, God dreams the world, dances it into creation. And just as light is the ground of the images on the screen, just as the mind of the dreamer is the ground of the dream images, the God which is not a Person is the ground of existence of this universe.

Action, dance, wave, and play suggest the energies of God. We previously saw the Hesychastic distinction between God's essence and His energies, and the analogy of fire's heat, light, and sound to fire itself. Rufus Jones drew a similar distinction in speaking of the thought of Clement of Alexandria.

> God, in His essential being, is transcendent, but
> dynamically He is immanent and near. The doctrine of
> an immanent God - God as Logos or Spirit, moving
> through all life and in immediate relation with the souls
> of men, is fundamental to Clement's thought . . . [H]e
> was . . . influenced . . . by the teaching of St. John and

> St. Paul. "In the beginning was the Logos; all things
> were made by Him." "In God we live and move and
> are." ([J02],48-9).

So God the Father is like fire, the thing-in-itself. And the Father's
Son, the Logos, Christ, is like the fire's energies; the Logos is the
dynamic energies of God which creates all things. The Spirit is those
same dynamic energies experienced inside one's self.

So our exterior and interior worlds are plays of the Uncreated
Light, as is the "supernatural" world. The God who is not a Person
assumes forms such as rocks, thoughts, and perhaps angels and Gods
who are Persons. All such entities are actions brought into existence
by an act, a play of the Eternal. They are all waves on an eternal
ocean of Uncreated Light. As Angelus Silesius wrote:

> It is as if God played a game
> immersed in contemplation;
> and from this game
> all worlds arose
> in endless variation. ([B05],55).

All worlds - the natural and the "supernatural" - are actions, plays of
Eternal Energy.

Voidness and Emptiness

Fists, tables, cars, fountains, whirlpools and Harvard University are
actions. They are component entities with an impermanent, unstable
type of existence. They come into existence from nowhere and then
vanish without a trace. And when they vanish, they don't "go"
anywhere, they simply cease to be. When someone stops singing, the
singing doesn't go anywhere, it simply ceases to be. When water is
running down a drain, a whirlpool exists. When the water has run out,
the whirlpool ceases to exist. When the water is turned off, a fountain
of water ceases to exist.

There is something unsettling about things like fists, fountains and
whirlpools which pop into and out of existence so easily. Such
chimerical entities seem to have a kind of existence which borders on
illusion. It's easy to feel whirlpools aren't, in some sense, fully real.
Realness, we feel, implies solidity and stability, and realness is what
we often prefer. Who, for instance, would advance money to a
business that just popped into existence yesterday and might pop out
of existence tomorrow, rather than to an established firm? If a home

could be built which might pop out of existence at any time, would anyone buy it?

Actions and component entities have a type of existence which *is* less than fully real. As we look deeply into component entities, down below the level where their components exist, we see that, in the ultimate, ontological sense, they don't really exist at all. They are void and empty of real existence. As Nicholson writes:

> Phenomena, as such, are not-being and only derive a contingent existence from the qualities of Absolute Being by which they are irradiated. The sensible world resembles the fiery circle made by a single spark whirling round rapidly. ([N11],82).

An example from the writings of Buddhadasa, a contemporary Buddhist monk, will illustrate how something which seems quite real is an appearance like the fiery circle. It might seem there are different kinds of water, such as rain water, well water, stream water, and river water. However, if we analyze each kind of water, we'll eventually find that there is really only one kind. If we disregard extraneous trace elements we find that each kind of water is identical.

> If you proceed further with your analysis of pure water, you will conclude that there is no water - only two parts of hydrogen and one part of oxygen. Hydrogen and oxygen are not water. . . . [W]ater has disappeared. It is void, empty . . . For one who has penetrated to the truth at this level there is no such thing as water. ([B13],88-9).

So water, which seems to possess no parts, has parts - one part oxygen and two parts hydrogen. Moreover, (if I correctly remember my college chemistry) these parts must be in proper relative relation to each other - the two hydrogen atoms each attached to the oxygen atom with a 105 degree angle between them - for water to exist. If a molecule composed of one oxygen atom attached to a hydrogen atom attached to another hydrogen atom could exist chemically, it wouldn't be water. Thus, water is a component entity. Or water may be called an action, for just as the fingers and thumb must remain in a certain position relative to each other for a fist to exist, the hydrogen and oxygen atoms must remain in a certain position relative to each other for water to exist.

Voidness and emptiness are the lot of all entities with only relative existence. So, in a sense the entire physical universe lacks a real existence. In a sense it's false and unreal.

> . . . [T]he physical world operates under one fundamental law of *maya*, the principle of relativity and duality. God, the Sole Life, is Absolute Unity; to appear as the separate and diverse manifestations of a creation He wears a false or unreal veil. That illusory dualistic veil is *maya*. ([Y02],310).

A word about a potentially confusing point: the Eternal is sometimes called empty or void but with an entirely different meaning. As Buddhadasa writes:

> The ultimately real is empty, not in the sense that it is vacuous, but in that it transcends any attempt to dichotomize or conceptualize it. ([B13],18).

Johannes Scotus Erigena, a 9th Century Christian philosopher, had a similar idea.

> Therefore so long as it is understood to be incomprehensible by reason of its transcendence it is not unreasonably called "Nothing" . . . ([E05],681A,308).

The meaning might be clearer if "No thing" had been used instead of "Nothing." "Void" or "Empty" would have also served. Erigena continues:

> [B]ut when it begins to appear in its theophanies it is said to proceed, as it were, out of nothing into something ([E05],681A,308).

A theophany is

> [a] manifestation or appearance of a deity or of the gods to man. ([F08],1389).

Erigena believed

> . . . every visible and invisible creature can be called a theophany, that is, a divine apparition. ([E05],681A,308).

Therefore, every rock, thought, angel, and God who is a Person is a manifestation of Uncreated Light.

Absolute Existence

Even water, though apparently a pure and simple substance, is actually an action, a component entity with relative existence. Is everything an action? Does everything have only relative existence? Is everything void and empty of real existence? Or is there something which isn't a component entity, which possesses full and real existence?

If something exists but has no parts, then it can't be a component entity and it can't have relative existence - there are simply no parts to be related. Such an entity could have full and real existence. Is there anything which isn't a component entity. Is there anything which has no parts? If we use "pure" in the sense of "unmixed," and "simple" in the sense of "composed of only one substance or element," then the question may be rephrased, is there anything that's pure and simple?

On the material level science has found only one entity that's pure and simple. What of thoughts and emotions, might not they be pure and simple? An obvious view is since thoughts apparently come into and go out of existence, they're actions. Therefore they possess only relative existence. There is, however, another view going back to Plato which sees thoughts and ideas, especially mathematical concepts, as pre-existing.

> According to Platonism, mathematical objects are real. Their existence is an objective fact, quite independent of our knowledge of them. . . . They exist outside the space and time of physical existence. They are immutable - they were not created, and they will not change or disappear. . . . [A] mathematician is an empirical scientist like a geologist; he cannot invent anything, because it is all there already. All he can do is discover. ([D04],318)

In *Infinity and the Mind*, mathematician Rudy Rucker proposes a similar concept, a "mindscape" where all thoughts already exist. In this view, when we think a thought our mind's eye sees that already-existing thought in the mindscape, just as we see an already existing rock as we walk across some landscape.

> Just as a rock is already in the Universe, whether or not someone is handling it, an idea is already in the Mindscape, whether or not someone is thinking it. ([R06],36).

When our mind's eye turns away or moves from a thought, we cease to think it. Just as when we walk far enough past the rock, we cease to see it. Yet both rock and thought continue to exist in the landscape and mindscape respectively. A similar idea, of course, could be proposed for emotions.

So thoughts and emotions may actually be pre-existing, unchanging entities, rather than actions that pop into and out of existence. Of course, they wouldn't be pure and simple if they were component entities. Are they?

Some thoughts do seem compound. The thought "I am hungry," for example, involves at least two components, the thought "I" and the thought "hungry." But what about the simple thought "hungry"? What about emotions such as love or fear? Are they simple entities? Are some thoughts and emotions pure and simple entities? I don't know, although my inclination is to consider only the Real as Pure and Simple.

If we suppose some emotions and thoughts are pure, simple, self-existent entities, then we seem to approach Plato's idea of eternal Ideas or Forms. Even though this idea has a respectable intellectual lineage, it doesn't appear in many of the systems of belief upon which the perennial philosophy is based. So we won't use it in this book.

So we'll assume thought and emotions have relative existence. Therefore, there's only one entity which isn't a component entity, the Ultimate Ground of Existence which is pure, simple, and one. Since the Source is one, It has no parts, and thus is not a component entity. If the Root doesn't have relative existence, what kind of existence does it have? What other kinds of existence are there? Absolute is frequently taken as an opposite of relative. What might the term "absolute existence" mean?

We've seen how relative existence implies parts in a certain relation, which implies a precarious type of existence since the parts may lose their special relation, causing the component entity to cease existing. Absolute existence, therefore, should imply something which has no parts, is pure and simple, and furthermore doesn't have an existence dependent on anything. In other words, absolute existence is existence pure and simple, self-contained existence with no dependencies on anything else, such as components and their relation. As we've seen, the Self-Existent has these characteristics, and so is in possession of absolute existence.

We've also seen component entities don't really exist below the level of their components. On the subatomic scale, water does not -

cannot - exist. This is why compound entities' existence is called void and empty. There's no level, however, below the level of the Root and Ultimate Ground of Existence. Therefore, the existence of the Source isn't void or empty. The Self-Existent fully and really exists. Thus It deserves the names the "Real," "Ultimate Reality," "Eternal Reality," and "Absolute Reality".

> The Brahman, the one substance which alone is eternally pure, eternally awakened, unlimited by time, space, and causation, is absolutely real. ([S01],254).

Discrimination

Our chain of reasoning has shown the inner and outer worlds we know are unreal, in a sense. Only the Eternal is fully real. In Hindu religious literature, such reasoning is called discrimination.

Today the word "discrimination" usually suggests bigotry and hatred. In contemporary society, the phrase "practicing discrimination" is an accusation rather than a compliment since it refers to the social evils of racial, sexual, or ethnic discrimination. For example, the first definition of "discriminate" in a dictionary is

> 1. to make a distinction in favor of or against a person or thing on a categorical basis rather than according to actual merit. ([R01],379).

However, an wider meaning appears next.

> 2. to note or observe a difference; distinguish accurately. ([R01],379).

In the past the word "discrimination" was often used in the second sense, as a compliment to a person's refinement and discernment. The discriminating tastes of the gourmet, for example, can distinguish a fine wine from an ordinary wine; a sharp business person can tell the difference between a legitimate deal and a scam; a critical reasoner can separate the valid argument from sophistry; a competent engineer can discriminate between solid ground able to support a heavy building and sandy, unstable soil which can not. Looking for a good used car, unmechanically inclined people push their mechanical discriminative ability to the limit.

In the religious sense "discrimination" refers to spiritual discernment. The ancient Christian monk, Evagrios, for instance, in his *Texts on Discrimination in respect of Passions and Thoughts* ([P13],VI,38), warns against "demons" such as avarice, gluttony, pride, anger, dejection, and unchastity. A religious seeker should learn to discriminate helpful passions and thoughts from unhelpful ones, but the highest type of religious discrimination is

> . . . the reasoning by which one knows that God alone is real and all else is unreal. Real means eternal, and unreal means impermanent. He who has acquired discrimination knows that God is the only Substance and all else is non-existent. . . . Through discrimination between the Real and the unreal one seeks to know God. ([G03],327).

Identity or Self

We've already discussed our own identity. Let's now investigate the identity of actions and component entities.

Do actions have an identity? If I fold my hand again have I made the *same* fist? It may seem I have since I'm using the same hand, but with other actions the answer isn't so obvious.

Consider the whirlpool created when water runs down a drain. The whirlpool is an action of the water just as fist is an action of my hand. Now plug the drain, come back the next day and unplug it. A whirlpool is created again. Is it the *same* whirlpool? It's hard to imagine how a whirlpool could have an identity. The water which composes it is always changing, always flowing. If you feel it is the same whirlpool, then what about the following? After I plug the drain, I move all the water to another sink or bathtub and open the drain. Is it the same whirlpool now? What if I move the water *and* mix in an equal amount of chlorine. Same whirlpool?

It can be difficult or impossible finding an identity in actions such as fists or whirlpools. What about more substantial actions? What, for example, about Harvard University? Does it have an identity?

We've seen Harvard University has been in existence since 1640. We've also seen the difficulty involved in trying to determine *what* has been in existence since that year. Just like a fountain or whirlpool, what we call "Harvard University" is a flow of students, professors, buildings, money, etc. Has any one thing persisted over those years that deserves to be called "Harvard University?" In other words, does Harvard University have an identity? If we try to find the enduring reality behind Harvard, the thing or things that were Harvard in 1640 and still are Harvard today, we fail. No one person or thing has been

Harvard University over the years. Its identity - such as it is - consists in the educational action of a multitude of components - students, professors, buildings, money, and degrees. It a strict sense Harvard University has no identity.

So attaching the idea of identity to actions is impossible. Actions lack a real identity, a real self. To be sure, for conversational and practical purposes, we use the terms whirlpool, fist, and Harvard University, and, in the practical sense, they exist. However, it's difficult or impossible to define their identity - to define exactly *what* exists - since they don't really exist in an ultimate sense. Assigning them a more than conversational, practical identity is impossible.

But what about simple, solid entities such as tables and cars? Does any component entity have an identity? Certainly we feel it's the same table, the same car, that existed when we last saw them. And certainly they have a conversational and practical type of identity. But it's a weak kind of identity since the very existence of component entities itself is so weak. Since a table's components (it's a component entity . . .) must maintain the *same* relation to each other (. . . possessing relative existence) for the table to keep existing, if we disassemble the table it ceases to exist as a table. So any identity the table had must cease too.

If the table is reassembled, is it the same table? If you believe it is, then what if we put the legs on different corners? Is it still the same table? Strictly, since the legs are now in different positions, it's a different table. In a more common sense view, however, the table with switched legs is the same since the same "stuff," the same top and legs, exist. (Common sense because if the legs of a table were rearranged, hardly anyone would claim it was now a brand new, freshly manufactured table - unless they were trying to dishonestly sell it for a higher price!) However, if the table were ground into sawdust its "stuff" would still exist (as sawdust), but the table would not. Would it be the same table if it were reduced to sawdust? No, since it wouldn't be a table at all.

Yet, the feeling may be that the table does have some sort of identity. "It" is really there, existing from one moment to the next, the same. Physicist Arthur Eddington discussed this question, not about tables but about elephants.

How do we know, he asks, if the elephant we saw a moment ago is the same elephant we see now? We can, of course, measure the elephant in all sorts of ways - weight, height, color, and others. Each of these measurements yields a pointer reading on a scale, a ruler, or some other measurement device. If it's the same elephant, the measurements should be approximately equal over a short time. However, might not an entirely different elephant have the same weight, height, etc.?

> Two readings may be *equal*, but it is meaningless to inquire if they are *identical*; if then the elephant is a bundle of pointer readings, how can we ask whether it is continually the *identical* bundle? ([E01],256).

Eddington concludes

> . . . the test of identity is clearly outside the present domain of physics. The only test lying purely in the domain of physics is that of continuity . . . ([E01],256).

On first sight this argument may seem unconvincing. After all, couldn't the elephant be marked in some unique way so we'd be sure the elephant we saw today was the same one we saw yesterday? The issue, however, is much deeper. It rests on the assumption the atoms which compose the elephant remain the same. This assumption has often been held. For example, Schrodinger wrote that all proponents of atomic theory from the Greeks to the nineteenth century believed

> . . . atoms *are* individuals, identifiable, small bodies just like the coarse palpable objects in our environment. ([S06],17).

However, as scientists investigated the deeper nature of atoms, they were forced to abandon the idea that an atom is

> . . . an individual entity which in principle retains its 'sameness' for ever. Quite the contrary, we are now obliged to assert that the ultimate constituents of matter have no 'sameness' at all. When you observe a particle . . . now and here, this is to be regarded in principle as an *isolated event*. ([S06],17).

Might the word "event" be replaced by "action"? Schrodinger continued:

> Even if you do observe a similar particle a very short time later at a spot very near to the first, and even if you have every reason to assume a causal connection between the first and the second observation, there is

no true, unambiguous meaning in the assertion that it
is *the same* particle . . . ([S06],17).

If individual atoms fail to have an identity, how can anything
composed of them - an elephant, for instance - possess an identity?
Buddha recognized compound entities lack an identity:

All compound things lack a self . . . ([C04],158).

He did, however, grant a kind of conditional identity to compound
entities, comparing their identity to that of a candle flame.

. . . [T]he flame of to-day is in a certain sense the
same as the flame of yesterday, and in another sense
it is different at every moment. ([C04],156).

In the sense of continuity, the flame now has descended from the
flame of a moment ago, the whirlpool now is the descendent of the
whirlpool of the past, "the child is father to the man" - but the child is
not the same as the man.

Previously, we saw the only absolute identity we possess is our
Consciousness which we equated with the Ultimate Ground of
Existence. Of course, if the Real didn't have an identity Itself, then it
certainly couldn't function as our identity. Does the Absolute have an
identity? First, the Root actually and truly exists, as opposed to
actions which have the temporary, unstable type of existence. Second,
the Source is simple, pure, and has no components or parts. Thirdly,
the Unconditioned is eternal and unchanging, so what It is today, It
was yesterday and will be tomorrow. So It remains the same under
different conditions. So the Root and Source has an identity.

This idea has found expression in mystical literature. For example,
the Islamic Sufis Abu Sa'id al-Kharraz and Abu Nasr al-Sarraj taught
only God has the right to say "I." ([E06],10). And, the Sufi Bayazid
wrote:

. . . the only real identity is God . . . God is the only
one who has the right to say "I am." ([E06],26).

Looking Back, Looking Ahead - II

In this chapter we saw that component entities: have an impermanent
kind of existence dependent on the relationship of their parts; can
cease to exist; in fact, are void and empty of real existence; and have
no real identity. The Absolute, on the other hand, has real, permanent,
independent existence, and a real identity. We've now completed Part

II, so we'll take some time to stop and see where we've been and where we're going.

Part I discussed the religious and scientific ways of knowing, the Ultimate Ground of Existence, and people who've had direct experience of It. It also discussed applying the scientific way of knowing to mystics' statements.

In the second part, we built a world view on mystical visions. We described the relationship of the outer, inner, and "supernatural" worlds to the Primal, the One. Of course, entirely different world views based on mystical visions could be constructed, as well.

It's worth observing that the world view we've seen isn't fully scientific. It couldn't be because it's the world view of a single individual while science is a group effort. Replication is an essential part of science. Scientific claims must be tested by others before they're accepted. So until our world view is tested by others it can't claim to be scientific. Until it's tested, it's only a tentative, first hypothesis, a starting point. And, of course, if it's to remain scientific, it must always remain open to question and criticism, subject to change and revision, capable of adaptation and improvement. It can never stagnate into dogma.

In the third and final part, we'll discuss practical consequences. The dominant questions will be: So what? Can these thoughts have practical consequences in how I live? How can these ideas and beliefs affect my life?

We'll go from world view to practical consequences in steps.

The first step will be describing the goals our world view contains.

A world view is a kind of map, and a map shows not only what is, but what is possible. If it shows mountains, then we may think of climbing them. If it shows a sea, we may think of sailing it.

Just like maps, world views differ. Some have wider scopes than others. Some world views have a limited scope in that they only discuss this world. They're silent about where we came from, what happens after death, and even our ultimate purpose here. In other world views, death is followed by heaven or hell; in others, by reincarnation; in still others, by destruction, the self just evaporates into nothingness.

Obviously, if someone's world view doesn't include an afterlife then getting to heaven or obtaining a good reincarnation won't be one of their goals. It can't be since it's not on their map. On the other hand, if a person's world view includes heaven, then they may value getting there. They may undertake some actions to ensure their place in heaven. Reaching heaven may be one of their goals.

Similarly, if a person's world view doesn't allow that gnosis - direct experience of and even union with the Eternal - is possible, then they aren't likely to value gnosis, or even know of its existence. But, since the map which is our world view does show gnosis, we can not only value gnosis but make it a goal, a life aim. We'll discuss goals, with emphasis on gnosis.

The second step will be deriving subordinate goals, things which aren't ends in themselves but means to gnosis. Someone whose goal is becoming a professional athlete might adopt the subordinate goal of exercising and practicing every day. The subordinate goal isn't an end in itself. Rather it's a means for developing the skill and strength required to achieve the main goal.

Similarly, a seeker of gnosis might adopt subordinate goals, certain attitudes and actions which help the journey to gnosis. The seeker will value these attitudes and actions not for their own sake, but as helps to direct experience of and union with the Eternal Light. We'll discuss various beliefs and actions which seekers of gnosis often value.

Identical world view and goals don't necessarily imply identical values. For example, among those who believe in heaven, some may be so eager for heaven now they pursue martyrdom or death in a holy war, while others are content to let death come in its own time. Similarly, our world view and the goal of gnosis don't necessarily imply only one set of values. Just as mystical declarations can support various world views, our world view and the goal of gnosis can support various values. In particular, we'll see two different values systems, the so-called negative way and affirmative way.

The third step will be picking out the values which have practical consequences. Some beliefs, actions, and attitudes which seekers of gnosis value have obvious practical consequences. For example, "Do unto others as you would have them do unto you." Others do not. For example, the major doctrinal difference between Western and Eastern

Christianity is that the West believes the Holy Ghost proceeds from the Father and the Son, while the East holds to the ancient belief that the procession is from the Father alone. Values which have practical consequences are called ethics. Thus, "Do unto others as you would have them do unto you" is an ethic, while "the Holy Ghost proceeds from the Father" or "the Father and Son" is not. Rather, it's a dogma. We'll discuss certain ethics which derive from our values.

The last step will be deriving specific morals from our ethics. The same ethic may lead to different morals. For example, two people might share the ethic that human life is sacred and should be protected. One, however, interprets this ethic as forbidding abortion but allowing war. The other feels the fetus is not fully human but believes in pacifism. Both are implementing the same ethic in their own way. Someone who refrains from abortion but fights in a war has different morals than the person who accepts abortion but refuses to go to war. But both may have the same ethic, a respect for human life. How they express this ethic in action - that is, their morals - disagree. We'll discuss certain morals which derive from our ethics, too.

So we'll be taking a "top-down approach." We'll begin at the top with our map, our world view, our mental picture of what is what. Based on our map, we'll describe potential goals. Based on our goals, we derive values. We'll pick out the values and principles which have practical consequences, the ethics. Morals are how we put our ethics in practice.

Part III: Consequences

11

- Goals -

Chapter Summary: This chapter explores various types of goals, with emphasis on mystical goals. Various afterlife possibilities (heaven, hell, reincarnation) are also discussed. Lastly, it discusses how various mystical goals can motivate a spiritual quest and lead to direct experience of God.

Deciding what's true is the business of a way of knowing. A world view's function is to create a map of the world and our place in it. But the number of true facts may be infinite. Which are important? Which are valuable? The number of places on a map may be numerous. Towards which are we traveling, or being carried? How are we to direct our lives? What goals should we choose?

We'll begin our discussion of these questions by examining various types of goals, various types of places on the map which is our world view. We'll discuss three kinds of goals: common goals, afterlife goals, and beyond-the-show-world (transcendental) goals.

Common Goals and the Afterlife

Common goals are common to people around the world, irrespective of their religion or philosophy. Common goals include physical goals such as food, clothing, shelter, and wealth; emotional goals such as love, friendship, and respect; and intellectual goals such as wanting to know, understand, and discover. Many examples of common goals may be given: someone drinks water to satisfy their thirst; someone works to earn money for life's necessities and pleasures; people marry for, among other reasons, emotional fulfillment; a student goes to college to get a degree. Some common goals remain completed for only a short time: water satisfies my thirst, but my thirst soon returns; the money I earn this week soon runs out. Other common goals remain completed much longer. An academic degree, for instance, lasts a lifetime; a building may easily outlive its builder.

Yet time eventually destroys the building, and death, seemingly, destroys the builder. Is death really a destroyer? Some people believe death does, in fact, bring utter destruction, that there is no afterlife, no continued existence beyond death. Even though religion offers the promise of eternal life, it actually gives, they believe, only emotional satisfactions - the assurance of justification in the sight of a God who probably doesn't exist, the fantasy of a pleasant afterlife.

Some reasons why this belief is not widely held are obvious. Many people find it impossible to believe our bodies, almost infinitesimal specks in the vastness of the universe, and our life spans, almost infinitesimal specks in the linesman of the universe, are all we are. Is it really true, they argue, we love, hate, hope, and strive, and none of it endures or matters? What kind of universe, they ask, could create such pitiful entities as ourselves, giving us enough intelligence to realize our own contingency, finiteness, and futility, and then utterly destroy us? Moreover, if death brings destruction, then death puts us beyond the consequences of our actions. Saint and sinner both cease to exist, with no reward for the saint and no punishment for the sinner.

I use "saint" and "sinner," by the way, in a loose sense. There are probably many thoroughly non-religious people who would reject the idea that a person who dies saving a bus full of children and a madman who dies bombing a hospital and nursery school both suffer the same fate - utter destruction - with the "saint" unrewarded, the "sinner" unpunished.

But if death isn't a destroyer then it must be a transformer, a deliverer into some sort of continued existence, some sort of afterlife. And if we continue to exist after death, then possibly our past actions still affect us, even as they do in this life. But how can the effects of past actions extend beyond death? One possibility is through the agency of a God who rewards and punishes. Another possibility is a more subtle mechanism where actions affect character and character affects destiny. In this view, we not only create our actions; our actions in a sense create us.

Afterlife Goals

If death indeed doesn't bring utter destruction, then some part of us must survive in some sort of afterlife existence. Those who believe in an afterlife existence may act with an eye to preparing for their afterlife state. That is, afterlife goals may motivate some of their actions here.

It's worthwhile noting that the usual association of the afterlife with religion isn't absolutely necessary. If we find ourselves born into this world by purely natural means then it's possible we'll find ourselves born into another, after death, by purely natural means too. Those who think of themselves as the body assume that if anything survives, it's the spirit. And so they naturally associate the afterlife with religion. But we've seen how the body is a entity with relative existence, lacking an enduring identity. Therefore, it's possible our real identity - however one conceives it - persists after death through purely natural means, without the agency of a God.

But since the afterlife is almost always discussed within the context of religion, we'll discuss it in that context too.

Many religions offer afterlife goals, such as the goal of everlasting life, of immortality, in the company of a God who is a Person. In some religions, such life begins immediately after death. In other religions, death is followed by reincarnation into another body; only after a long series of lives does the soul reach its final state, life with God.

Afterlife goals include the attainment of heaven, the avoidance of hell, and the avoidance of a bad reincarnation, such as rebirth as an unfortunate person or even an animal. My family's religion offered the afterlife goals of heaven, hell, and purgatory. Even at an early age, however, I didn't find these ideas completely believable. As an 8 year-old boy sitting in a Roman Catholic catechism class, I decided if only those who were baptized and believed in Jesus could get into heaven, then, for example, someone who lived in China 5,000 years ago was forever excluded from heaven - through no fault of their own but simply because they had been born 3,000 years too early.

Later, I doubted the Catholic teaching that dying with an unforgiven mortal sin resulted in eternal hell. At the time, deliberately eating meat on Friday was a mortal sin. I couldn't believe some young boy who knew it was Friday but ate a hot dog anyway might die and spend all eternity tortured in hell. (Nevertheless, I never ate hot dogs on Friday!)

As I grew and saw more of the world, the very existence of heaven and hell seemed more and more doubtful. Nonetheless, I could still admire the intelligence of this explanation of why the good suffer and the evil prosper. The good sometime do evil and must suffer for it on earth so when they die they may go straight to heaven. The evil sometime do good and must be repaid on earth so when they die they may go straight to hell. The explanation's cleverness pleased me. But I had lost belief in the existence of heaven and hell. Most people, it seemed, led moderately good and moderately bad lives, lives deserving of neither the eternal bliss of heaven nor the eternal punishment of hell.

Consider for instance "Pete," who died when he was seventy. Pete had never been very religious. He had an average disposition, sometimes cheerful and sometimes moody. As a father, Pete was often attentive and loving - when he wasn't too drunk. He'd committed adultery once, but felt guilty afterwards and had managed never to do it again. Yet many women were the object of his fantasies. Pete was mostly honest at business, but he paid his employees as little as possible. He wrote highly imaginative income tax returns. There were some people in town Pete didn't like, and he usually let them know it - if they weren't too powerful or influential. Over his life, Pete maintained a few close friendships. These friends as well as Pete's wife and children grieved when he died. They missed him.

Does Pete deserve the eternal bliss of heaven in the company of God and angels? Or the eternal torment of hell in the company of Satan and demons? Or neither?

Such doubts may have motivated the Roman Catholic idea of purgatory, a place where souls go who die with unforgiven non-mortal sins. Since these souls are not sinless, they do not yet deserve heaven. But since their sins are not mortal, they don't deserve hell either. So they go to purgatory. Eventually when they're cleansed of their less-than-mortal sins, they advance to heaven.

For me, however, the addition of the after-death alternative of purgatory didn't help. Purgatory was only for those who had died with unforgiven non-mortal sins. The young boy who knowingly ate the hot dog and died still went to hell. Forever.

Afterlife and Personal Identity

Eventually after I'd come to some understanding of personal identity, I began to doubt that an individual person could dwell in heaven for all eternity. Believers in a God who is a Person often interpret immortality as perpetual existence for their own personal identity. However, if personal identity does not consist of the Eternal - if it's changing and transitory, and forever distinct from the Eternal - then which relative personality or personalities get "frozen" into immortality?

A woman, "Anna," is 12 years old when her 30 year-old mother dies. Anna herself lives to see ninety. During those 90 years, Anna is many different people: infant, child, adolescent, student, woman, wife, mother, attorney, Sunday school teacher, bridge player, grandmother, etc. At various times of her life, Anna is naive, sophisticated, generous, stingy, serious, playful, frivolous, patient, quick-tempered, trusting, suspicious, etc. Suppose a few traits such as jealousy, pride, sloth, or impatience appear in most, or all, of the people Anna is. That is, suppose Anna has a few characteristic vices.

Which of Anna's personalities, which of her traits, are given immortality in heaven? Just one or, in some way, all of them? Is her characteristic jealous streak, her tendency to pride and arrogance, her slothfulness, or impatience incorporated into her heavenly personality? Certainly, most people don't picture heaven with prideful, arrogant, slothful, or impatient residents. Are, then, only Ann's good qualities included in her heavenly personality?

By the way, I'm assuming qualities such as honesty, patience, kindness, charity, and love are compatible with life in heaven, and qualities such as lust, hatred, greed, envy, sloth, and malice are not. Many religions where God is thought of as a Person have such a morality. Later I'll discuss morality for followers of the God who is not a Person.

For 78 years, Anna looks forward to seeing her mother in heaven. But when Anna reaches heaven, will she have to assume the body and personality of her 12 year-old self to meet her long deceased mother? What if Anna would rather have the body and personality of her forty-fifth year? How could she relate to her mother, whose body and personality were set at 30? Moreover, Anna's own children and grandchildren look forward to seeing her in heaven. Her children look forward to seeing Anna as she was in middle age; her grandchildren hope to see sweet, old granny Annie with the silver hair. When they all are in heaven, which body and personality does Anna have?

How can Anna's different earthly personalities, the frivolous and naive child, the serious and sophisticated adult, fuse to form one single person, a person purified of Anna's characteristic faults and vices? Isn't an essentially new person created if Anna's purified and fused heavenly personality has only Anna's good elements, and lacks the bad? Doesn't the transformation produce an essentially new person, only distantly related to the many people Anna was on earth?

Further, if we assume Anna is somehow transformed into a purified and fused person then we may ask: does a purified and fused person remain the same forever, throughout all eternity? If it does, it would forever lack any good qualities that Anna never acquired on earth. Or, at least, they'd be underdeveloped. For example, suppose Anna never learnt patience. Then even though her heavenly personality might never be obviously impatient, would it have perfect loving patience? How would a personality which was never patient on earth acquire perfect patience in heaven? It would have to be able to change, to evolve.

Suppose heaven's fused and purified personalities can evolve, can acquire or bring to perfection any good qualities they lacked on earth. Eventually, after they've acquired and perfected all the virtues, they would be very similar to each other, if not identical. The originally distinct mother and child would have evolved into two, essentially identical persons. Purified of all vices and possessing all virtues, any heavenly person seeing another would merely see its own reflection. Anna and her mother would have evolved into two identically perfect and perfectly identical persons.

Reincarnation

Proponents of reincarnation believe the process of purification and evolution just described occurs over a series of lives, on earth or elsewhere. Slowly, life after life, we move closer to our ultimate destination: experience of, and eventual union with, the One. Only when we've reached the level of purity enjoyed by the Absolute, they say, may our consciousness merge with the One; for only then is the purified soul identical to the One. Thus the merging doesn't change the Unchangeable, or add to the One. Here, too, the assumption is the One possesses all the moral virtues in the highest degree and we too must acquire them before we may merge with It.

Although reincarnation is often thought to be an Eastern belief, it's been held in the West as well. In Plato's *Phaedo*, for example, Socrates says

> . . . the living come from the dead, just as the dead come from the living . . . the souls of the dead must exist in some place out of which they come again. (*Phaedo* 72A, [D07],VI,424).

A more contemporary expression occurs in the modern fable *Jonathan Livingston Seagull*.

> Most of us came along ever so slowly. We went from one world into another that was almost exactly like it, forgetting right away where we had come from, not caring where we were headed, living for the moment. Do you have any idea how many lives we must have gone through before we even got the first idea that there is more to life than eating, or fighting, or power in the Flock? A thousand lives, Jon, ten thousand! And then another hundred lives until we began to learn that there is such a thing as perfection, and another hundred again to get the idea that our purpose for living is to find that perfection and show it forth. . . . [W]e choose our next world through what we learn in this one. Learn nothing, and the next world is the same as this one, all the same limitations and lead weights to overcome. ([B01],53-4).

The idea of reincarnation initially attracted me. I'd heard about cases like the ones in ([S24]) *Twenty Cases Suggestive of Reincarnation*, and been impressed. (Years later, *Children Who Remember Previous Lives* ([S23]), by the same author, appeared.) A representative case is as follows.

A young child, Sarah, has memories of a previous life in another town, with another family. In some cases, Sarah remembers being a child in her former life; in other cases, a woman or a man. Sarah describes her former town, house, and family in detail. Her parents eventually take her to the town. Sarah immediately finds her former house and family; they are as Sarah remembered them. The person

Sarah claims to have been indeed existed but died before Sarah was born.

Rebirth on earth seemed a fitting fate for most souls, an appropriate reward for moderately good and moderately bad lives. At first, it also seemed to explain why some are born to good circumstances and some are born to bad: their births are simply a reward or punishment for actions in previous lives. Certainly, that seemed much fairer than supposing some infants begin their one-and-only life in loving families, nurtured materially, emotionally, morally, and intellectually, while others begin their one-and-only life in hostile, deranged families, enduring material, emotional, moral, and intellectual deprivation and abuse.

But seeing birth circumstances as simply a reward or punishment for behavior in past lives doesn't fit the facts. For certainly, some born to good circumstances have unprincipled, selfish, even degenerate natures. Conversely, others born to harsh circumstances have principled, loving, altruistic natures. These natures, the result of lifetimes of good and bad character development, should imply commensurate birth opportunities, but often do not. Moreover, seeing birth circumstances as a straightforward reward or punishment means an infant suffers blindness, physical deformation, or mental retardation because of actions in some past life. In effect, it says the baby deserves their fate; that innocents are, in fact, not innocent.

But if life circumstances aren't a straightforward reward or punishment for past actions, then what are they? Do they occur with any rhyme or reason? Perhaps not. As we've seen, pain and suffering are an inherent part of our world. If we suppose multiple life spans, then it's probable that, in some life or another, we'll find ourselves in unfortunate circumstances. Thus birth circumstances need not reflect our evolution. Given enough lives, all of us will, in our experience of the variety this universe has to offer, experience unhappy, undesirable lives.

So perhaps chance determines why someone is born into a particular situation. There is, however, another possibility. It's possible our experiences are meant to teach us something, to help us on our journey home to the Source. So, someone may be born into wealth not as a reward for previous good deeds, but as an opportunity to learn that wealth is ultimately unsatisfactory as a replacement for experience of the Absolute. After a number of such lives, this individual's desire for wealth will be quenched. A much more evolved individual, on the other hand, may be spending this life learning to transcend an unhealthy body, or learning patience and forgiveness as a member of a despised minority. However, I don't intend to promote passivity and acceptance of injustice; such an individual might just as easily be learning to fight for justice.

Whether our experiences are meant to teach us something or happen at random, it is up to us to make the most of them. How we respond leaves its record in our character, and character marks how far a consciousness has progressed in its journey toward conscious reunion with the Eternal.

Between Death and Birth

Even if a person remembers parts of what appear to be a past life, they usually have no recollection or description of the time between death and rebirth. *Life After Life* paints a partial description based on the near-death experience of 150 people. A composite near-death experience is as follows.

> A man . . . hears himself pronounced dead . . . He begins to hear . . a loud ringing . . . and . . . feels himself moving . . . through a long dark tunnel. . . . [H]e sees his own body from a distance . . . Others come to meet and to help him. . . . [A] loving, warm spirit . . . - a being of light - appears before him. . . . He is overwhelmed by intense feelings of joy, love, and peace. . . . [H]e somehow reunites with his physical body and lives. ([M16],23-4).

Life After Life discusses similar experiences from the Bible, the writings of Plato and Swedenborg, and ([T05]) *The Tibetan Book of the Dead*, which has the most complete description. *The Tibetan Book of the Dead* claims to describe not only the immediate time after death, but the complete journey of the soul from death to rebirth as well. A summary follows.

The newly disembodied consciousness encounters ([T05],89) "the fundamental Clear Light . . . the Unborn *Dharma-Kaya*", the ([T05],95) "Radiance of the Clear Light of Pure Reality" which is ([T05],104) "subtle, sparkling, bright, dazzling, glorious, and radiantly awesome." An editorial footnote describes ([T05],12) this

Light as "The Uncreated, the Unshaped, the Unmodified" and as ([T05],11) containing "the essence of the Universe." Our Consciousness, unborn and undying, is that Light.

> Thine own consciousness, shining, void, and inseparable from the Great Body of Radiance, hath no birth, nor death, and is the Immutable Light . . . ([T05],96).

If the soul can recognize and unite with the Clear Light, it will escape from Maya's show, i.e. it will attain liberation. However, most souls are unprepared to behold, much less unite with, the "radiantly awesome" Clear Light of Reality. As Huxley writes:

> Following Boehme and William Law, we may say that, by unregenerate souls, the divine Light at its full blaze can be apprehended only as a burning, purgatorial fire. An almost identical doctrine is to be found in *The Tibetan Book of the Dead*, where the departed soul is described as shrinking in agony from the Pure Light of the Void . . . in order to rush headlong into the comforting darkness of selfhood . . . ([H10],55-6).

Thus, most souls journey through various levels of phenomenal existence, encountering peaceful and wrathful deities and a judgement, and eventually are reborn, on earth or elsewhere.

What constitutes an "unregenerate" soul? What prevents a consciousness from uniting with the Real? A footnote in *The Tibetan Book of the Dead* discusses this question.

> In the realm of the Clear Light, . . . the mentality of a person dying momentarily enjoys a condition of balance, or perfect equilibrium, and of oneness. Owing to unfamiliarity with such a state, which is an ecstatic state of non-ego, . . . the consciousness-principle of the average human being lacks the power to function in it; *karmic* propensities becloud the consciousness-principle with thoughts of personality, of individualized being, of dualism, and, losing equilibrium, the consciousness-principle falls away from the Clear Light. It is ideation of ego, of self, which prevents the realization of *Nirvana* . . . and so the Wheel of Life continues to turn. ([T05],97).

"Karmic propensities" refers to our habits, which are the results of past actions. Nirvana means union with the Real. We've already seen how attachment to dualistic perception of the universe veils the One.

What creates attachment to dualistic perception? Pleasure. Therefore, the yin aspect of pleasure is that desire increases and we become more enmeshed in the drama. And what creates detachment to dualistic perception? Pain. Therefore, the yang aspect of pain is that desire decreases and we draw away from the drama, we become less enmeshed in it. Therefore, pleasurable and painful entities actually have equal amounts of yang and yin.

By the way, there's a more basic dualistic perception than perception of the universe: perception of our own relative selves, our ego. Indiv*iduality* is still a kind of duality.

A Third Kind of Goal
After many lives, we tire of the show. The dancing Light which is Maya fails to amuse. Component entities are transitory; we begin to seek something which is permanent. Component entities don't really exist below the level of their components; we begin to seek the truly Existent, the Real. Component entities contain a measure of pain and are unsatisfying, at least partially; we begin to seek That which is perfectly fulfilling. Tired of perceiving the individual qualities of various thoughts, emotions, and physical objects, we seek to experience their Isness. Knowing component entities can never satisfy our thirst for the Eternal, we seek to transcend dualistic perception, to undo the flip that occurred in Eden. We wish to behold the One, the Source, the Ultimate Ground of Existence. We seek return to the Kingdom of Heaven, the Pure Land, Eden.

Now, desire has been born for experience of the One, the Eternal, the Ultimate Ground of Existence. Now, longing has begun for God, for religion in the root meaning of the word - re-joining or re-fastening. Now,

> [t]he race is precisely the flight from creatures to union with the uncreated. ([M11],89).

For

> [n]o man shall ever know
> what is true blessedness
> Till oneness overwhelm
> and swallow separateness. ([S13],53).

Thus,

> [t]he wise have one wish left:
> to know the Whole, the Absolute.
> The foolish lose themselves in fragments
> and ignore the root. ([B05],51),

Once the desire is born in someone for gnosis, for knowledge of the Eternal, either first-hand or unitive, a quest has begun which will eventually dominate their entire life.

> "Here," says Ruysbroeck on the soul which has been lit by the Uncreated Light, "there begins an eternal hunger . . . If God gave to such a man all the gifts which all the saints possess, and all that He is able to give, but without giving Himself, the craving desire of the spirit would remain hungry and unsatisfied." ([U01],265).

Is gnosis a third type of goal? It certainly isn't a common goal. It is an afterlife goal? In a sense it is, since union with God, once achieved, continues after this life. But, as Nicholson writes,

> . . . the whole purpose of Sufism . . . is . . . a recovery of the original unity with The One, *while still in this body.* ([N11],16).

Thus, gnosis is not an afterlife goal; it's a goal which can be achieved in the here and now. It is, in fact, the fourth state of consciousness open to us in *this* life, where the others are the awake, dreaming, and dreamless sleep states. Therefore, gnosis is a third kind of goal, a beyond-the-show-world or transcendental goal.

Of all the goals available to us, gnosis is the highest.

> To surmount *maya* was the task assigned to the human race by the millennial prophets. To rise above the duality of creation and perceive the unity of the Creator was conceived of as man's highest goal. ([Y02],311).

Gnosis - and perhaps only gnosis - fulfills religion's promise. For it grants: salvation, since we are saved from Maya's illusion; awakening, since we awaken from the dream world of Maya; enlightenment, since the world we perceive is filled with Light; Self-realization since we realize our true Self; and liberation, since it liberates us from yang and yin. Liberation from Maya's drama gives transcendence, the "peace which surpasses all understanding" of Christian scriptures.

Entirely and Partially Transcendent Goals

Gnosis is a purely transcendental goal when it's motivated by a pure desire for the Eternal, a desire to know God simply for the sake of knowing God, with no other motive. Both motive and goal are grounded beyond the show world. So a desire to know and unite with the Center born of a pure love for the Ground which is our basis, the Eternal which is our Father, the Source which is our Mother, is an entirely transcendental goal. The Uncreated Light is loved for Itself, with no ulterior motive.

But pure love of That which we do not yet know and have not yet experienced is difficult. Therefore, the goal of gnosis, although ultimately leading beyond this world, is often grounded in it.

How can gnosis, something which is beyond the world of appearance, the world of maya, be grounded in it? By being a secondary goal, or even an unanticipated by-product of some other goal. There are religious goals, afterlife goals, and even worldly goals which lead to gnosis, some quickly, some slowly, some inescapably, and some only occasionally.

A few such goals are: escape from the transitory, from duality, from suffering, or from desire; realization of Self; immortality; spiritual rebirth; universal love; and following God's will by emulating and obeying a religious saint or Incarnation. All these seemingly distinct goals may lead to gnosis. Just as the distinct goals of political power, prestige, social status, fame, or high salary might all lead someone to seek their country's presidency, apparently different goals lead to the same end, gnosis.

Let's examine some of these goals.

Desire for Unchanging Reality

Achieving gnosis means transcending duality.

> Until duality is transcended and at-one-ment realized,
> Enlightenment cannot be attained. ([T06],206).

Knowing the One implies recovering unitary vision. So, someone with a pure love of the Unchanging might wish to transcend duality because it's an obstacle.

> If you dare
> call Him "Father"
> and live this in reality

286

> You must become a newborn child
> and overcome duality. ([B05],139).

In this case, both goal and motivation are still transcendental; escaping duality is a means, not an end.

Sometimes, however, the desire is not so much to gain the Eternal as to be rid of duality, with its pain and imperfections, its fleeing entities with only relative existence. Like shifting sands and ocean waves, everything around us changes. It's transitory; it passes away. Moreover, the entities possessing relative existence which compose our world lack a real identity. Their very existence is dependent, precarious, and temporary. In fact, they don't exist at all below the level of their components.

Lay not up your treasures where rust corrupts and thieves steal, advised (Mt6:19) Jesus. Don't build your house on shifting sands, he (Mt7:24-7) said, but on rock. As someone lost too long in the desert will desperately seek an oasis, someone lost too long in Maya will seek the pure Light of the Real. As someone lost at sea desperately reaches out for a boat, or a piece of wood, someone afloat in a world of fleeting, changing entities will reach out for the Eternal.

When escaping duality, when escaping an endless round of existence which is less than perfectly fulfilling, is the main goal then the goal is grounded in this world and is, therefore, less than purely transcendental. Nonetheless, since escape from duality necessarily implies perception of the One, such a goal will inescapably lead to gnosis.

Escape from Suffering

Often, however, the goal is escape not from both yang and yin, but merely from yin, from the disagreeable, from suffering. People are usually quite willing to be rid of the painful, the disagreeable, the annoying. Giving up the pleasurable, the agreeable, the pleasing, is quite another matter.

But yang and yin are inseparable and their perception - dualistic perception - is the root cause of suffering.

> The conception of duality is the root of all suffering; its
> only cure is the perception of the unreality of all
> objects and the realisation of myself as One, pure
> Intelligence and Bliss. (II,16,[A10],8).

We ignore the One and, as a direct consequence, live in the world of duality, a world where suffering is inevitable.

People who pursue the goal of escape from the yin aspects of existence, from suffering, pain, and distress, may someday realize yang and yin, pleasure and pain, are inseparable. At this point, no doubt, some will abandon their quest. Others, however, will decide to escape from yin *and* yang.

Therefore, the goal of ending suffering, of escape from yin, *may* lead to gnosis. If pursued far enough, sooner or later it may lead to a turning away from duality, which in turn will lead to gnosis. Therefore, even though the goals of escape from yin and yang, or from yin alone, are this-world goals, they can lead to gnosis.

Escape from Desire

But the world we live in is *composed* of yang and yin; can they be evaded? No, not until we perceive the Eternal. Only when we reach the One do we leave the "two" behind. But the desire for, the regard of, and the attachment to them can cease. They remain; but we lose our concern with them. Our concern is with the Eternal. The Bhagavad-Gita describes a person who has achieved such desirelessness as close to gnosis.

> He neither longs for one thing
> Nor loathes its opposite;
> The chains of his delusion
> Are soon cast off. ([S18],56-7).

Therefore, someone who wishes to experience the Real, or even someone who only wishes to escape yin and yang, may come to see escape from desire as a means. And even someone who wishes only to escape suffering may eventually wish to escape desire, too. In fact, this is what happened to Buddha.

In a very sheltered childhood, Buddha knew little of pain and suffering. One day, however, he realized how painful life could be. Specifically, he realized sickness comes to many, old age comes to those who don't die young, and death comes to all. Driven by his vision of a world full of pain and suffering, he set out in search of a solution. How to escape life's pain? was his question.

By giving up desire, was his answer.

But how to give up desire? One way is realizing the dual nature of entities, seeing that yang and yin are inseparable, that in the long run nothing is more or less desirable than another. Another way is the practice of asceticism, which we'll discuss later. Still another is the dangerous path of Tantra, which we'll also discuss.

Of course, the lack of unitary awareness is the root cause of desire.

> Desire returns
> as soon as we ignore
> the divine essence
> at our core. ([B05],105).

So perhaps the only way to genuinely extinguish desire is to achieve gnosis.

In any case, vision of the One often does leave someone desireless; it often gives them

> . . . a general condition of indifference, liberty, and
> peace, and elevation above the world, a sense of
> beatitude. (Delacroix, *Etudes sur le Mysticisme*, p.370
> in [U01],330).

Unitary awareness removes us from the world of yang and yin, pleasures and pains, desire and aversion.

And even a brief experience of gnosis may leave in the mystic's heart the same "Certitude. Certitude. Feeling. Joy. Peace." whose memory Pascal cherished. Not surprisingly, someone filled with such certitude, peace, love and joy may have no desires.

Realization of Self

Sometimes, an aspirant, especially one who's introverted, isn't motivated by the transitoriness or imperfection of the exterior world, since they aren't very concerned with the external world in the first place. Rather, they're concerned with themselves. They look inside and ask: Where did I come from? Where am I going? Where is my enduring identity? Who or what am I?

Who am I? This question, the basis of "the question of personal identity," is also the basis of the spiritual method (refer [S19],3-14) of the Hindu sage, Ramana Maharshi, who recommended the aspirant always keep it in mind. Why? Because full realization of our true Self, the Eternal, leads to gnosis - in fact, *is* gnosis.

> In those who have cognised the Self, illusion is
> dispelled, and the light of pure consciousness shines
> through them; their distress is at an end and they live
> in bliss. (XVIII,6,[A10],40).

How is someone to discover who they are? Through gnosis, in the long run. There is, however, a preliminary step.

Someone who has realized who they are has gained some knowledge. Let's call this positive knowledge. And someone who realizes they don't know who they are has zero knowledge. Can a person have *less* than zero knowledge? Yes! They can have "negative" knowledge; they can think they know something when in fact they do not.

If I realize I don't know which country Paris is in, I can go to a dictionary or encyclopaedia and find out. But if I think Paris is in Italy, I may never discover its true location. Negative knowledge can be more of a hindrance to knowledge than plain ignorance. The story of Socrates illustrates this.

The oracle at Delphi declared Socrates the wisest of men. "How can this be?" thought Socrates. "I know nothing. Yet the oracle cannot lie." Socrates eventually realized he was wise because he knew the extent of his ignorance. Everyone else he met thought they knew or understood things they in fact did not.

So the first step in discovering our real, enduring self is realizing that what we think of as our self - our body, our emotional or intellectual centers, our changing personality - is not really us.

Moreover, it seems identification with Self and identification with self, the ego, are mutually exclusive. As the Sufi Nifari wrote:

> When thou regardest thyself as existent and dost not
> regard Me as the Cause of thy existence, I veil My
> face and thine own face appears to thee. ([N11],85).

A commentary explains:

> If a man regards himself as existing through God, that
> which is of God in him predominates over the
> phenomenal element and makes it pass away, so that
> he sees nothing but God. If, on the contrary, he
> regards himself as having an independent existence,
> his unreal egoism is displayed to him and the reality of
> God becomes hidden from him. ([N11],85-6).

If identification with Self and identification with self are indeed mutually exclusive, then identification with Self must bring a kind of "death" to the self. A person who identifies more and more with Self must identify less and less with self. In a sense they "die" to ego, to self.

> Those only who do not believe, call me Gotama, but
> you call me the Buddha . . . And this is right, for I have
> in this life entered Nirvana, while the life of Gotama
> has been extinguished. ([C04],160).

How complete and actual is "dying to self"? Some Sufis believe the death is *to* the self, not *of* the self.

> When they say, "Die before ye die," . . .[they] do not
> mean to assert that the lower self can be essentially
> destroyed, but that it can and should be purged . . .
> ([N11],41).

Other mystics teach an absorption in God so total that their individual self is obliterated.

The Art of Dying

Identification with Self leads to detachment from self. Some mystics teach the inverse relations holds too - that dying to self leads to awareness of Self. For instance, the Sufi Jalaluddin Rumi wrote:

> Become pure from all attributes of self,
> That you may see your own bright essence ([N11],70).

And Nicholson writes:

> . . . in realising the non-entity of his individual self the
> Sufi realises his essential oneness with God . . .
> ([N11],155).

In fact, he claims:

> The whole of Sufism rests on the belief that when the
> individual self is lost, the Universal Self is found[,] . . .
> [T]hat ecstasy affords the only means by which the
> soul can directly communicate and become united
> with God. Asceticism, purification, love, gnosis,
> saintship - all the leading ideas of Sufism - are
> developed from this cardinal principle. ([N11],59).

Dying to self, by the way, may be the object of religious practices such as humility and obedience.

Sometimes, lessened identification with self is seen as a kind of death. The Quaker John Woolman seems to have experienced gnosis in this way. Woolman's journal records a dream where he heard an angel-like voice say "John Woolman is dead." Interestingly, a question of personal identity seems to have provoked his experience.

> In time of sickness . . . I was brought so near the gates of death that I forgot my name. Being then desirous to know who I was, I saw a mass . . . of human beings . . . . I was mixed with them, and . . . henceforth I might not consider myself as a distinct or separate being. ([F01],59-60).

It was then he heard the voice. Later, he realized

> "John Woolman is dead," meant no more than the death of my own will. ([F01],61).

A Quaker publication suggests:

> He . . . realized that the dream had shown him the death of his individual will and his submergence into the divine unity. ([P08],28).

Woolman lost his identification with his ego when he ceased to consider himself "a distinct and separate being." He felt he'd died. Yet he still lived - not as a separate ego but submerged in "the divine unity."

Thomas Kelly is another Quaker who may have had a similar experience. He wrote:

> It is an overwhelming experience to fall into the hands of the living God, to be invaded to the depths of one's being by His presence, to be, without warning, wholly uprooted from all earth-born securities and assurances, and to be blown by a tempest of unbelievable power which leaves one's old proud self utterly, utterly defenseless . . . [A]s Moses knew, no man can look on God and live - live as his old self. Death comes, blessed death, death of one's alienating will. ([F01],64-5).

And the Quaker George Fox noted that a certain part of the self seems incompatible with consciousness of Eternal Light, and must die if consciousness of the Light is to live.

> . . . [T]here did a pure fire appear in me . . . (ch.1,[J05],14)

wrote Fox. He then described the parts of the self which were in conflict with that Fire.

And that which could not . . . endure the fire . . . I
found to be the groans of the flesh (that could not give
up to the will of God) . . . and could not give up self to
die by the Cross, . . . that which would cloud and veil
from the presence of Christ, that which the sword of
the spirit cuts down and which must die . . .
(ch.1,[J05],14-15).

Immortality

Those who have died to their self have mastered the "Art of Dying" to
self. They've undergone a death which renders them immortal, for
they have nothing to fear from the death of the physical body or of the
relative selves.

I do not face my end in fear
for, knowing death must come,
I let my Me die long ago
and watched desire disappear. ([B05],122).

Thus a person, rather than seeking some form of immortality for their
relative selves, can realize a form of immortality by transcending their
own ego and identifying with their true Self, their Ultimate Ground of
Existence, which already *is* immortal.

His life is everlasting: whoever sees it is thereby made
everlasting. ([N11],7).

The idea of transcendence of the finite, contingent selves, of the
ego, is found in a short story by Jorge Luis Borges, an imaginative
Argentinean writer. His story *The God Script* tells of an Aztec priest,
Tzinacan, long imprisoned by the Spaniards. One day, Tzinacan
experiences

. . . union with the divinity, with the universe . . . God
has been seen in a blazing light . . . ([B11],172).

Now, Tzinacan believes he can shatter his stone prison, destroy the
captors who tortured him, evict all Spaniards from Mexico, rebuild
the sacred pyramid, and rule the entire land. Yet he knows he never
shall, for he no longer identifies with Tzinacan.

Whoever has seen the universe . . . cannot think in
terms of one man, of that man's trivial fortunes or
misfortunes, though he be that very man. ([B11],173).

Many religions have the idea that ego transcendence gives a kind
of immortality and deathlessness. Swami Paramananda, for example,
wrote

When we thus realize Him as the underlying Reality of
our being, we transcend death and become immortal.
([U03],116),

while a Buddhist monk writes

. . . the truth of nonselfhood or emptiness, makes a
man immortal because it makes him free of the 'self'
idea. When there is no self, how can there be death?
([B13],17).

Ego transcendence, by the way, can occur in meditative states.
The Taoist philosopher Ch'i, for instance,

. . . looked strangely dazed and inert, as though only
part of him were there at all. "What was happening to
you?" asked his disciple . . . [S]aid Ch'i; "when you
saw me just now my 'I' had lost its 'me'." ([W01],116).

We've been using "ego" to indicate our self image. It's also
sometimes used in a wider sense to refer to our body, heart, and
thinking centers, to all but our Self. Therefore, gnosis may be
described as separating the Self from the self, the relative selves, the
ego. Ch'i achieved this state when his "I" lost its "me."

Yet, our ego exists in the external world - at least, in other
people's external world. Therefore, separation of Self from ego may
also be described as separation of Self from world. And since the
emotions are sometimes considered a function of the heart, and
thoughts a function of the brain - both bodily organs - gnosis is
sometimes pictured as separation of "I" from body.

According to Plato, Socrates thought that separating "I" from body
was the true aim of philosophy. Calling the process "purification"
Socrates said

. . . purification is nothing but the separation of the
soul from the body, . . . the habit of the soul gathering
and collecting herself into herself from all sides out of
the body; the dwelling in her own place alone . . .
(*Phaedo* 67C, [D07],VI,418).

He thought

. . . the true philosophers . . . are always occupied in
the practice of dying . . . (*Phaedo* 67C, [D07],VI,418).

Physical death separates Consciousness from physical sensations
and, perhaps, emotional and intellectual sensations, for a time. The
"practice of dying" of Socrates was, I believe, a meditative exercised

which also temporarily separated Consciousness from physical sensations and, perhaps, emotional and intellectual sensations.

Some people, it seems, are born with a Consciousness detached from the physical, emotional, and intellectual facilities. For example, a Hindu saint said

> My consciousness has never associated itself with this
> temporary body. Before I came on this earth . . . 'I was
> the same'. As a little girl, 'I was the same.' I grew into
> womanhood; still 'I was the same.' When . . . this body
> married, 'I was the same.' And . . . now, 'I am the
> same.' Ever afterward, though the dance of creation
> change around me in the hall of eternity, 'I shall be the
> same.' ([Y02],524).

The New Birth

Since some form of the individual still exists, dying to self may be experienced in a positive manner, as the birth of a new self. Thus, the art of dying, paradoxically, brings Life. As St. Symeon the New Theologian wrote:

> A man who has attained the final degree of perfection
> is dead and yet not dead, but infinitely more alive in
> God with Whom he lives for he no longer lives by
> himself. ([W11],132).

And Meister Eckhart described his death to the temporal, finite ego as an "eternal birth."

> To this end I was born, and by virtue of my birth being
> eternal, I shall never die. It is of the nature of this
> eternal birth that I *have been* eternally, that I *am* now,
> and *shall be* forever. What I am as a temporal creature
> is to die and come to nothingness, for it came with
> time and so with time it will pass away. ([M11],231).

Moreover, Rufus Jones wrote Eckhart teaches

> . . . by a Divine birth the soul may rise to a mystical
> insight, which is above knowledge and which is *union* -
> an experience beyond subject-object. So only does
> the soul escape from the show-and-shadow world. He
> only can arrive at reality who can rise to the *Ever-
> present Now* in which all things are together.
> ([J03],232).

Rebirth imagery is common in mystical writings, as well as religious literature in general. Rebirth may refer, as it does here, to the final union with the Eternal, the last step of the path. Or it may symbolize the initial awakening, the first step on the path. Eckhart's eternal birth is, of course, to be understood in the second sense. These words of Angelus Silesius, however, might apply to either birth:

> Were Christ a thousand times
> reborn in Bethlehem's stall
> and not in thee, thou still
> art lost beyond recall. ([S13],20).

Universal Love

The ancient Greek language distinguished ([M07],975-6) four different types of love: epithemia, eros, philia, and agape. In contrast, English uses the single word "love" for essentially different things. Writes Huxley:

> . . . love, unfortunately, stands for everything from
> what happens when, on the screen, two close-ups
> rapturously collide to . . . when a John Woolman or a
> Peter Claver feels a concern about Negro slaves,
> because they are temples of the Holy Spirit - from . . .
> when crowds shout and sing and wave flags . . . to . . .
> when a solitary contemplative becomes absorbed in
> the prayer of simple regard. ([H11],83).

Let's examine a few types of love. First, there is the "love" of objects which please us, which give us pleasure. A child loves chocolate ice cream, a man loves to watch sports, a woman loves a dress. In this love, the thing is loved not for its own sake, but only for the pleasure it brings. Should the dress become torn or soiled, it's discarded.

Second, there's affection between people, the love found in friendship. This love also depends on pleasure; people usually become friends because they enjoy each other's company or have common interest. Yet it doesn't depend entirely on pleasure; sometimes friends argue and cause each other pain but remain friends nevertheless.

Third, there's the love between parent and child, or between very close friends. This love depends even less on pleasure. It's more self-sacrificing and can withstand much trial. A parent may care for a

296

disabled or mental-disturbed child for years. A son or daughter may tend an elderly, senile parent who doesn't even recognize them. In this love, someone may care more for the welfare of the other than for their own comfort or welfare. In an extreme case, they may even give their life for the sake of the other.

Erotic love can have elements of all three kinds of love. When it's mostly the first kind of love, the kind based solely on pleasure, it's more lust than love. In fact, many people wouldn't call it love at all. For them, erotic "love" needs an element of friendship, at least, to be genuine love. Of course, erotic love may go beyond friendship and reach the closeness of the third type of love.

The lower forms of love are based on pleasure derived from the object or person. The higher forms of love are not, but are still based on the other person. That is, the other person is loved because of who they are. All people in general aren't loved equally, regardless of their characteristics. Universal love is love which is independent of not only pleasure but person, too. It's love which shines like the sun, on everyone, universally, depending upon nothing. It's disinterested, not in that there's no interest in the other person but in that nothing the other person is or does will diminish it. Huxley calls ([H11],81) disinterested love charity and writes:

> . . . "charity" has come, in modern English, to be synonymous with "almsgiving," and is almost never used in its original sense, as signifying the highest and most divine form of love. . . . [C]harity is disinterested, seeking no reward, nor allowing itself to be diminished by any return of evil . . . [S]ince charity is disinterested, it must of necessity be universal. ([H11],82-3).

The founder of a medieval religious order advocated such love when she recommended

> [t]he sisters should not have particular friendships but should include all in their love for one another . . . ([B04],131).

How can someone develop charity, that is, universal, disinterested love? One method is to

> . . . take the whole universe as the expression of the one Self. Then only our love flows to all beings and creatures in the world equally. ([P12],610).

That is, if we love the Eternal and if we see all persons as Its manifestations, then we'll naturally love all people; we'll love them for the Root which is their basis. How can we develop love of the eternal? By directly experiencing It, by gnosis. As Nicholson writes:

> Gnosis and love are spiritually identical; they teach the same truths in different language. ([N11],101).

Following a Spiritual Teacher

Many people who have none of the preceding goals wish to live according to God's will. Often they try to obey the teachings and emulate the life of some religious founder or saint. This goal can lead to gnosis.

For example, Gotama's direct experience of the "Unborn, Unoriginated, Uncreated, Unformed" was what made him a Buddha.

> Realization of . . . the Voidness, the Unbecome, the Unborn, the Unmade, the Unformed, implies Buddhahood, Perfect Enlightenment . . . ([T05],97).

Therefore, out of a desire to follow Buddha, a Buddhist might aspire to realization of the Eternal.

Similarly, Jesus spoke of the "Kingdom of Heaven," and the "Pearl of Great Price," the possession of which was worth every earthy thing. These, I believe, were references to gnosis. Moreover, unitive gnosis, by uniting a mystic with the Perfect, allows them to fulfill the command of Jesus:

> Be ye therefore perfect, even as your Father which is in heaven is perfect. (Mt 5:48,[H08],996).

Spiritual teachers, however, often speak of God in personal terms; Jesus, for example, spoke of the Father. We may wonder if emulating or following the teachings of someone who taught or revealed a God who is a Person leads to gnosis.

We've seen Gods who are Persons are like blocks of ice in the infinite, eternal ocean of the Ultimate Ground of Existence, and moreover that only unitive knowledge of the Absolute in Its pure form is ultimate knowledge. Therefore, we might suppose following the teachings of Christ, Krishna, Mohammed, or some other spiritual teacher would eventually bring one to experience and knowledge of God. At first such experience and knowledge might be only of some God who is a Person. But since the Real underlies such a God, deep and intimate experience and knowledge of that God might naturally lead to the ultimate knowledge of Godhead, to That which manifests in the forms of Gods who are Persons. Therefore, the goal of practicing the teachings of some spiritual teacher, even if they teach a God who is a Person, may ultimately lead to gnosis.

Summary
We've examined various goals in this chapter, with emphasis on the goal of gnosis and various other goals which lead to it. The next chapter discusses various principles and values, ethics and morals which someone might adopt who has the goal of gnosis.

12

- Values -

Chapter Summary: This chapter discusses various value and moral systems, with emphasis on systems appropriate to someone who desires gnosis. Various spiritual goals and motivations are also discussed.

Values and morals govern much of our conduct - what we do and what we refrain from doing. How does someone who has the goal of gnosis conduct their life? What do they value? What do they avoid? This chapter discusses some of the values and morals, some the attitudes and actions, which seekers of gnosis often embrace - that is, it discusses some mystical value systems. It's not intended as a handbook for mystical life, however. Anyone seriously interested in such a life should refer to their own religious tradition, or to a few of the many recognized mystical texts in the bibliography.

We'll discuss value systems which contain only common goals, then systems which contain afterlife goals as well. Finally we'll discuss mystical value systems.

Common Value Systems
Some value systems contain only common goals. The people who follow such systems have no afterlife or mystical goals. Their hopes and aspirations are entirely for the things of this life, for physical, emotional, and intellectual satisfactions.

Of the people with common value systems some are naturally charitable, humble, forgiving, or chaste. They're kind and considerate, not for the sake of any ulterior goal, but simply because they feel that's how one ought to act. As we'll see, such people are unknowingly progressing toward gnosis. Others who follow common value systems are ruthless. Being kind, charitable, etc., isn't part of their value system, and they consider people who are to be sentimental and foolish. For those who are ruthlessly pursuing one or a few common goals morality is simple: do whatever is necessary to win. Their morality is much like the one described by Niccolo Machiavelli, medieval author of *The Prince*.

The Prince is a classic manual of values - of "morals," if you will - for someone interested in the single-minded, utterly unscrupulous pursuit and welding of political power. Machiavelli recommends such a person not limit themselves by morals.

> For there is such a difference between the way men
> live and the way they ought to live, that . . . anyone
> who determines to act in all circumstances the part of
> a good man must come to ruin among so many who
> are not good. Hence . . . he must learn how to be not
> good, and to use that ability or not as is required.
> ([M01], 141).

For example, says Machiavelli, it's not necessary to keep one's word.

> A prudent ruler . . . cannot and should not observe
> faith when such observance is to his disadvantage
> and the causes that made him give his promise have
> vanished. ([M01],148).

And while it's best if a ruler can win the love and fear of others, if only one can be won then fear is the better choice.

> . . . [S]ince men love as they please and fear as the
> prince pleases, a wise prince will evidently rely on
> what is in his own power and not on what is in the
> power of another. ([M01],147).

Machiavelli doesn't regard virtues as completely useless, however. Their appearance, at least, does have its uses.

> It is not necessary . . . for a prince really to have all the
> virtues . . . but it is very necessary to seem to have
> them. ([M01],149).

Machiavelli's recommendations make sense if a single common goal - in his case, political power - is all someone desires. If acting morally isn't a goal itself and if no other goal demands moral action, then there is no reason to be moral.

But what single goal is worthy of ruthless pursuit? Is any goal so valuable that for its sake all other considerations - morals, consideration for others, common decency - may be ignored? Fables such as the tale of King Midas dramatize the risk of achieving a poorly chosen goal. There seems to be a real danger in "putting all one's eggs in one basket," in pursuing any single goal, no matter how well chosen. Moreover, an unprincipled pursuit usually earns enemies. So even if one's own conscience permits enjoyment of a goal unscrupulously obtained, one's enemies may not. Is there any single goal worth ruthless pursuit and not liable to be lost to others? Probably not.

Perhaps for these reasons, perhaps for others, most people have many goals. First of course is surviving, fitting in with others, "getting by." A common strategy is accepting the reigning ideology. If one lives in a communist country, then one is communist. If one lives in a capitalist country, one is a capitalist. If one lives in a fascist country, even perhaps a Nazi-like country bent on genocide of a certain sub-group, then one is a good fascist and, if necessary, participates in genocide. (I don't mean to condone or recommend this strategy. I merely want to acknowledge its existence and widespread use.)

After survival comes enjoyment, the acquisition of desirable people such as a loving spouse, children, and congenial friends; of desirable things such as possessions and wealth; and of intangibles such as fame and creative or intellectual accomplishment, respect and power. People famous and unknown have devoted their entire lives to the pursuit and enjoyment of such things. Some of the famous evidently found them unfulfilling, as their frustrated lives or even suicides demonstrate. Others, no doubt, were well satisfied with their achievements - for a time. But even if that time reached to the end of their life, death sooner or later separated them and their possessions.

For death awaits all, regardless of how much or little they've won here. Therefore, many people have afterlife goals as well as common ones.

Afterlife Value Systems

Afterlife value systems include a goal beyond common goals and beyond the present life. Paradise and another life here, that is, heaven or a favorable incarnation are afterlife goals. Many people have some form of afterlife value system. Since our time here is limited, it's not surprising that there's widespread concern for what happens afterwards. Understandably, many people are willing to devote some small part of their time and effort to the next world. Conveniently, some religions promise heaven or a favorable incarnation in return for a minimal investment of time and money. An hour or two a week, a small donation, observance of moral and ritualistic rules which usually aren't too taxing, and one is right with God, yet available for the somewhat moderated enjoyment of this world.

Some of the rules and values which allegedly lead to heaven or a better incarnation seem natural: charity, love of neighbor, honesty. Others seem quite arbitrary. For example, there is a great diversity in religious rites and observances. In particular, rules about food vary widely. For example, ham is allowed in some religions, but forbidden in others. Beef is allowed in some religions, forbidden in others, and allowed in others only if slaughtered in the ritually correct way. Some religions allow, even demand, the use of alcohol; others strictly forbid its use.

Liturgies also differ. Some are somber and austere. Others, performed with candles and incense in dark churches, seem close to magic.

> . . . [I]nvocative arrangements of the Names of God . . . Sacred numbers, ritual actions, perfumes, purifications, words of power, are all used . . . by institutional religion in her work of opening up the human mind to the messages of the suprasensible world. In certain minor observances, and charm-like prayers, we seem to stand on the very borderland between magician and priest. ([U01],163-4).

The Bottom-up Approach to Morals

How are religious values - rules, rituals, and morals - determined? What are they based on? Religions often derive their morals and rules of conduct in a bottom-up manner. In a bottom-up approach to morals we begin with the practical rules of conduct called morals and then figure out the implied ethics, values and world view. We start with morals and derive the rest. But where do the morals come from in the first place? Usually, from some God who is a Person. Bottom-up approaches are common to systems where morals are simply the will - the commands - of some God who is a Person, or the dictates of some impersonal entity such as Reason or Natural Law. To act in accordance with God's (or Nature's) will is to act morally. To act otherwise offends God, and therefore is immoral and sinful. Knowing what is moral and what is not - i.e., knowing God's will - isn't a problem since there are scriptures and established churches to make it known.

One problem which does arise, however, is the following: is God free to will anything at all into rightness or wrongness? or are there standards of right and wrong even God must respect? In other words, is something wrong simply *because* God happens to forbid it? or does God forbid it *because* it's already wrong, harmful or evil?

Suppose we choose the first alternative and define "good" as whatever God wills. Then saying "God is always good" is merely a tautology - it's true by definition, just as if we define "dozen" to mean twelve, and then say "a dozen always had twelve things." It's true, but has little significance. It's just a kind of game with words. Moreover, if whatever God wills is good, then war, murder, sadism, torture, and rape are good when God wills them. You may feel that God never actually does will war, murder, sadism, torture, and rape. The millions throughout history who have fought holy wars, burnt heretics, and conducted inquisitions, however, would disagree. Some of them sincerely believed they were doing God's will. Didn't the invading armies of Europe's "holy" crusades and the religious leaders who organized it shout "God wills it"?

On the other hand suppose we choose the second alternative and decide there are certain standards of right and wrong which even God must respect. Then God can will only what is already inherently good. In this case, God seems the discoverer of good rather than its creator. How can such a God be omnipotent?

So is anything God wills good, or can God only will what's already good? It's a dilemma that's more theoretical than practical since regardless of the answer, right and wrong are forever etched into sacred scriptures in a bottom-up moral system. All a believer need do is follow them, with no explanation or justification needed or given. For example, in Exodus Yahweh commands (Ex 20:1-17) the Israelites to obey the Ten Commandments. They are to blindly follow what Yahweh commands *because* Yahweh commands it. With no explanation.

Another question which arises in a bottom-up moral system is the following: if moral principles really are the dictates of some universal God (or Reason or Natural Law) then they should be universal too.

But cultures have different, sometimes vastly different, morals. The following retells a story found in the *History* of Herodotus.

> Darius . . . found . . . the Callatians . . . customarily ate the bodies of their dead fathers. . . . [T]he Greeks practiced cremation . . . One day . . . he summoned some Greeks . . . and asked them what they would take to eat the bodies of their dead fathers. They were shocked . . . and replied that no amount of money could persuade them to do such a thing. Then Darius called in some Callatians, and . . . asked them what they would take to burn their dead fathers' bodies. They Callatians were horrified and told Darius not even to mention such a dreadful thing. ([S02],12).

Why do moral codes differ? An unassuming solution is that all moral codes are imperfect and still struggling toward the one, objective, true moral code. A more common answer says that one's own existing moral code perfectly embodies the true, objective, universal moral code, and all other moral codes are wrong. Another solution is that there is no single perfect moral code. What one calls sin, another may call virtue. If so, then morals are subjective, either to individual persons or to whole societies. Things are good or evil according to society's or the individual's taste. There are obvious problems with this approach. Many people feel the murder of innocents, the genocide of entire ethnic groups, sadism, etc., are objectively and universally wrong, not merely not to "taste."

The Value of Religious Practices
What's the value of religious practices? of morals? of food taboos? of rites and rituals? The religious see them as part of the optimum way to live. They believe the practices are valuable because God wants us to follow them and because a reward in the afterlife follows.

Skeptics on the other hand often view religious practices as mere superstition and ignorance. Or if they take a more charitable view, they admit the practices reinforce one's sense of belonging to a community, and give the believer peace of mind and the assurance they're right with God.

But many people, religious or skeptic, agree that religious values such as love, humility, charity, honesty, concern and respect for others, contribute to social harmony. In fact, many people - especially if they don't believe in any sort of existence beyond death - see teaching the community's commonly-held values as religion's main purpose. But as we'll see, some religious practices have a value beyond social harmony and integration: they can lead to gnosis. And that perhaps is their most important function, as well as religion's.

Yet while religion can be a path to gnosis, it doesn't seem to go all the way: it can't actually give the experience of the Ultimate Ground of Existence. Religion can't

> . . . extract finality from a method which does not really seek after ultimate things. This method may and does teach men goodness, gives them happiness and health. It can even induce in them a certain exaltation in which they become aware, at any rate for a moment, of the existence of the supernatural world - a stupendous accomplishment. But it will not of itself make them citizens of that world: give to them the freedom of Reality.
>
> "The work of the Church in the world," says Patmore, "is not to teach the mysteries of life, so much as to persuade the soul to that arduous degree of purity at which God Himself becomes her teacher. The work of the Church ends when the knowledge of God begins." ([U01],164).

Religion can make us aware of the existence of the door, show us its location, and encourage us to knock. The knocking, however, is up to us. Once we decide to knock for ourselves, our value system includes a mystical goal: we have a mystical value system.

Mystical Value Systems

Mystical value systems are value systems which include the goal of direct experience of and, perhaps, union with God. In some, God is a Person who has left explicit instructions for how to achieve gnosis. In such a system, morality is based on the will of that God. What about systems where God isn't a Person? What basis can be given for morality in such a system, especially if good and evil are said to be ultimately illusions?

Of course, a moral code could be offered with no basis or theoretical justification, as a collection of rules which in some undefined sense "should" be followed. It could be presented as something which has been found over time to promote a pleasant, harmonious life. In art, crafts, and manufacturing there are "rules of thumb," rules which have no theoretical basis but nonetheless are widely followed. Such rules, like rules handed down from on high, aren't intellectually satisfying. To be a science rather than an art, a scientific religion would require a theoretical foundation for its moral code.

We'll base our mystical value system's values and morality on the idea of helps and hindrances.

Helps and Hindrances

"Norma", a 15 year-old high school girl, has a goal: she wants to be an Olympics ice skater. Norma skates at least an hour every morning before school, and feels missing a practice is a kind of "sin." The word "sin" comes from words which signify ([P01],148-9) falling short or missing the target, and for Norma missing a practice falls short of her ideal of daily practice. Of course, missing an hour's skating isn't a moral failing; so "hindrance" would be a better word than "sin." Conversely, skating each morning isn't a moral virtue, but it is a "help" toward the goal of Olympics participation.

Similarly, certain thoughts and acts help our journey to gnosis; others delay it. Therefore, a value may be placed on thoughts and acts depending on whether they aid or impede our journey towards knowledge of or union with the Eternal. That is, we can construct a moral system based not on the supposed commands of some God who is a Person, but on whether the act, action, or thought in question has proven in general a help or hindrance to gnosis.

The Vedantist sage's Shankara derived morals in a similar way. Shankara taught that all acts belong to the realm of Maya. Yet he divided them into those which bring us closer to seeing the One (which we are calling helps), and those that re-enforce Maya's illusion (hindrances). He called these two kinds of maya

> . . . *avidya* . . . and *vidya* . . . . Avidya is that which causes us to move away from the real Self, or Brahman, drawing a veil before our sight of Truth; vidya is that which enables us to move towards Brahman by removing the veil of ignorance. ([V01],111).

Helps and hindrances to gnosis (or any other goal) still exist, even when there's no objective God who is a Person whose will defines virtue and sin. They exist independent of any God who is a Person. Something is a help or hindrance not because of some Divine command, but rather because of the very nature of the universe.

When we exercise, the result is written into our body in the form of stronger muscles, increased flexibility, or more efficient cardiovascular function. When we fail to exercise, the result is also written into our body in the form of weaker muscles, decreased flexibility, or less efficient cardiovascular function. Our body is a living record of our past physical activity. Even though Norma's actions, either practicing or not, aren't recorded in some heavenly book by a god who gives or withholds Olympics success, her actions nonetheless leave their record in her body and her level of skill. Norma carries with her the consequences of her past actions.

In a similar way, our character and state of consciousness results from - is a living record of - our past thoughts, emotions and actions. And as we'll see, our character and state of consciousness is a measure of our progress towards gnosis.

Reward and Punishment

In the biblical system of good and evil, Yahweh directly distributes rewards and punishments. He punishes disobedience, eventually, and rewards virtue, sooner or later. Does the system of helps and hindrances still have rewards and punishments? In one sense, yes; in another, no.

Since all entities, including our thoughts and acts, have equal yin and yang, each is equally rewarded and punished. Getting out of a warm bed early on a dark winter morning to practice is punishing; the reward is increased skill. Staying in bed and resting is rewarding; the punishment is eroded skill. In this sense, therefore, it doesn't matter which course is taken since practicing or not practicing both have their own rewards and punishments, their own yang and yin aspects.

On the other hand for a person with a direction and a goal, there are rewards and punishments. For Norma, increased ability is rewarding, and deteriorated ability is punishing. She chooses to see only the rewarding aspect of practice, and dismisses the punishing aspect as simply the price required for increased ability, as "paying dues." Similarly, she chooses to see only the punishing aspect of missing a practice.

The dual aspects underlie why some good, virtuous actions are such a pain, while some sinful, forbidden actions are so much fun. (A quip I once heard: "Everything I like is either immoral, illegal, or fattening.") Some virtuous actions have their yin aspect "up front," and some sinful actions have their yang aspects in the forefront. Later, the yang aspect of virtuous actions becomes apparent, as does the yin aspect of sinful actions. Perhaps this phenomena underlies the Christian "As you sow, so you shall reap" principle and the similar Hindu "Law of Karma."

Concerning Theoretical Explanations

Before using the idea of helps and hindrances to derive our mystical value system - that is, before discussing a theoretical foundation for many of the acts, thoughts, and beliefs which mystics value - let's discuss a point about theoretical explanations.

Our theoretical explanation for mystical practices and attitudes will explain why the practices and attitudes are helps in the quest to gnosis. It won't necessarily explain why mystics valued these practices. Some mystics may have adopted the practices and attitudes for entirely different reasons than our theory gives.

There's an analogy in nutrition. Suppose a scientist finds a primitive society's most popular recipes are also the most nutritious. The scientist would say the recipes are popular because they are nutritious. The local people, however, might know nothing of nutrition. They might believe the recipes are popular because of tradition, or because some god or seer commanded it.

In *Diet For A Small Planet* Frances Lappe writes:

> The proteins our bodies use are made up of 22 amino acids, in varying combinations. Eight of these amino acids cannot be synthesized by our bodies; they must be obtained from outside sources. ([L03],66),

that is, from food. Moreover, the eight essential amino acids must be present in the proper proportion for the body to use them. In other words, proteins are component entities, and their components (the amino acids) must have the proper relation to create a protein. Given a hundred *b*'s, *o*'s, and *d*'s, but only ten *y*'s, the word "body" can be made only ten times. The remaining ninety *b*'s, *o*'s, and *d*'s are useless. Similarly, if one essential amino acid is deficient, the body can't use the others.

Few foods all by themselves have the perfect balance of amino acids. Suppose beans have ninety *b*'s and *o*'s, but only ten *d*'s and *y*'s. And suppose wheat has only ten *b*'s and *o*'s, but ninety *d*'s and *y*'s. Then eating beans and wheat together will be much more nutritious than eating either alone, since their amino acids complement each other.

So if a scientist finds that popular recipes tend to have ingredients with complementary amino acids, for example that bean and wheat dishes are popular, the scientist would say the recipes are popular because they are more nutritious. The local people might know nothing of amino acids and complimenting protein, and say they like the recipes for entirely different reasons.

Similarly, mystics may have had many different reasons for adopting the practices and attitudes we'll discuss. They might have known little or nothing of the God which is not a Person, and little or nothing of the idea of the "end of drama."

The End of Drama
A familiar picture of the avid seeker of God is the contemplative hermit who lives devoid of almost all possessions, in a cave, the desert, or atop some lonely mountain. The stereotype is not entirely fanciful; in the past seekers have lived that way. Some do even today. Let's attempt to understand what might motivate someone to adopt such a life.

For someone whose goal is vision of, and eventual union with, the One, with Isness, the habit of seeing "the many," dualistic seeing, is a hindrance. This habit in turn is a result of our attachment to Maya's drama, of our involvement with duality and consequent ignorance of the One.

Watch a movie and try to continuously remember the images on the screen are nothing but light. If the movie is exceptionally boring, you may succeed. If, however, the movie has lots of attraction - lots of adventure, if that is what you like, or romance, if that's your preference - you'll become absorbed in the drama, forgetting that all you're seeing is light projected onto a screen. Even when the movie turns frightening or sad, you'll probably remain absorbed, feeling the appropriate fright or sadness. Only if the movie turns very disagreeable - frightening, sad, or boring beyond endurance - will the thought return "it's only a movie, it's only a play of light."

So it is with us and the Eternal. When life is going well, when there are lots of interesting things occurring, it's a rare individual who seeks the Eternal, the Light behind the show. Even when life is going badly, many do not turn to the Reality behind the illusion. For what prevents turning to the Root is not whether life is pleasant or unpleasant, but how deeply the individual is absorbed in life's drama. Attachment to the show, the appearance, the illusion, - the world - hinders our perception of the Reality standing behind it, the Eternal Substance.

How might someone completely dedicated to the goal of vision of and eventual union with the One conduct the movie which is their life? One straightforward way is to reduce involvement with duality, the drama which diverts perception of the One to perception of the many. Reduction of drama may be accomplished by withdrawal from the world and society to a life of solitude and quiet. Of course, desires and passions wouldn't be satisfied in such a life, so they would have to be somehow transcended, even uprooted. Poverty would eliminate most of the things which absorb our attention; solitude would eliminate the people which absorb our attention. Obviously, solitude would also imply abstinence from sexual intercourse if not complete chastity.

Emotions, thoughts, and fantasies are just as much a movie and drama as the exterior world. Someone completely dedicated to union with the One, therefore, might seek to control, still, and quiet them, too. Indeed,

> [i]t is a common teaching of mystic writers that
> introversion is effected by a successive silencing of
> the faculties . . . till . . . the very being of the soul . . .
> comes into immediate relation with the Ultimate
> Reality which is God. ([B17],33).

Moreover, awareness must be freed from consciousness of the ego. Our body, emotions, and thoughts exist in the world of Maya. That is, our very own self, the ego, is a perishable entity with only relative existence. Our egos are waves on the Eternal ocean of existence, but are not themselves the ocean.

Eventually, when Awareness has been freed from all that's relative, It becomes aware, not of any impermanent entity with only relative existence, but of Itself.

> . . . the unitive state is the culmination of the simplifying process by which the soul is gradually isolated from all that is foreign to itself, from all that is not God. ([N11],149).

Thus, for the sake of perception of the Eternal in Its pure state an aspirant may practice detachment of Consciousness from the show world. They may reduce to a minimum their involvement in, and perception of, any and all created entities, even their own self. Such practices constitute an apparent rejection of creation, which is viewed negatively as a veil, a hindrance to vision of the One. Not surprisingly therefore this way to pure vision of the Source has been called the Negative Way.

It will be helpful to investigate the practices of the Negative Way in more detail. Since withdrawal from society and created entities, the practice of silence and solitude, poverty and chastity, the control of emotions and the mind, all motivated by the desire for direct experience of the Root, constitute (surely, not coincidently) the main values of the strictest cloistered and hermetical monastic traditions, we'll turn to them for our illustrations.

In the religions in which it exists, monasticism is often considered the most radical, demanding, and direct way to vision of and union with the Eternal. Monasticism has three forms: life within a monastic community (cenobitic), life with the companionship of a few others (cloistered), and the life of solitude (eremitical). Of course, monastic communities with a scholarly or humanitarian function can't always observe solitude and silence. But many monastic situations have these characteristics: detachment from the world and society, fasting and abstinence, poverty, the practice of solitude, silence, and continual prayer, the battle against passions and desires, the fight to control the heart and mind, indifference to created entities, all motivated by the search for vision of, and eventually conscious union with, the Real.

Renunciation and Monastic Withdrawal

An obvious step for reducing attachment to drama, and the first step in many monastic traditions, is reducing physical involvement with the external world to a minimum.

312

> . . . [I]t was the universal conviction of the ancients . . .
> separation from the world constituted the climate . . .
> essential to the pursuit of monastic life. ([P09],193).

For example, *The Ladder of Divine Ascent* ([C09]), a famous
Orthodox Christian work on the monastic life describing thirty stages
of the soul's journey back to God, begins with withdrawal to a quiet
place. Historically and stereotypically, monastics have withdrawn to a
cave, mountain, desert, forest or cloistered monastic cell.

The life of Julian of Norwich, who lived during the ([J06],22)
"Golden Age of the English Recluse," offers a vivid illustration.
Julian was an anchoress, a religious seeker who pursued the search for
God almost entirely alone.

> The anchorite's ideal was a renunciation of the world
> so complete that all thought of the world and its cares
> would be totally banished from his mind, leaving him
> free to fill his mind and heart with God alone.
> ([P09],386).

To become an anchoress, Julian first petitioned her bishop. Then,
a Mass, probably a Mass of the Dead, was said for her.

> In the Exeter rite the whole service strongly resembled
> an actual burial service . . . ([J07],xxxviii).

Next, Julian was led to the anchorhold, a enclosed suite of rooms
about the size of an apartment, perhaps provided with an open air
courtyard.

> . . . [T]he In Paradisum was chanted as the postulant
> walked into the anchorhold, and the prayers for the
> commendation of a departed soul were said over the
> prostrate body of the newly-enclosed . . . The
> occupant of the anchorhold was now officially 'dead to
> the world'. ([J07],xxxviii).

Dead to the world but ([J06],23) "alive unto God."

An anchoress didn't necessarily maintain absolute physical
solitude, however; she might have a servant who purchased food, for
example. Moreover, she could even have an occasional visitor.
Nonetheless, each anchoress was considered sealed in their
anchorhold for life.

> Some ran away, of course; some went mad; the great
> majority were faithful unto physical death . . . What did
> they do? Fundamentally, they prayed . . . Julian
> cannot have been unique in the quality of the prayer

life she lived: many another found that he was in fact
alone with God, and was raised to great heights of
prayer. ([J06],24)

Not all monastics are anchorites, of course. Many live retired from
the world in the company of a few like-minded brethren. A famous
example is Mount Athos, an Eastern Orthodox group of monastic
communities. ([S12] describes Mount Athos in words and pictures.)

Hermit monks came to Mount Athos as early as the
ninth century. The first monastery, Great Lavra, was
founded by St. Athanasios in 963 . . . By 1400, 19 of
the 20 monasteries active today had been completed.
. . . Some 1,500 monks now inhabit the Holy
Mountain. ([N01],740).

The monks live in Mount Athos's ancient monasteries; the hermits in
its small huts and caves.

In the West, the Carmelites, Carthusians, and other orders still
follow a strict monastic life. Originally hermetical, Carmelite life
eventually became more communal. Then, Teresa of Avila, the
([C10],42) "greatest mystic of her day," founded a reform order,
closer to the original ideal, and closer too to the life Julian led. Teresa

. . . insisted upon enclosure . . . and limited the
opportunities for nuns to receive visitors . . .
([B04],132).

She also

. . . stressed voluntary poverty and the ascetic lifestyle
it entailed . . . ([B04],131-2).

Like Julian, Carmelite nuns and the hermits of Mount Athos live
in a world with few enticements to their attention. They've reduced to
a minimum their involvement with physical entities having only
relative existence. Such material poverty is an integral part of
monasticism.

The abandonment of material goods was an essential
ingredient of the renunciation involved in the monastic
vocation from the very beginning. ([P09],247).

Voluntary poverty is often recommended by the enlightened.
Jesus, for example, advised his followers:

Lay not up for yourselves treasures upon earth, where
moth and rust doth corrupt, and where thieves break
through and steal: But lay up for yourselves treasures
in heaven, where neither moth nor rust doth corrupt,

and where thieves do not break through nor steal: (Mt
6:19-20,[H08]),

and:

... go *and* sell that thou hast, and give to the poor ...
(Mt 19:21,[H08]).

For

... [i]t is easier for a camel to go through the eye of a
needle, than for a rich man to enter into the kingdom
of God. (Mt 19:24,[H08]).

Why should this be so? Should a Christian view created things as
inherently evil? Hardly. After all, Genesis 1:31 teaches God surveyed
all He had created and saw it was good. But involvement with and
attachment to such objects seems to prevent attachment to the Eternal.

For where your treasure is, there will your heart be
also. (Mt 6:21,[H08]).

Poverty and Detachment

Poverty involves more than physical

... detachment from all that is worldly and unreal ...
True poverty is not merely lack of wealth, but lack of
desire for wealth ... ([N11],36-7).

Detachment, the giving up of desire for, involvement and concern
with, and attachment to any created entity, is the second step in *The
Ladder of Divine Ascent*. As hesychastic monks have found,

... the first step ... in the long process of returning to
God is to cut oneself off from all extrinsic attachments
and then to tie the self to God in one's heart. But
before there can be attachment to God there must be
detachment from the world. ([M02],52).

Before we can clearly see the movie as a play of light, we must
first become unattached to seeing the various forms - the people and
objects - the play of light creates. Similarly, before we can clearly see
the world as a play of the Uncreated Light - and so come to see the
Uncreated Light Itself - we must first become unattached to seeing the
various forms, the people and objects, the play of Uncreated Light
creates.

The Eternal Light is Reality, but Its play creates entities with
relative existence, entities which are, in comparison, unreal.
Detachment from the play gives us discrimination, enabling us to
distinguish the Real from the unreal. For example, as Teresa

> . . . severed her attachments to things of the world, so
> her experience of . . . God deepened. ([B04],121).

Religious teachers, therefore, often advise detachment. Buddha
suggested his followers cease desiring any entity with only relative
existence.

> For that which is impermanent, brother, you must put
> away desire. . . . For that which is suffering, brother,
> you must put away desire. . . . For that which is no
> self, brother, you must put away desire. . . . ([B08],65).

Jesus, too, seems to have taught a similar detachment.

> . . . if any man will sue thee at the law, and take away
> thy coat, let him have *thy* cloak also. (Mt 5:40,[H08]).

More generally, he recommended

> Take no thought for your life, what ye shall eat, or
> what ye shall drink; nor yet for your body, what ye
> shall put on. (Mt 6:24,[H08]).

In short,

> Take therefore no thought for the morrow . . . (Mt
> 6:34,[H08]).

Passionlessness And Thoughtlessness

The world's exterior show isn't the only play of the Eternal to which
we are subject. And desires with respect to it aren't our only desires.
Inside of us is a drama of emotions and thoughts, memories and
fantasies. These too are impermanent entities having only relative
existence; their perception therefore is also a hindrance to perception
of the Eternal. So, to completely free Awareness, we must detach It
from the internal drama, from passions and other emotions, from
thoughts, memories, and fantasies.

Thus, "in the earnest exercise of mystic contemplation,"
"Dionysius" recommends one should

> . . . leave the senses and the activities of the intellect
> and all things that the senses or the intellect can
> perceive . . . ([D08],191).

Similarly, hesychasm recommends one should ([M02],76) "contain
the mind in the heart, freed of all imaginings." For

> . . . the mind, in order to reach true contemplation,
> must begin by emptying itself of all thoughts, whether
> they be good or bad. ([M02],113).

In fact, Patanjali, in his *Yoga Sutras*, defines yoga as

. . . the control of thought-waves in the mind.
([H09],11),

Passions of course also exist in the drama, and so must be abandoned, too. Thus, Islam's Sufis seek *fana*, a term which includes both passionlessness and thoughtlessness, and sometimes refers to gnosis as well. Fana is

1. A moral transformation of the soul through the extinction of all its passions and desires. . . . 2. . . . passing-away of the mind from all objects of perception, thoughts, actions, and feelings through its concentration upon the thought of God. . . . 3. The cessation of all conscious thought. ([N11],60-1).

Detachment from the Ego, Detachment from World

To withdraw from the drama, an Awareness must be freed of everything external or internal which is not the Eternal. The play of the Eternal includes more than objects and people, more than feelings and thoughts. It also includes our relative selves, the ego. For as we've seen, our selves are changing entities with only relative existence. They are part of the drama. Awareness, the Self, must be detached from the ego, the self, if It is to be free of the drama. Therefore, religious teachers often condemn pride and arrogance, recommending instead detachment from ego in the form of humility, meekness, and self-surrender. Jesus, for example, taught

Blessed *are* the meek . . . (Mt 5:5,[H08]),

and

. . . whosoever shall smite thee on thy right cheek, turn to him the other also. (Mt 5:39,[H08]),

and

And whosoever shall compel thee to go a mile, go with him twain. (Mt 5:41,[H08]).

Moreover, he held the person who serves others the greatest in the kingdom of heaven.

Similarly, Teresa of Avila

. . . rejected the principles of honor and lineage as incompatible with the religious life. For Teresa, obsession with one's reputation was a particularly insidious example of attachment to "things of the world. ([B04],127).

So

> [I]ike many of the great religious reformers, Teresa
> replaced honor with its reverse, humility, as the value
> most appropriate to the spiritual life. ([B04],130).

In her convents, wrote ([B04],127) Teresa, "All the sisters must be equals." And the author of *The Ladder of Divine Ascent* recommended

> . . . patience in annoyances, unmurmuring endurance
> of scorn, disregard of insults, and the habit, when
> wronged, of bearing it sturdily; when slandered, of not
> being indignant; when humiliated, not to be angry . . .
> ([C09],13).

Detachment from ego, in the form of humility and even-temperedness, helps dying to self and the consequent new birth of awareness of Self. It gives independence from external circumstances, making one a "king" of the interior world.

The usual image of a king is someone who has power and control of the external world. On a king's command, buildings are constructed, people are knighted, wars are fought, etc. But such a person might have little or no control of their own interior world; they might be unable to resist anger, lust, greed, gluttony, etc. A king of the interior world, on the other hand, has power and control over their own interior world.

The ancient Stoics seem to have held the ideal of the interior king. For them

> . . . sovereignty over oneself ceased to be a civic
> virtue and became an end unto itself. Autonomy
> secured inner peace and made a man independent of
> Fortune . . . This was preeminently the Stoic ideal . . .
> ([H07],v1,36).

And Buddha had an analogous idea. He considered various negative states to be "defilements" and suggested ([B10],180-2) one should "cleanse the mind of obstructive mental states" such as ill-will, sloth, and torpor, restlessness and worry, dejection, and coveting for the world.

The Negative and Affirmative Ways

One criticism of the negative way is that it's life-denying. Is it? Yes, if life is identified with drama, the picture show of events, feelings, and thoughts that absorbs most of us. In fact, that's it purpose. Drama-denying, however, is more accurate, for far from denying life, the goal of the negative way is the life which never ends, conscious union with the Eternal.

Another criticism is that it's not always available. Some religions don't have a monastic tradition. For some people, a retired, private life, in effect, a private cloister, may be the only option. Indeed, many people have pursued the negative way in that manner.

> . . . [M]en and women have built their own cloister in
> the midst of the worldly activity around them.
> ([N03],v4,263).

Many people have lived in the world yet practiced renunciation, withdrawal, poverty, detachment, passionlessness, desirelessness, and mental stillness. So monasticism isn't the only way to God. As Parrinder observes, there have been

> . . . noted lay men and women living in the world yet
> famed for their mystical devotions and writings.
> ([P05],187).

The most serious criticism of the negative way, however, is it's open to only the few. Many people, myself included, aren't able (or, at least, aren't willing) to make the radical life changes it demands. Complete dedication to beyond-the-show-world goals is itself beyond most of us.

So, what about us? How can we move towards gnosis? For us, there's another way, the affirmative way.

In the negative way, creation is denied and withdrawn from. It's considered an obstacle, a veil, a hindrance to perception of the Eternal. Siddhartha, protagonist of Hermann Hesse's novel of the same name, once held this attitude.

> "This," he said, handling it, "is a stone . . . Previously I
> should have said: This stone is just a stone; it has no
> value, it belongs to the world of Maya . . ." ([H05],145).

Siddhartha saw that the play of the Uncreated Light, creation, veils the Eternal. Yet it embodies the Real, too. Like a diamond which holds and reflects light, the world around us holds and reflects the Eternal Light. Eventually, Siddhartha realized this.

> But now I think: This stone is stone; it is also animal,
> God and Buddha. . . . I love it just because it is a stone
> . . . I see value and meaning in each one of its fine
> markings and cavities, in the yellow, in the gray, in the

hardness and the sound of it when I knock it . . . There
are stones that feel like oil or soap, that look like
leaves or sand, and each one is different and worships
Om in its own way; each one is Brahman. ([H05],145).

Since the universe is a manifestation of the Real, knowing the
universe can lead to knowledge of the Real. So the show world, the
play of light, can be a bridge to the Center. Some mystics, in fact, see
that as its primary purpose. In the words of Eckhart:

The world . . . was made for the soul's sake, so that
the soul's eye might be practiced and strengthened to
bear the divine light. ([M11],161).

For the divine light

. . . is so overpowering and clear that the soul's eye
could not bear it unless it were steadied by matter . . .
so that it is led up to the divine light and accustomed
to it. ([M11],161).

Indeed, as Ghazzali wrote,

Allah hath Seventy Thousand Veils of Light and
Darkness: were He to withdraw their curtain, then
would the splendours of His Aspect surely consume
everyone who apprehended Him with his sight.
([A03],76-7).

Mystical value systems which view creation positively, as a help
in reaching gnosis, are part of the so-called "affirmative way." The
affirmative way seeks to come to awareness of the Real not by
denying Its manifestations, but by learning to see the Eternal Light
behind all Its varied appearances.

Since it accepts the Eternal's dance, Its drama, the affirmative way
better suits someone who is in the world. It accepts the everyday
world we live in, and demands no rejection and separation, no cave,
mountaintop, or hermetical retreat. It looks for experience of God in
the people and things around us. It replaces renunciation and
withdrawal with a worldly life whose aim is gnosis, a life which is "in
the world but not of the world." In place of poverty there's the
moderated and charitable use of things; in place of detachment there's
an acceptance of occurrences as God's will; in place of chastity there's
a restrained indulgence in sexuality, often only within wedlock.

Attitudes towards Others

The affirmative way looks for experience of God in the people and things around us. It sees each living and non-living entity as a manifestation or embodiment of the Eternal Light, what Johannes Scotus Erigena called a "theophany."

How should someone treat a theophany? If the theophany is another person, one way is pacifism, an absolute refusal to use violence against them.

Some religions are pacifist. Eighteenth century Quakers ([F01],3), for instance, gave up political control of Pennsylvania, a state which they founded, rather than vote for war. Today, Quakers still believe that

> . . . we must all seriously consider the implications of our employment, our investments, our payment of taxes, and our manner of living as they relate to violence. ([F01],35),

and that

> . . . war is wrong in the sight of God. . . . We would alleviate the suffering caused by war. We would refrain from participating in all forms of violence and repression. ([F01],34-5).

In this, they follow Jesus' command to

> [l]ove your enemies, bless them that curse you, do good to them that hate you, and pray for them which despitefully use you, and persecute you; That ye may be the children of your Father which is in heaven: for he maketh his sun to rise on the evil and on the good . . . (Mt 5:44-45,[H08]).

Some religious groups even refuse to use violence against animals; they are vegetarian. Buddha, for example, described the monk as one who has abandoned

> . . . the slaying of creatures . . . the taking of what is not given . . . the unchaste life . . . falsehood . . . slanderous speech . . . bitter speech . . . idle babble . . . injury to seed-life and plant-life . . . highway robbery, plundering and deeds of violence. ([B06],221-2).

And Patanjali, in his *Yoga Sutras*, makes "not injuring" the first principle of "Abstinence," which in turn is the first of the eight "limbs" of yoga.

(1) "Abstinence" includes (a) not injuring, (b) not lying,
(c) not stealing, (d) not being sensual, and (e) not
being possessive. (XXX,[Y01],96).

Of course, pacifism and vegetarianism aren't the only ways to treat living theophanies. But many religious and mystical value systems do recommend patience and forgiveness of injury and insult, and a positive concern manifest in schools, hospitals, orphanages, humanitarian and poverty relief efforts, and social action groups.

The Rightful Use of Things

In some systems, non-injury is practiced, in so far as possible, toward animals and even plants. For example, the hermit whose food is predominately dairy, fruit and nut is someone who refuses to injure animals and plants unnecessarily. In other systems, no obligation toward animals and plants is seen. Rather, they are freely exploited for the benefit of the human race. Even in these systems, however, a proper concern for other people may imply a certain treatment of animals and plants.

For example, we've seen that about 16 pounds of plant feed are needed to produce a pound ([L03],9) of animal flesh for the table. In addition, the animal consumes much fresh water between birth and slaughter. Therefore, a desire the conserve water and food for people, rather than any direct concern for animals, might persuade someone to avoid animal flesh.

So, concern for people may determine the proper treatment of animals and plants. The proper use of things may be derived in a similar way. The proper use of inanimate objects helps us and other people, while improper use hurts, not the objects, but ourselves or others. Therefore, even though the affirmative way doesn't demand personal poverty, it does demand the ethical, charitable use of things, moderation in one's own personal possessions, and an interest the welfare of others. Humanitarian and poverty relief efforts are examples that come to mind. Religions often engage in such efforts.

There is, however, another type of humanitarian relief effort which religions often neglect.

It's one thing to relieve poverty and sickness; it is quite another to attempt to understand and eliminate their cause. This realization has come recently to the Roman Catholic church in Latin America. Once,

[p]riests had . . . often preached resignation to "God's
will" in a way that could reinforce the belief that the
present distribution of wealth and power comes from
God. ([B03],31).

Moreover,

. . . morality focused on sins of marital infidelity or
drinking, or treatment of other individuals, and was
little concerned about the impact of social structures.
([B03],66).

Now, however, in some quarters material poverty is understood to be

. . . an evil, as the result of the oppression of some
people by others. Poverty that dehumanizes human
beings is an offense against God. ([B03],32).

Such ideas are sometimes labeled "Liberation Theology."
Liberation theology attempts to eliminate the causes of poverty by
restructuring society. It teaches

[p]eople do not simply happen to be poor; their poverty
is largely a product of the way society is organized.
([B03],5).

Therefore, it criticizes economic systems that

. . . enable some Latin Americans to jet to Miami or
London to shop, while most of their fellow citizens do
not have safe drinking water. ([B03],5).

Such thinking, however, is not entirely new. For example,
believing that all people were equal in the sight of God ([F01],3),
Quakers centuries ago worked for what were then unpopular causes,
such as

. . . the abolition of slavery and of war, the welfare of
Negroes and Indians, temperance, prison reform and
the rights of women. ([F01],5).

Their motives, perhaps, were similar to those of liberation theologians
today who've decided

. . . the causes of poverty were structural and would
require basic structural changes . . . [S]uch changes
would come about only through political action.
([B03],15).

Such theologians envision

. . . a government that feeds the hungry, clothes the
naked, teaches the ignorant, puts into practice the

> work of charity, and love for neighbor . . . for the
> majority of our neighbors. ([B03],18).

There's one danger to the spiritual seeker in political action, and indeed all acts, that should be mentioned. In the affirmative way, actions are meant to aid the journey to gnosis. When properly performed,

> [f]ar from being an obstacle to spiritual growth, the
> giving of oneself in the service of others out of charity
> fosters the interior life of the soul. ([N03],v1,99).

But if actions, the means, become more important than experience of the Eternal, then the goal becomes political rather than spiritual. The object shifts from changing one's own inner world to changing the outer world. Someone who started out as a spiritual seeker becomes a political activist. Their action increases, rather than reduces, their attachment to drama.

The Two Ways Compared; Transcendence and Immanence

The negative and affirmative ways regard the world differently. One views it as a veil of the Eternal, a hindrance to gnosis. The other views it as an embodiment of the Eternal, an aid to gnosis. These two views have their roots in two different ways of thinking of the Real: as either immanent or transcendental. Let's examine these two ideas.

When we first introduced the idea of Ultimate Ground of Existence, many chapters ago, we started with a table and progressed to wood to molecules to, eventually, the table's Eternal Substance. Approached in this way, the Real is immanent, inherent, and indwelling in the table and, indeed, in all entities. In so far as It's the world's Ultimate Substance, the Eternal is the world and the world is the Eternal.

Yet the Unformed transcends the table, too. The table is brown, perhaps; the Unborn isn't brown. In fact, It's very different from anything we know. Grass is green, the Unconditioned isn't green. Water is wet, the Uncaused Cause isn't wet. Lead is heavy, the Unformed isn't heavy. The Real transcends the physical, emotional, and intellectual spheres. Therefore, the Center goes beyond and is not limited by the world. In this sense, the Eternal isn't the world and the world isn't the Eternal.

Light may be thought of as particle or wave. The God who is not a Person may be thought of as Person or not. Similarly, the Real may be considered immanent in the world or transcendent to the world. Shankara illustrated the situation as follows.

In India, cobras are greatly feared since their bites are often deadly. Imagine a rope left coiled along a village path. It's twilight. Someone on the path "sees" a snake and becomes fearful.

The immanent reality and the ground of existence of the "snake" is the rope. Therefore, in a sense the rope is the snake. Yet the rope transcends the "snake," goes beyond the "snake," and is very different from any genuine snake. In this sense, the rope isn't the snake.

Even as the "snake" is the rope misperceived, the world is the Eternal misperceived. And even as we may regard the "snake" as actually a rope, or as something very different from rope, we may regard the world as the Eternal Substance, or as something very different. And, finally, even as the "snake" is illusory but its ground is real, the world is illusory but its ground is real, in fact, the Real.

Ramana Maharshi's once declared [T03],16) that 1) the Eternal is real, 2) the world is unreal, and 3) the Eternal is the world. The world is unreal, he says, but the Eternal, which is real, is the world. Is, then, the world real or not? The statements may seem to contradict themselves, but when understood in the light of Shankara's illustration, they're no more contradictory than 1) the rope is real, 2) the snake is unreal, and 3) the rope is the snake.

Perhaps, "the Eternal appears as the world" and "the rope appears as a snake" is clearer. For world and snake are unreal in that they exist only in appearance.

How does all this concern the negative and affirmative ways? The negative way seems based on a transcendent view of the Real. Since the Real is very different from anything we perceive, it says, perception of those things must be abandoned before perception of the Real can arise. The affirmative way, on the other hand, seems based on an immanent view of the Real. The Real is here, right before us, if we could only see. Therefore, there is no need to deny the world around us. Rather, seeing the world properly will reveal its Eternal Basis.

Are then the two ways equally effective for reaching gnosis? Perhaps not. For it seems as long as the "snake" exists there some measure of illusion and unreality. Even if the "snake" is known to be a rope, even if it's seen as such, as long as even the appearance of snake remains, the rope is not fully seen as it is, clearly, without illusion.

Rufus Jones observes that in the affirmative way,

> . . . the seeker follows after the "beneficent progression of God," and gathers up what light he can from the revelations and manifestations, as God unveils Himself by going out of His Hiddenness. ([J03],105).

He continues:

> The discovery of the truth through manifestations is . . . the affirmative way. . . . [I]n the *outgoing* of God we can discover the attributes which in the Godhead "at home" are swallowed up in the unity of His perfect self. ([J03],108).

Not only are the attributes of God swallowed up in Godhead. For if we follow the outgoings of God far enough, we too are swallowed up in the unity of His perfect Self. But at this point we've lost sight of creation, and see only God. The affirmative way has turned into the negative way.

So, by itself, the affirmative way seems to have a certain limitation.

> The *affirmative way* never carries the seeker beyond "reflections" of the ultimate reality. ([J03],108).

Only the negative way goes beyond reflections to Godhead. It seems the affirmative way is a way of preparation which leads one eventually to the negative way, and eventually the pure contemplation of, and finally union with, the One.

Summary

We've examined different kinds of values systems, with emphasis on mystical value systems, on systems which include the goal of gnosis. We've seen that many of the negative way's practices, that is, many monastic actions and attitudes, follow naturally if someone is trying to abandon the drama of life to reach the Reality behind it. We've also briefly discussed the affirmative way, with emphasis on a few of its possible practices, namely, pacifism, vegetarianism, and action for social justice.

Many more practices from both ways can be discussed. We'll see more in the next chapter.

13

- Paths And Pitfalls -

Chapter Summary: This chapter discusses the spiritual path, an idea that pre-supposes goals and values. Various types of actions and attitudes are explored.

There are many more attitudes and actions which seekers of gnosis value. So we'll need a more detailed scheme than negative and affirmative way to discuss them, just as we needed a finer classification scheme than extrovert and introvert to discuss personal identity. This chapter uses the idea of path to place thoughts and actions into an overall scheme, to help organize some ideas we've already seen, and to introduce some new ones. It also discusses a few pitfalls on the road to gnosis.

The Path

A world view is a kind of map. It tells us who and where we are in relation to the world around us. And it shows us other places we could go, that is, it describes various goals we could work towards. The goals we choose express where we are going, or at least where we'd like to go.

A map also shows the intervening terrain. With a goal but no map, we might set out in the right direction, but encounter so many obstacles we turn back. But with a good map, we can intelligently choose a path which avoid pitfalls and obstacles.

Our values express the terrain we'd like to visit, and the terrain we'd like to avoid. Someone who values the affirmative way might choose a life in the normal world; someone who values the negative way might join a cloistered monastic community. Each would be taking a different path towards the same goal.

Paths appear in many religions. Eastern Christianity's *The Ladder of Divine Ascent* ([C09]) for example describes a path back to God where each step on the path is like a rung on a ladder. And the Hindu sage Patanjali describes eight "limbs" of yoga:

> (1) Abstention . . . (2) Devotion . . . (3) Posture . . . (4) Relaxation of Breathing . . . (5) Retraction of the Senses . . . (6) Fixation of Attention . . . (7) Fusive Apprehension . . . (8) Full Integration of Consciousness . . . ([Y01],94).

These limbs are actually milestones on the path. Abstention and devotion are moral practices, a list of do's and don'ts. Posture, relaxation of breathing, retraction of the senses, and fixation of attention refer to meditative practices which lead to gnosis. The last two limbs correspond to different levels of gnosis, different types of knowledge of God.

In this chapter a path loosely based on the eightfold path of Buddha (refer, for example, [B07],74) will be useful. We'll discuss right world views, right goals, right attitudes and acts in general, and right attitudes and acts specifically directed toward gnosis. These steps are similar to Buddha's right views, aims, speech, action, livelihood, and effort. The next chapter continues with meditative exercises which lead to gnosis, and gnosis itself, steps similar to Buddha's right mindfulness and right concentration.

Right Views, Right Goals
The world view we've been developing is a "right view" in that it includes a "right goal," a transcendental goal, the goal of gnosis. Other - sometimes vastly dissimilar - religious, philosophical, or metaphysical world views contain a similar goal. Since these systems contain a "right goal" we'll call them each a "right view." Of course, the systems don't completely agree with each other and may contain errors. So, "right view" doesn't necessarily mean a perfectly correct view, it means a world view which offers the goal of union with God.

For our purposes purely secular world views aren't "right views." Even though their rejection of superstition and ignorance may place them closer to objective truth and reality, they offer no right goal, no ideal of transcendence, no goal beyond the world of people and physical objects, emotions and thoughts. Science as it exists today isn't a right view because it offers no goal beyond the physical, emotional, and intellectual spheres. If it had the goal of direct experience of the Eternal, however, it would become a right view.

Is a right view - a comprehensive world view which offers the goal of gnosis - necessary for an aspiring mystic? No. Sometimes, just having a right goal and a few associated beliefs are sufficient. For example, believing it's possible to directly experience our Eternal Basis is enough to convince some people to undertake the arduous struggle for gnosis.

Most people, however, need some additional motivating beliefs. For example, believing that the world is soon to end has been a powerful motivation; based on it many people have eagerly sought That which will not pass away. We'll discuss a few motivating beliefs. We'll see some are sanely based on truth and insight; others are sane beliefs exaggerated to an unhealthy pitch; still others are unhealthy, or are based on falsehood and delusion.

Right Actions, Right Effort

Ideally, a right view should offer more than the goal of gnosis; it should offer some means of achieving it. Science unfortunately doesn't even acknowledge the possibility of direct experience of or union with the Eternal, so it naturally offers no means of achieving gnosis, of transcending the physical, emotional, and intellectual spheres. Our world view offers a means of achieving gnosis, as do other world views. Right actions and right efforts constitute some of the means to gnosis.

A right action is an attitude adopted or an action performed for a goal other than gnosis which nonetheless contributes to, or at least doesn't hinder, the journey toward gnosis. For example, we may work primarily to earn a livelihood but if our work helps our journey toward gnosis, then it's right action. Right effort, on the other hand, is an act undertaken or an attitude adopted specifically to help our return to the Eternal. Examples of right effort include prayer, fasting, vigils, meditation, charity, humility, and compassion.

The distinction between right action and right effort is sometimes hard to apply since some actions may be performed either for their own sake, or for the sake of gnosis, or both. Someone can fast to lose weight, to purify the body, to deepen prayer or meditation, or for all those reasons. Many, if not most, situations and occurrences can be used to move towards gnosis. Almost daily, opportunities for patience, kindness, resignation, love of neighbor, etc. present themselves. In fact, the perfection of the affirmative way is to use everything as a stepping stone on the road to union with the One, to live one's entire life as a prayer.

Because the distinction between right action and right effort can be troublesome, we'll often discuss acts and attitudes without deciding whether they are primarily, or only partially, concerned with gnosis. That is, we won't always decide whether something is a right effort, or a right action.

Let's begin with attitudes towards inanimate entities, towards things and events.

Attitudes Toward Things and Events

How might an aspiring mystic regard things and events? What attitudes might they adopt towards these nouns and verbs of the external world?

Many religions teach a disillusioned view of the things to which we are so attached. "Disillusioned," that is, in a positive sense.

To one who wishes to know the truth, disillusionment - losing illusions - is desirable. Conversely, someone who wants to avoid disillusionment seems to feel our illusions are a precious shield against a truth too horrible to behold.

A disillusioned view of component entities is easily derived from what we've already seen: they are "empty," that is, they lack enduring substantial identity and fully real existence; they are transitory and have an existence dependent on certain conditions; and they lack the ability to satisfy us completely, that is, they contain an element of imperfection, of suffering. Moreover, since they contain equal yang and yin, they're only apparently desirable or undesirable, not actually.

Yet some entities certainly seem very desirable. As an antidote to this desire, religious teachers often emphasize, sometimes quite strongly, the unalluring, yin aspects of component entities. Buddha, for example, said:

> Body . . is impermanent. Feeling, perception, the
> activities . . . consciousness is impermanent. So . . .
> the well-taught . . . disciple is repelled by body, is
> repelled by feeling, by perception, by the activities. He
> is repelled by consciousness. Being repelled by it he
> lusts not for it: not lusting he is set free. ([B08],20).

He repeated the same statements replacing "impermanent" with "suffering" and "non-self."

I find "is repelled by" a bit too strong and prefer "is detached from." The original version seems to border on that exaggerated depreciation of and aversion to things which often passes for genuine disillusionment.

Genuine disillusionment is a dispassionate, detached attitude based on insight. It's quite different from and often confused with another attitude which I'll call "sour-grape disillusionment," after a famous Aesop fable. In the tale, a fox wants some grapes hanging from a tree, but can't jump high enough to reach them. After many attempts, he gives up and walks away in disgust and disappointment. "They're probably sour anyway," he decides.

Sour-grape disillusionment is based on disappointment rather than insight. The fox would still like to have the grapes, if he could. He depreciates them only because of disappointment. Genuine disillusionment, on the other hand, is based on insight. Once someone fully realizes the yin inherent in pleasurable things and the yang inherent in painful things, or once they fully desire the Eternal above all else, then they've lost interest in grapes, no matter how sweet.

People who adopt a sour-grape attitude toward the world still very much desire its pleasures, which for some reason seem out of reach. They depreciate the world only because of disappointment. Often, their disappointment arises from injuries, disasters, and calamities they've suffered or witnessed. For example, many Europeans who endured the horrors of the Bubonic plague adopted sour-grape disillusionment toward the world.

Another example is found in ([M02]) *Russian Hesychasm*. During the 14th and 15th centuries C.E. the Russian people faced a series of calamities. Perhaps as a result, many people held the

> . . . conviction of the world and all that is found in it as seditious and evil. ([M02],27).

Religious writings echoed the theme.

> The ascetical tracts certainly painted the world in its worst colours . . . ([M02],27).

There was another reason to depreciate the world then. Based on biblical prophecy it was widely held ([M02],26) throughout Russia the world would end in 1492.

That the world is evil and will soon end are two powerful reasons for giving it up. They led some Russians to adopt the monastic life, since

> [w]ith their thoughts so often centred on the destruction of this present world, the monastic life held out special appeal as the best preparation for the Heavenly Jerusalem. ([M02],27).

The impending end of the world has been a popular religious belief, although obviously wrong. In the past, many religions preached the world would end soon. Early Christianity, of course, was partially based on this belief. Similarly,

> [t]o Muhammad the Final Judgment seemed a near reality, and he constantly urged his followers to abstain from material pleasures in order to lay up treasure in Heaven. ([A08],26)

Some religious sects even today expect the world to end soon. I once saw a person carrying a poster which announced this belief. The poster cited a scriptural prophecy. It occurred to me the scriptural prophecy had been penned over two thousands years ago. I couldn't understand why I should believe a "prophecy" that had been consistently wrong for over two thousand years.

Believing the sour-grape disillusioned view of the world and believing it's going to end soon can powerfully motivate the search for gnosis, for That which is perfectly fulfilling and never ends. Moreover, the two beliefs naturally give rise to detachment from things. After all, if the world is evil and soon to end, then we'll soon lose our possessions. So clinging to them is futile. And if the world is a dirty, evil place then there's no use desiring anything in it.

Yet they are inferior motivations. First, they aren't based on truth. At least, there is no reason to expect the world to end soon. Whether it's a dirty, evil place anyone may decide for themselves. They are inferior too since when the world doesn't end as scheduled, a believer may abandon the search for the Eternal and even decide all religious and mystical activities are worthless. A much superior motivation is the type of disillusionment based on accurate, deep insight into the actual nature of things.

An attitude akin to sour-grape disillusionment is fatalism, the view that everything is preordained and already fated, that effort is useless, for what shall happen shall happen. Fatalism says we are helpless pawns at the mercy of powers beyond our control, and must resign ourselves to the inevitable. Believing the world will soon end may encourage fatalism, since if the end of the world is preordained then perhaps all other occurrences are, too.

Is the future already determined? The movie we see at a theater certainly is; events are fixed before the film begins to run. Is the movie which is our life preordained as well? Is the future already immutably fixed? Or do we have free will; are we free to choose our actions?

These questions obviously concern predestination and free will, issues to which entire books have been devoted. In this book, I just assume we have free will. If we do, then value systems are useful as guides to action, and we can in fact choose our goals, actions, and attitudes. If we don't, then value systems and goals are useless; we'll perform whatever actions are predestined. But if we don't have free will, then I can't help assuming we do.

Predestination, by the way, isn't the only position which denies free will. Another is presented in a book I read as a child, *What is Man?* by Samuel Clemens who wrote as Mark Twain. Clemens attempted to prove we are machines, incapable of making real choice, always choosing the path of least pain. In his view, the future wasn't preordained, but the choice we'd make in any circumstances was.

Whether they believed in free will or predestination, many mystics recommended equanimity, an even-tempered, balanced acceptance of good and bad fortune. Equanimity can be based on the insight that both yang and yin aspects inhere in events, or on resignation to God's will. In either case, it's a help to gnosis. For it promotes a peaceful, calm state of mind detached from the world's drama. Equanimity becomes unhealthy, however, when it turns into fatalism.

Things and events are part of the external world, part of what is often beyond our control. Our own physical, emotional, and intellectual actions, on the other hand, are usually under our control, at least partially. We'll discuss them next.

Physical Actions and Efforts

Physical methods are methods which work with the body, for example, hatha yoga or asceticism. One ascetic practice many religions recommend, or even require, is fasting.

> For Symeon there could be no serious prayer life
> without fasting. ([S26],31).

Fasting is said to deepen prayer and meditation, as well as purify and revitalize the body. The bibliography contains a few references ([C17], [P06], [R08]) on fasting. Many monastic traditions recommend not only occasional fasting, but a permanently restricted diet, as well as vigil, long hours of prayer, and other ascetic practices. Buddha, for example, said a monk should be

> . . . moderate in eating. . . . should take food reflecting
> carefully, not for fun or indulgence or personal charm
> or beautification, but just enough for maintaining this
> body . . . ([B10],180).

The negative way uses asceticism to help detach Self from body. Asceticism also aims at a dying to self, a detachment of Self from self, the ego. At least one monastic author believes the two are related.

> It is self-deception to think that we can eradicate our
> self-love without doing violence to our flesh, which is
> the favorite breeding ground of egoism. ([P09],232).

He believes ascetical practices are necessary (at least, for monks), since while they

> . . . do not constitute perfection, . . . the experience of
> centuries has taught that they are an efficacious
> means to obtain it, and that it can scarcely be
> achieved without them. ([P09],232).

Of course, the body and ego may resist. Therefore asceticism can become ([P09],192) "ascetical combat."

Asceticism's ultimate aim is equanimity and indifference to pleasure and pain. Buddha listed some practical consequences of such equanimity and indifference when he described a monk as someone who is

> . . . able to endure heat, cold, hunger, thirst, the touch
> of mosquitoes, gadflies, wind, sun and creeping
> things, abusive language and unwelcome modes of
> speech; he has grown to bear bodily feelings which . .

. are painful, acute, sharp, severe, wretched,
miserable, deadly. ([B10],182).

Equanimity can lead to gnosis since someone who sees the yin
aspects of pleasurable things and the yang aspects of painful things
sees "pleasurable" and "painful" things equally. They're less
enmeshed in dualistic vision and therefore closer to gnosis. Similarly,
insensitivity to pleasure and pain means insensitivity to dualistic
vision, which means some measure of release from duality. Such a
person is closer to gnosis.

There's a pitfall which sometimes traps the would-be ascetic. They
allow their desire to discipline the body to turn into a hatred of the
body. Shankara seemed to fall into this trap when he wrote the body is

. . . a bundle of bones held together by flesh. It is very
dirty and full of filth. ([S11],57),

and, therefore, should be regarded as

. . . impure, as though it were an outcast. ([S11],75).

Similarly, while many early Christian saints of the Egyptian desert
were models of holiness, some exhibited a "ascetical one-upmanship"
that seemed very unhealthy. Therefore, St. Benedict warned his
monks against

. . . an inhuman kind of asceticism which would
destroy the very faculties that must be perfected. All
ascetical practices are nothing more than means . . .
([P09],230).

Rather than ends in themselves, they are means to gnosis. Like St.
Benedict, Buddha condemned the "self-tormentor."

He is a plucker out of hair and beard . . . He remains
standing and refuses a seat. . . . He is a "bed-of-
thorns" man, he makes his bed on spikes. . . . He lives
given to these practices which torment the body.
([B06],219).

Though fasting and asceticism play a role, in the affirmative way
physical acts which affect other people are more common. Examples
are humanitarian efforts which either alleviate suffering, such as the
care of the sick or elderly and the feeding of the hungry, or help
people in some other way, such as education or public service. Such
efforts not only help the recipients, but help the giver become more
concerned with others and less concerned with self, that is, to die to
self. And if self-centeredness is the problem and Self-centeredness the

cure, then whatever helps us become less egocentric is also a help to gnosis.

The affirmative way also offers a physical method which is the opposite of asceticism. This method, formalized in India under the name of "Tantra," teaches that the world's people and objects are to be loved and quite literally embraced, rather than rejected. It teaches one can rise above desire and reach gnosis through the fulfillment and satiation of desire. In particular, Tantra ([M15],40-1) teaches that sexual desires shouldn't be suppressed, but indulged.

Move toward union with God while enjoying all sorts of pleasures along the way? Tantra sounds too good to be true, and it probably is. Some mystics denounce the Tantric path as no path at all but an excuse for debauchery; others seem to grant it some measure of legitimacy by condemning it as a path which is extremely dangerous. Many of Rajneesh's disciples ([M15],38-42) followed the path of Tantra; some of them probably wished they hadn't.

There's a phenomena which passes for asceticism though it's not. Once some measure of union with the One has been achieved, the mystic may naturally transcend duality, and therefore be naturally insensitive to dualistic pairs.

> Having . . . transcended the influence of the pairs of opposites, the Sage, free from desire, does not feel pleasure or pain in anything he experiences.
> (III,14,[A10],13).

At this point, the mystic isn't ascetic. They aren't resisting their natural inclinations, rather they are following them. But clear vision of the One has robbed the Two of its attraction, so their inclination is to be unattracted to component entities with relative existence. Similarly, they don't have to help and love others to detach their Self from their self. Rather they have a natural regard for other people, who they see as embodiments of the One they love.

Drugs

Physical methods also include the use of certain "mind-manifesting" substances, that is, of certain drugs. During the 1960s, these drugs were well-known and widely used in some circles. What wasn't well-known, however, was that some people used the drugs to achieve mystical experiences.

Aldous Huxley observes

> . . . it is a matter of historical record that most
> contemplatives worked systematically to modify their
> body chemistry, with a view to creating the internal
> conditions favorable to spiritual insight. ([H10],155).

Their methods included ascetic practices such as fasting, vigils, self-flagellation, continuous psalms, and breathing exercises. He thought certain drugs could bring about similar changes in body chemistry, and recommended ([H10],156) them over the older methods. In fact, Huxley himself experienced mystical insights under the influence of mescaline, a drug derived from the peyote cactus and used sacramentally in the religious services of some Native American tribes. *The Doors of Perception* contains a record of his mescaline experience.

> The Beatific Vision, *Sat Chit Ananda*, Being-
> Awareness-Bliss - for the first time I understood, not
> on the verbal level, not by inchoate hints or at a
> distance, but precisely and completely what those
> prodigious syllables referred to. ([H10],18).

Sat Chit Ananda is a Hindu phrase for the experience of Brahman, the God which is not a Person.

Can drugs really bring about direct knowledge of God?

The religious use of drugs (refer, for example, [F04] or [N05],v14,201) is ancient. Thousands of years ago, Hindu sages sang the praises of soma, Jews and Christians drank consecrated wine, and Aztecs used sacred psychedelic mushrooms which they called "flesh of the gods." More recently, William James found nitrous oxide greatly stimulated his mystical consciousness. And philosopher Alan Watts had drug induced religious experiences which he described in *The Joyous Cosmology*. Watts seemed to experience the God which is not a Person.

> . . . . I see . . . a face which reminds me of the Christos
> Pantocrator of Byzantine mosaics, and I feel that the
> angels are drawing back with wings over their faces in
> a motion of reverent dread. But the face dissolves.
> The pool of flame grows brighter and brighter, and I
> notice that the winged beings are drawing back with a
> gesture, not of dread, but of tenderness - for the flame
> knows no anger. Its warmth and radiance - "tongues of
> flame infolded" - are an efflorescence of love so

endearing that I feel I have seen the heart of all
hearts. ([W03],78).

Timothy Leary was another proponent of drug-aided religious
experiences. While still a Harvard psychology professor, Leary

. . . ate seven of the so-called sacred mushrooms . . .
During the next five hours, I was whirled through an
experience which could be described in many
extravagant metaphors but which was, above all and
without question, the deepest religious experience of
my life. . . . I have repeated this biochemical and (to
me) sacramental ritual several hundred times . . .
([L04],13,4).

Leary also administered psychedelic drugs to others and studied their
experiences. He found

[s]ubjects speak of participating in and merging with
pure (i.e., content-free) energy, white light . . .
([L04],24).

He regarded such energy as God.

God does exist and is to me this energy process . . .
([L04],275).

And he believed psychedelic experience was best described in
mystical terms.

[T]he panoramas and the levels that you get into with
LSD are exactly those areas which men have called
the confrontation of God. The LSD trip is the classic
visionary-mystic voyage. ([L04],260).

Thus, much of Leary's drug experience was religious, at least to
him. Even R. Zaehner, who criticized mystical claims for drugs,
recognized Leary believed LSD could bring about

. . . direct *religious* experience of the eternal being that
pervades all that is ephemeral and transient.
([Z01],76).

Zaehner described this as experience of

Brahman. . . 'boundless being' . . . the base of all
becoming. . . . It is this 'principle of eternity' rather than
any personal God that the takers of psychedelic drugs
claim to experience. ([Z01],43).

So Leary and others believed psychedelic drugs could help bring
about direct experience of God, not of some God who is a Person, like
Jesus or Krishna, but of the God which is not a Person.

Eventually Leary and some associates moved into a mansion, which they regarded as a religious community,

> . . . a religious center. About 30 people are devoting their lives and energies to a full-time pursuit of the Divinity through the sacrament of LSD. ([L04],293).

One of the people who lived there, Richard Alpert, later known as Baba Ram Dass, describes his spiritual journey in ([A04]) *Be Here Now*. His book greatly influenced the spiritual journey of myself and many others.

Leary advocated using LSD to bring about mystical states of mind.

> Drugs are the religion of the twenty-first century. Pursuing the religious life today without using psychedelic drugs is like studying astronomy with the naked eye . . . ([L04],44).

But some people felt Leary's estimation of LSD was not entirely accurate.

> . . . Huxley, Alan Watts, and others . . . in their various writings imposed upon the psychedelic experience essentially Eastern ideas and terminology which a great many persons then assumed to be the sole and accurate way of approaching and interpreting such experience. ([M05],260).

LSD and other psychedelic drugs enjoyed some popularity in the 1960-1970's during the "Hippie" era partly due to the efforts of Huxley, Watts, Leary and Alpert. Many people found the drugs didn't live up to their reputation as doorways to mystical experience. Psychedelic drugs induced all kinds of experience from beatific to hellish and anything in between.

What had gone wrong? An answer may lie in the drug experiences of the poet, Allen Ginsberg, who had a non-drug religious experience in 1948 reading a poem by the mystic William Blake, and spent 15 years vainly trying to recapture it with LSD. His efforts not only failed, but were counterproductive.

> Ginsberg found that . . . self-programming could create formidable psychic tensions often resulting in awful bummers. ([L06],110).

On one trip, he

> . . . felt faced by Death, my skull . . . rolling back and forth . . . as if in reproduction of the last physical move

> I make before settling into real death - got nauseous,
> rushed out and began vomiting, all covered with
> snakes . . . I felt like a snake vomiting out the universe
> . . . ([B15],56).

Ginsberg eventually saw the attempt to control psychedelic trips was futile.

> It's just like somebody taking acid and wanting to have
> a God trip and straining to see God, and instead,
> naturally, seeing all sorts of diabolical machines
> coming up around him, seeing hells instead of
> heavens. So I finally conclude that the bum trip on
> acid as well as the bum trip on normal consciousness
> came from attempting to grasp, desiring a
> preconceived end . . . ([L06],112).

Therapists who used psychedelic drugs in their practice before 1966 while it was still legal seemed to have reached a similar conclusion. For they

> . . . simply sought to help subjects relax and remain
> open to the experience without defining what was
> supposed to occur. ([L06],109)

But psychedelic drugs had induced religious experiences in Huxley, Watts, and Leary. How could the same drugs give such different experiences?

It was a well-known in the 1960's that "set and setting" influenced the drug experience. "Set" referred to a person's "mind set," their basic attitudes, character traits, education, etc. "Setting" referred to their environment while under the drug's influence. A night club could induce a very different experience than a beach. Huxley had a long-standing interest in spiritual matters, as witnessed by *The Perennial Philosophy* written over a decade before his mescaline experiments. Alan Watts had a masters in theology, a doctorate in divinity, and was a professional philosopher. I believe I once read Leary had considered the priesthood in his youth. That such people would have religious drug experiences is perhaps not too surprising.

And when they tried to describe their religious experiences, Eastern terminology would be a natural choice because Eastern religions have so much more to say about the God which is not a Person than Western religious thought. As Zaehner wrote of Leary:

> Among the Eastern religions, then, he is drawn
> principally to the pantheism developed in ancient India

by the authors of the Upanishads and to Taoism in
China. In neither is God as a person relevant.
([Z01],73)

About 1964 Leary re-wrote the *Tibetan Book of the Dead* as *The Psychedelic Experience* ([L05]), an LSD tripping manual describing what was supposed to happen under the influence of LSD.

Leary now presented turning on as a process of
initiation into a great brotherhood of free souls
christened by the mind-blowing apprehension of the
Clear Light during the peak of an acid trip. ([L06],109).

It's one thing to say mystical experience, or any other kind of experience, may occur under the influence of a drug; it is quite another thing to say it is *supposed* to happen. Based on Leary's writings, many people took psychedelic drugs to gain spiritual insights. The result seems to have been a repetition of Ginsberg's experience on a much larger scale.

It is as if (Leary) . . . polluted the stream at its source
and gave half the kids in psychedelic society a bad set
to start out with. Almost every acidhead I talked to for
years afterwards told me he had, as a novice, used
The Tibetan Book of the Dead as a "guide" - and
every one of them reported unnecessary anxiety,
colossal bummers, disillusionment, and eventual
frustration and exasperation, for which, in most cases,
they blamed themselves, not Tim or the book.
([K04],29).

There's an historical analogue to the story of psychedelic drugs and mystical experience. The ancient world had beer and wine but no "hard" liquor. When the distillation process was discovered about 1,000 years ago, some people were certain the thousand years of peace and plenty foretold in the Bible (Rev 20:1-7) when Christ would reign over the earth had arrived.

Emotion Actions and Efforts

Now let's turn to right actions and efforts which concern the emotions.

Because the negative way seeks to reduce an Awareness's attachment to emotions, it places restrictions on social interaction. For example, some Christian monks are advised against ([P09],197) excessive talking, a habit which has been found to interfere with meditation, contemplation, and prayer. Similarly, Teresa recommended that

> [a]s much as they can, the sisters should avoid a great deal of conversation with relatives . . . ([B04],129)

since

> . . they will find it difficult to avoid talking to them about worldly things. ([B04],129).

And Buddha recommended the monk turn away from talk

> . . . which is low, of the village, of the ordinary folk, . . . not connected with the goal, which does not conduce to . . . detachment nor to . . . calm nor to super-knowledge nor to self-awakening . . . ([B10],156-7),

more specifically, talk about ([B10],157) thieves, great ministers, armies, battles, food, drink, clothes, relations, vehicles, towns, women, men, and streets. He recommended, instead, talk which is

> . . . austere, . . . which conduces to . . . detachment, stopping, calm, super-knowledge, self awakening . . . ([B10],157).

Eliminating excessive talking is but one example of the withdrawal from society the negative way often involves. Some Christian monks are also advised against ([P09],237) pursuing friendships and associations with secular people. And, of course, cloistered and hermetical monks withdraw from society entirely.

Sometimes, the seeker finds social withdrawal difficult, especially from family. Teresa, for example,

> . . . recognized that for many the most difficult aspect of "detachment" . . . involved severing ties with family members. ([B04],128).

On the other hand, sometimes withdrawal is easy, even desperately desired. For at one point along the path, a seeker with an intense love of God may

> . . . desire to shun like poison his wife and children and other relations, worldly connection with whom deflects him from the divine Lord; ([S01],366).

Ramakrishna once had this state of mind.

> I could not then bear the very atmosphere of worldly people, and felt when in the company of relatives, as if my breath would stop and the soul leave the body. ([S01],366).

And perhaps Jesus referred to this state when he said:

342

> If any *man* come to me, and hate not his father, and
> mother, and wife, and children, and brethren, and
> sisters, yea, and his own life, also, he cannot be my
> disciple. (Mk 14:26,[H08]).

The companionship of the opposite sex is another area where
social withdrawal can be difficult. Since the affective and sex drives
are so strong, someone pursuing the negative way may have difficulty
giving them up. Drastic means, therefore, are often advised.
Ramakrishna, for example, taught ([G03],874) a male monk should
keep his distance from women, shouldn't talk to one, and shouldn't
even look at picture of one. He also taught a man could conquer
passion and lust ([G03],601,701) by "assuming the attitude of a
woman". He himself, at one point, dressed like a woman! And some
hesychast monks were taught to avoid, not only the sight of women,
but ([M02],90) "youthful, beardless, and effeminate faces," as well.

Reduced involvement with emotions and withdrawal from society
may seem less than loving. Ideally, however, they're motivated by
love for the Eternal, rather than hatred of people. Ideally, they're
based on insight, not on a negative view of people. A few pitfalls are
probably obvious.

One pitfall is when involvement with emotions remains, but
negative emotions replace positive ones. Underhill describes a mystic
who viewed

> . . . with almost murderous satisfaction the deaths of
> relatives who were "impediments." ([U01],216).

Hating, as much as loving, still involves the hater in duality.
There's still emotional involvement. And moving from a loving
concern of relatives to hatred can hardly be considered a step towards
gnosis. To avoid this pitfall, a path would lead from the loving
concern of the affirmative way to a transcendent indifference of the
negative way, around, not through, the pitfall of hatred and aversion.

Another pitfall is when other people are seen as evil, and derided,
scorned, or shunned. Underhill describes a mystic whose idea of
chastity included ([U01],216) shutting "himself in a cupboard for fear
he should see his mother pass by". If avoiding the opposite sex is to
remain healthy, it should be understood that the danger is one's own
lust, not the other person. To a man who wishes to practice celibacy, a
woman, particular a beautiful woman, is certainly a hindrance. But to
her, he may be a similar hindrance.

So the hindrance is not the other person but lust. Why? Lust, along with anger, envy, hatred, and other acknowledged vices, make entities with only relative existence seem quite real. Vices enmesh us more in the drama; this, in fact, is what makes them vices. So, a person's attitude becomes unhealthy when they see another person, rather than their own anger, lust, envy, hatred, greed, etc., as the problem. This attitude is similar to sour-grape disillusionment; it's based on an overly negative view of other people rather than insight.

A healthy attitude may be easier to maintain in a help/hindrance system. A good/evil system, on the other hand, seems more liable to promote unhealthy attitudes. If I'm trying to be good and find someone else an impediment, then it's natural to feel that they must be evil. After all, if something is good, its opposite is evil. If I am of one gender and those of my gender are good, so this thinking goes, then those of the other gender must be evil; if my race is good then other races must be evil.

Since the affirmative way doesn't deny them but employs the emotions in the journey to gnosis, often directing them toward some God who is a Person, it's less likely to lead to emotional pitfalls, in particular unhealthy attitudes toward other people. Since its followers are often motivated by love of some God who is a Person, their attitudes tend to be healthier. In Christianity, for example, one is encouraged to live a virtuous life for the sake of Jesus who died for us. Love of Jesus is supposed to promote love of people in general.

However, the love of God can lead to the pitfall of hatred of those who don't worship the same God, or who worship in a different way. Certainly, religion has motivated many wars, although political and economic factors often contributed too.

Or such love may create a self-righteousness toward those who are not on as good terms with God as one supposes themselves to be. Or it may lead to a neglect of others. For doesn't every expensive religious meeting place and every gold religious object represent money which could have been better spent relieving hunger, disease, pain, and suffering?

Intellectual Actions and Efforts
Studying philosophy, metaphysics, or theology are intellectual acts which can help bring about experience of the Absolute. If the study is

primarily undertaken to bring one closer to gnosis, then it's right effort. On the other hand, if the study is directed toward purely intellectual enlightenment, or to winning academic position or honors, then it's only right action. In either case, understanding the nature of things, ourselves and the world around us, can motivate a search for the Eternal.

Moreover, some groups see a lack of true understanding of ourselves and the world around us as the main cause of suffering. In this view, we need to overcome not sin but ignorance "of the way things are." Buddha, for example, declared:

> If ye could see things as they are, not as they appear,
> ye would no longer inflict injuries and pain on your
> own selves. ([C04],200).

And Buddhadasa declares ([B13],111) suffering is the result of acting inappropriately which in turn is the result of not understanding the "true nature" of things.

> . . . [W]e are ignorant of the true nature of things; thus
> our behavior results in suffering. Buddhist practice is
> designed to teach us how things really are. ([B13],111)

Therefore it's designed to eliminate suffering as well.

Some ancient, non-orthodox Christians called "Gnostics" also

> . . . insisted that ignorance, not sin, is what involves a
> person in suffering. . . . Both gnosticism and
> psychotherapy value, above all, knowledge - the self-
> knowledge which is insight. . . . [L]acking this, a
> person experiences the sense of being driven by
> impulses he does not understand. ([P01],149).

Of course, understanding presupposes a capable mind. Buddhists believe deep insight and understanding of the true nature of things comes easier to the calmed and controlled mind. Conversely, a turbulent mind and short attention span hinder understanding and insight. Television has been criticized for promoting a fickle, uncontrolled mind, incapable of sustained concentration. I don't know if Buddhist monks are advised against it, but at least one Christian spiritual director warns his monks that too much television, newspapers, magazines, and radio can ([P09],237) "deaden their sense of the supernatural." He also advises the curbing of indiscriminate intellectual curiosity and recommends instead a ([P09],237) "careful selection of worthwhile objects of knowledge."

Intellectual discrimination should avoid the pitfall of overly depreciating learning and curiosity. For if curbing is taken to an extreme, then all secular learning may be scorned. As a result, the study of arts and sciences will probably decline. Since a scientific religion is a science as well as a religion, it would probably not thrive in an environment hostile to other sciences. In contrast, traditional religion may flourish in such an environment. It may even promote it. Dampier writes ([D01],65) that early Christians little valued secular learning for it own sake. In time,

> Christian thought became antagonistic to secular learning, identifying it with the heathenism which Christians had set out to conquer. . . . [I]gnorance was exalted as a virtue. ([D01],65).

Perhaps as a result, Europe eventually lost much of its knowledge. During those "Dark Ages," the Christian religion peaked in power and prestige. And Thomas Kempis, fourteen hundred years into the Christian era, could write in an otherwise highly valuable Christian book

> . . . there are many matters, knowledge of which brings little or no advantage to the soul. Indeed, a man is unwise if he occupies himself with any things save those that further his salvation. . . . Of what value are lengthy controversies on deep and obscure matters, when it is not by our knowledge of such things that we shall at length be judged? . . . [W]hat concern to us are such things as *genera* and *species*? ([K03],29-30).

The Best Motivation

Someone has a detached attitude toward things because they believe the world is soon to end; another, because they seek detachment from things with relative existence, and attachment to the Real. Someone avoids society out of hatred for other people; another to pursue the negative way in solitude. Someone practices celibacy because they believe the other sex, or sexuality itself, is inherently evil, a doorway to hell; another, because they find celibacy a help to union with the One. Someone practices asceticism because they believe the world of matter, of body and things, evil; another, because they seek release from the drama. In each case, the action is similar, but the attitude is quite different.

Excessive and erroneous beliefs can motivate actions and efforts helpful to union with the One. But anyone seeking the truth should, it seems, avoid them as a matter of principle. Moreover, even though they offer short-term gain, they may also entail long term loss. When the world fails to end as predicted, for example, the believer may lose faith entirely, in all things religious or spiritual.

Yet, might not excessive or erroneous beliefs be better than nothing? Is it better to tread the spiritual path motivated by delusion than to live content, exclusively in the world of relative existence, lacking any goal beyond the physical, emotional and intellectual? I've met many people who apparently think it is. For they practice a religion half-heartedly. "You need something," they say. Or "Kids need something." It seems that a spiritual path followed under delusion or with skepticism is widely preferred to the natural world seen clearly.

Yet there is another alternative. Someone may practice helps to gnosis because one finds them appealing, because they are naturally drawn to them. Moreover, this is clearly the very best motivation.

That mystics have embraced practices such as silence, solitude, fasting, or chastity may seem odd, unnatural, even perverse. That they have *eagerly* embraced such practices - passionately desiring to achieve passionlessness, becoming deeply attached to unattachment - may seem entirely unbelievable. (And paradoxical, as well. The paradox is acknowledged by the mystics themselves who explain with analogies such as "a thorn is used to remove another thorn, then both are thrown away" or "a boat is used to cross the river, then the boat is abandoned." Said Ramakrishna, "If you must desire, desire God.")

Mystics often actively seek out apparently "unnatural" and "disagreeable" attitudes and activities. Why?

What's unnatural for one person may be natural for another. Consider the following illustration. "Tom", an 8 year-old boy, has a 12 year-old brother, "Frank." Tom has recently noticed some peculiar changes in Frank. Formerly, Frank, like Tom, saw girls as mostly a nuisance. Tom and Frank once shared a mutual disgust of "icky" movie love scenes. Now Frank's eyes glaze over at such scenes. Frank walks Nancy home from school every day and even carries her books.

Tom doesn't understand what's happening to Frank; he hopes it's not contagious. What's happening, of course, is puberty. It happens to almost everyone. And it's not contagious even if you wish it was.

Suppose Tom wanted to be just like his big brother. Suppose he tried to make his eyes glaze at movie love scenes. Suppose he walked little Sue home every day, and forced himself to talk to her an hour each night on the phone, just like Frank talks to Nancy.

It wouldn't work, of course. Aping the actions of someone undergoing puberty wouldn't change Tom a bit. He just can't feel it yet. As Ramakrishna said:

> One cannot explain the vision of God to others. One cannot explain conjugal happiness to a child five years old. ([P12],628).

Let's suppose, however, that puberty only happened to a few people; suppose it could happen at any age; suppose it was contagious, that it could be caught either from someone who had it, or by merely adopting appropriate attitudes and values; and suppose once you had it you could loose it if it wasn't nurtured. Then puberty would have much in common with spiritual awakening.

Until the first birth into spiritual life, many things the mystics said and did, many of their attitudes toward God, themselves, and the world, are as mysterious to us as Frank's actions are to Tom. During our first awakening, however,

> . . . the eye is opened on Eternity; the self, abruptly made aware of Reality, comes forth from the cave of illusion like a child from the womb and begins to live upon the supersensual plane. Then she feels in her inmost part a new presence, a new consciousness - it were hardly an exaggeration to say a new Person . . . ([U01],123).

The second eye is opened of which Angelus Silesius wrote:

> Man has two eyes.
> One only sees what moves in fleeting time,
> the other
> what is eternal and divine. ([B05],43).

Curiously, it's not uncommon for someone undergoing spiritual awakening to experience very advanced mystical states for a while. It's as if novice piano students often played like a master during their first few weeks of lessons, but later reverted to a beginner's level.

After awakening, the spiritual sight of the newly awakened, would-be mystic is

> . . . weak, demanding nurture, clearly destined to pass
> through many phases of development before its
> maturity is reached . . . ([U01],123).

The mystic stands at the beginning of the path to God.

The "Path"

We've seen making gnosis, direct experience of Reality, a goal
implies certain values. These values, in turn, imply a path. And
someone seriously wanting such experience will order their lives
accordingly. Yet, "path" and similar concepts such as "way," and
"method" shouldn't be taken too literally. We've already seen
practicing religion can't guarantee gnosis. Indeed, nothing we can do
can compel gnosis. For experience of the Unconditioned is itself not
conditioned, not obtainable through so many prayers or so many fasts,
much less through taking a drug. It seems to happen freely, unforced.
The experience is obtained only as a free gift.

Nonetheless, many things promote experience of the Eternal.
These things are like knocking at a locked door. The knocking
doesn't, in and of itself, unlock and open the door. It does, however,
show we want the door opened. Until we knock, there is no reason for
anyone to open the door.

Until someone opens the door, we can only repeat our knocking -
and wait. Often aspirants have had to knock for quite a long time
before the door opened. Some, no doubt, gave up, deciding there was
no one behind the door, that the door would never open. Yet, as we've
seen, experience of the Eternal is possible, and, moreover, is
experience of our very Self. This being the case, why should such
experience be difficult to obtain?

We previously saw Alan Watts' mythological, Vedantist
description of creation. Here's his answer, also expressed in the form
of myth.

> Now when God plays hide and pretends that he is you
> and I, he does it so well that it takes him a long time to
> remember where and how he hid himself. But that's
> the whole fun of it - just what he wanted to do. He
> doesn't want to find himself too quickly, for that would
> spoil the game. That is why it is so difficult for you and

me to find out that we are God in disguise, pretending not to be himself. ([W02],14).

14

- Union -

Chapter Summary: This chapter discusses the goal of spiritual paths: union with God. Various meditative methods that lead to such union are also discussed. Then, union itself and its consequences are explored.

Right views, goals, actions, and efforts eventually lead to what Buddhist's call right mindfulness and right concentration. For us, right mindfulness will mean certain mental states conducive to gnosis. And right concentration will indicate first-hand experience of or union with the Eternal - gnosis.

Right Mindfulness

The process of achieving right mindfulness or right concentration is variously called contemplation or meditation since "meditation" and "contemplation" have opposite meanings in East and West.

In the West, "contemplation" refers to gnosis, to right concentration. For example, in *Western Mysticism* Edward Butler writes:

> Contemplation at its highest limit is identical with the mystical experience, and involves . . . an experimental perception of God's Being and Presence. ([B17],213).

"Meditation" refers to some sort of mental activity, for example reflecting on a biblical incident, a theological concept, or a doctrinal statement. (See [P09],421 for an example of this use of "meditation.") It's meant to lead to contemplation.

In Eastern religious literature, the terms are reversed. "Meditation" refers to experiential perception of God, while "contemplation" means mental reflection. Basil Pennington, a Roman Catholic monk, describes ([P11],29) this confusing situation and decided to use "centering prayer" or simply "centering" for what's called "meditation" in the West and "contemplation" in the East.

The term "centering" is quite compatible with our world view since experience of the Ultimate Ground of Existence is also experience of the Center. The term "centering prayer," however, is less general since prayer is religious. A secular person could think of drawing closer to the Center as a metaphysical, philosophical, but entirely non-religious process. For such a person, there's nothing inherently religious about consciousness becoming aware of itself (not "Itself" since they don't identify consciousness and deity). Similarly, a scientist might maintain becoming more directly aware of energy (again, not "Energy") is entirely secular. For such individuals, "centering" is the appropriate term. Yet, centering can be religious, and religions have often acknowledged its value.

So, for some people, "centering prayer" is the more appropriate term. And for others, "centering meditation" is better. And for still others "centering" is preferable. I'll generally use "centering" to indicate all three. So "centering" indicates right mindfulness, a mostly mental process which seeks to promote experiential perception of what can be viewed religiously as God, or non-religiously as our own true Self.

Centering

Centering seeks to still the thoughts, emotions, and senses that so often occupy our Consciousness so that Awareness may become aware of Itself. The Christian saint Albert the Great described centering when he wrote:

> When St. John says that God is a Spirit, and that He must be worshiped in spirit, he means that the mind must be cleared of all images. When thou prayest, shut thy door - that is, the doors of thy senses. Keep them barred and bolted against all phantasms and images. Nothing pleases God more than a mind free from all occupations and distractions. Such a mind is in a manner transformed into God, for it can think of nothing and love nothing, except God; other creatures and itself it only sees in God. He who penetrates into himself, and so transcends himself, ascents truly to God. . . . Leave thy body and fix thy gaze on the uncreated Light. *Let nothing come between thee and God.* ([JO3],219).

How to still the body? In normal circumstances it's helpful if the body is fit. Aches and pains demand attention and capture awareness. So one aim of Hatha yoga is a fitness which prepares the body for centering. Of course, other types of exercise would also serve this purpose. If the body is fit, then sitting still and quiet in a comfortable position should allow awareness of the body to lessen.

Yet aches and pains can be a powerful motivation for lessening body awareness. Someone with an ill body might find it harder to lose awareness of body, but be much more motivated to do so.

How to still the emotions? Roman Catholicism defines ([N09],42) the "seven deadly sins" or "capital sins" as pride, covetousness, lust, anger, gluttony, envy, and sloth. Gluttony and sloth keep us aware of the body; the others, the emotions. So one task is to reduce the pride, covetousness, lust, anger, and envy which may surface in centering. Resolving such emotions is not only good psychologically but aids centering. How can we reduce them? That's one of the aims and results of a moral life. That's why moral virtues are one of the first steps of the path.

Another task is reducing positive emotions which may surface in centering. Love of spouse or child, for example, should be put temporarily aside for a higher love, love of God. Later we'll see how Ramakrishna found even the love of a God who was a Person a barrier to the highest level of centering.

If the body and emotions are quiet, then all remaining is to quiet the thoughts. The work of preparing the body and emotions are done in daily life. Although the work of controlling the mind can occur in daily life, too, it's usually done during centering. Therefore, the main goal of centering is often viewed as quieting the mind so that various levels of gnosis become accessible.

Many books describe centering exercises. For example, *The Relaxation Response* ([B02]), written by an M.D., explains the health benefits of centering and describes, among other techniques, the technique ([B02],162-3) of "breathing mindfulness."

Breathing mindfulness is a common centering exercise which may be used for religious or non-religious purposes. It seems of Buddhist origin, and Buddha recommended it highly. For example, in the *Anapanasatissuta* (the "discourse on mindfulness when breathing in and out"), he says

> [m]indfulness of in-breathing and out-breathing,
> monks, if developed and made much of, is of great
> fruit, of great advantage. ([B10],124).

He described it as follows. A person

> . . . sits down cross-legged, holding his back erect,
> arousing mindfulness in front of him. Mindfully he
> breathes in, mindfully he breathes out. ([B10],124).

Forms of breathing mindfulness appear in many religions. The Jewish Shneur Zalman, for example, believed

> . . . God's invisibility to be an illusion wrought by man's
> ignorance. ([E04],132)

and therefore thought the

> . . . best means for breaking the spell was consciously
> to motivate his soul by means of the breath . . . Only
> sustained contemplation could awaken the divine
> intelligence that resides in every soul. ([E04],132-3).

One type of breathing mindfulness is as follows. Sit in a quiet place and turn your attention to your breathing. Concentrate exclusively on your breaths. Count them, either on the inhalation or the exhalation. Or count both, the inhalation as one, the exhalation as two, etc. If you find yourself thinking of anything but your breath, go back to one. If you reach ten, also go back to one. Continue.

Surprisingly, breathing mindfulness, as well as other forms of centering, can give a mild form of ecstasy. The mind can find rapt attention on one fixed focus peaceful and soothing. Of course, this procedure can also be extremely boring, especially to a mind accustomed to frequent stimulation, particularly of the kind offered by television. Someone with a short attention span might find breathing mindfulness very difficult or even impossible. If they persevere, however, concentration and attention span increase.

Mantra

A Quaker publication first describes the centering process:

> . . . [S]ilence is not an end in itself but a way toward
> worship. . . . In the normal course of everyday life the
> mind is filled with flotsam from external stimuli. This
> must be calmly put aside . . . [W]orries . . . should not
> stand in the way of submerging your individual self in
> the one eternal Self. ([P08],21),

and then describes the usefulness of mantra:

> The way to this union is through prayer. Some Friends
> use a short and oft-repeated prayer or a mantra as an
> aid to concentration. ([P08],21).

Concentrating on the breath can be difficult, especially at first.
The breath is subtle and content free; the mind may demand
something it can more easily grasp, something more definite.
Therefore, a meaningful word or phrase is sometimes used as the
focus instead of, or along with, the breath. If the word or phrase is
"ah," "peace," or even "wealth, power, and fame" then the centering is
non-religious.

Often, however, the phrase is specifically religious. In the Hindu
religious tradition it's called a "mantra", and beads similar to the
Catholic rosary are used to count repetitions. The Eastern Orthodox
Hesychastic tradition uses the Jesus Prayer - "Lord Jesus Christ, Son
of God, have mercy upon me, a sinner" - often synchronized with the
breath or heartbeat. (A good introduction to this method is given in
[W04].) Hesychastic monks incessantly repeat this phrase, seeking to
fulfill the recommendation of Jesus "that men ought to always pray"
(Lk 18:1,[H08]) and Paul to "Pray without ceasing." (1Thes
5:17,[H08]). In fact, in Christianity

> [t]he ideal of the first monks was to carry on incessant
> prayer. ([P09],386).

Pennington notices ([P11],33) similarities in the dhikr centering
method of Islam's Sufis, the Buddhist nembutsu method, and the Jesus
Prayer. He also says centering was the principal kind of prayer in
Western Christianity ([P11],29) for a thousand years, until about the
14th century. Then, vocal liturgical prayers became dominant.
Liturgical prayers are closer to what most people think of as prayer. In
Teresa's time, academic theologians ([B04],142) favored vocal prayer
and opposed her insistence ([B04],144) on "mental prayer," i.e.,
centering. Such theologians, however, were themselves opposed by
([B04],143) "people who disdained technical theology as dry and
detached from the spirit of religion" and by ([B04],142) "people who
exalted the knowledge gained through direct religious experience and
prayer."

Merging

In centering, one aims for the direct experience of one's own
Awareness. As such, centering's focus is inward. Awareness is

intentionally directed away from the external universe. Indeed, centering, by its very name, implies coming closer to something which is immanent.

In contrast, one might aim for direct experience of the Ground of the external universe. Here, the focus is outward. Awareness is directed to the external world. Of course, we already experience the external universe, but we don't usually experience its transcendental Basis. Such experience is the object of merging.

How is merging practiced? Someone begins by seeking the transcendental Basis of some particular external object. Underhill describes how this might be done. First, the object is chosen. It

> . . . may be almost anything we please: a picture, a statue, at tree, a distant hillside, a growing plant . . . ([U01],301).

Next, we are asked to

> [l]ook . . . at this thing which you have chosen. Wilfully yet tranquilly refuse the messages which countless other aspects of the world are sending; and so concentrate your whole attention on this one act of loving sight that all other objects are excluded from the conscious field. Do not think, but as it were pour out your personality towards it: let your soul be in your eyes. Almost at once, this new method of perception will reveal unsuspected qualities in the external world. First, you will perceive about you a strange and deepening quietness . . . Next, you will become aware of a heightened significance, an intensified existence in the thing at which you look. . . . It seems as though the barrier between its life and your own, between subject and object, had melted away. You are merged with it, in an act of true communion: and you *know* the secret of its being deeply and unforgettably, yet in a way which you can never hope to express. ([U01],301-2).

Teresa may have practiced such merging, for she writes she

> . . . found gazing at fields, water, or flowers a great help, for they spoke to me of the Creator, and served as a book in bringing me to a state of Recollection. . . . I used at times to feel . . . the presence of God . . . ([C10],42-3).

Merging may lead to illumination, one of the first and most common forms of gnosis. At this stage, the world is lit up; the mystic perceives a glow and knows it is of God. For such a mystic,

> [n]othing appears to him any longer as purely profane.
> . . Creatures . . . become sacraments which proclaim
> the presence of God everywhere . . . ([P09],425).

In fact, such experience may be an unrecognized, first-hand experience of the Uncreated Light. Illumination is like seeing a glow but not recognizing the fire from which it comes. However, if recognition is present, if not only the glow, but the fire itself is seen, then we have first-hand experience.

Benefits of Centering and Merging

In addition to helping achieving right concentration, centering and merging have other benefits. Let's begin with an analogy.

Many tasks require a steady hand; if a person's hand is unsteady and always shakes, that person can't do calligraphy, sculpture, and surgery is certainly out of the question. A steady hand, a hand which can be held fixed on one point at will, is prerequisite to these activities and others.

Many people have a steady hand, but an unsteady mind; they can't concentrate steadily at will. True, their mind is sometimes concentrated when something grabs their interest, a thrilling novel or show, perhaps. But they can't steady and concentrate their mind where and when they please.

But just as a person might develop a steady hand by practicing holding it fixed, a person may developed a more steady, concentrated, "one-pointed" mind by practicing holding it fixed on some object of centering or merging.

Not only is a quiet and controlled mind a help for various activities, it can also reduce suffering.

Imagine a young child, a boy of six, whose parents take him to a physician to be vaccinated. They don't mention the vaccination needle. After an examination, the physician quickly rolls up the little boy's sleeve and gives the shot. The boy yells and starts to cry, but soon, with his parent's encouragement and perhaps some candy, he feels better. He's suffered a brief sharp pain and a few minutes soreness.

Now imagine instead the boy is brought to the physician's office, and shown the needle he'll receive the next day. Perhaps he sees another child vaccinated. At home, he can't eat and has trouble going to sleep. The following morning he feels terrible. He cries and begs his parents not to take him to the physician. The ride to the office isn't pleasant for him or his parents. When it's over his parents, as before, give him words of encouragement and candy.

The little boy has suffered much more than in the first scenario. For twenty-four hours, his thoughts and emotions have given him much torment and little peace.

Imagine now an ability rare for adults and even rarer in children - the ability to control the mind and heart. If the little boy had this ability, he would have been able to put the thought of the needle out of his mind for twenty-four hours. He would still have had to experience the pain of the first scenario, but would have saved himself from the added torture, the worry and fear, of the second. The ability to control his thoughts and emotions would have saved him from pain. Even an imperfect, limited ability to partially control his thoughts and emotions would have saved him from some pain.

So, control of thoughts and emotions can reduce pain, while uncontrolled thoughts and emotions (worry, fear, etc.) may themselves cause us much pain. In fact, much if not most of the pain people in advanced countries suffer - people who have enough to eat, sufficient clothing and housing - is this type of pain. Controlled thoughts and emotions could greatly reduce it.

It's easy to feel such control is not within the reach of the normal person. Yet the normal person clearly has the potential. It's not uncommon to cut yourself while playing a sport, basketball or soccer for example, but not realize it until you notice the blood. Your intense concentration on the game made you insensitive to the moment's pain. If we could produce such intense concentration at will, might we not be able to lessen pain?

It's interesting, by the way, that religious beliefs such as "God always provides" have effects similar to those obtainable with control of thoughts and feelings. The beliefs give some measure of control over thought and emotion, since genuine belief in Divine Providence banishes worrisome regard for the future. Also, religious forgiveness and atonement provide a powerful antidote to regret and worry over the past. Thus, some religious beliefs produce results similar to centering or merging practice: they effectively limiting negative, useless, painful thoughts and emotions. (Of course, religion sometimes promotes negative, useless thoughts and emotions, too.)

Right Concentration: Self-Referential Awareness

Ideally, a quiet body, emotions, and mind lead to gnosis. Unfortunately, they can lead to pitfalls instead. Rolt writes:

> There is a higher merging of the self and a lower merging of it. The one is above the level of personality, the other beneath it; the one is religious the other hedonistic; the one results from spiritual concentrations and the other from spiritual dissipation. ([D08],34).

Underhill describes the lower merging as a ([U01],322) "half-hypnotic state of passivity", a "meaningless state of 'absorption in nothing at all'", and ([U01],324) "vacant placidity". She associates these evils with "Quietism."

Quietism, a unbalanced mystical theory which appeared three to four hundred years ago in Europe, overemphasized (refer [U01],471-2) the passive nature of centering and merging and carried it over into all aspects of life. In Quietism, it seems, a quiet body, heart, and mind lead to a Consciousness aware of nothing at all. Of such quietness Ruysbroeck wrote:

> Such quietude . . . is nought else but idleness, into which a man has fallen, and in which he forgets himself and God and all things in all that has to do with activity. This repose is wholly contrary to the supernatural repose one possesses in God; for that is a loving self-mergence and simple gazing at the Incomprehensible Brightness . . . ([U01],322).

Quietism was suppressed, perhaps overly so since the effort, writes Huxley, resulted not only in its elimination but also ([H11],66) "to all intents and purposes the extinction of mysticism for the better part of two centuries" in Europe.

So how can a quiet body, emotions, and mind lead to gnosis? We've seen the Real is immanent in the world, yet transcends it. Therefore, some mystics experience the God Which is not a Person immanently, as their very own, true and deepest Self, while other mystics experience It transcendently, as external to, and different from, their own self or selves. Gnosis may be experience of That which is immanent or That which is transcendental. Since we've identified Consciousness with the Eternal, we'll describe right concentration, gnosis, mostly in terms of immanence. We'll present it primarily as an immanent, inward experience where Consciousness becomes aware of Itself.

Usually, Consciousness is aware of something external to Itself, of body, emotion, or thought, all entities with only relative existence. Imagine someone with a large measure of detachment from thoughts, emotions, and physical sensations who is sitting still in a quiet place. Their senses are unstimulated, so their bodily awareness is minimal. Their emotions, too, are still; they're at peace, neither loving or hating. Their mind is calm and quiet, with little or no thought. Moreover, though the practice of humility and selfless action they've turned away from their relative selves. What remains for their Consciousness to be aware of?

Itself.

For a Consciousness aware of only Itself there's no duality. It has temporarily become unconscious of the body, heart, and mind. It knows only Itself. The triad of knower, knowing, known disappear since Consciousness is all three. A Consciousness aware of only Itself has escaped from duality and ceased to be aware of entities with only relative existence. Such an Awareness is in a self-referential state. Self-referential states and processes naturally lead to infinity, physically, mathematically, and logically.

An example of a physically self-referential state or process occurs when an audio speaker's output is (usually inadvertently) fed back into a microphone. A feedback loop is established where output becomes input becomes output, etc. - each time passing through the amplifier and getting louder. If the amplifier's electronics allowed, this self-referential physical process would eventually produce infinite volume. Instead, the amplifier quickly reaches the limits of its electronics. The resulting, high-pitched squeal is familiar to anyone who has ever set up a music or public address sound system.

Self-referential processes also lead to infinity logically. As Rudy Rucker writes in *Infinity and the Mind*:

> The philosopher Josiah Royce maintained that a
> person's mental image of his own mind must be
> infinite. His reason is that one's image of one's own
> mind is itself an item present in the mind. So the
> image includes an image that includes an image, and
> so on . . . The old can of Pet Milk, for instance, bore a
> picture of a can of Pet Milk, which bore a picture of a
> can of Pet Milk, etc. ([R06],38).

As another example: We said long ago that when a knower transcends the triad of knower, knowing, and known, when they merge with the known, then only One remains. But does only the One remain? Rather, aren't there two elements, knowing and the now united knower/Known? A common answer to this objection is that since God is One, God's self-knowledge is not different from God. Thus, there are not two elements, but One. God and God's self-knowledge are identical.

If the answer is accepted, then it follows that God is, in some way, a self-referential process. Why? Because if God's self-awareness is itself the same as God, then we have awareness aware of itself, that is, self-referential awareness.

Rene Descartes

Rene Descartes, the 17th century French philosopher and mathematician who gave the world the Cartesian coordinate system of high school Algebra, may have experienced self-referential consciousness. Descartes believed the soul or mind resided in or near the pineal gland. Let's see why this suggests self-referential consciousness.

First of all, which is it, soul or mind? We consider these as two separate functions: the mind, the intellectual function, and the soul, the consciousness function. Descartes may not have made that distinction. The caption to a figure ([T02],137) taken from the 1677 edition of his *De Homine*, for example, shows the *soul* near the pineal gland. However, another reference ([N05],v5,601) has Descartes dividing the world into matter and *mind*. So perhaps Descartes used the same word for what we consider two separate functions. If so, either soul or mind might be a valid translation.

Second, where is the pineal gland? An atlas of anatomy ([C07],25,95) shows it about level with the ear tops, slightly rear of center. So, Descartes believed consciousness or mind resided somewhere in that area. Of course, he may not have believed it actually resided *in* the pineal gland.

Finally, why? What could have suggested to Descartes that soul or mind resides in that part of the body? Perhaps, it was a "centering" experience like the following.

Sometime when you're alert, not fatigued or sleepy, sit up straight in a quiet place and close your eyes. Become aware of the space around you. Attach these labels: front, behind, above, below, left, and right. For example, picture the floor or ground and think "below"; picture the ceiling, or the sky, and think "above"; picture the tree to your left and think "left." Now begin to zero in on the center point by thinking of closer things: if you are conscious of your left ear, think "left"; if you feel the air in your nostrils, think "front." If you feel your jaw, think "below." Try to get as close as possible to the center point which divides front from back, left from right, and above from below. This point seems to be the base of consciousness, the place from which you now look out at your body and the world beyond. It also happens to be close to the pineal gland.

The Location of Awareness

If consciousness actually resided in or near the pineal gland, then our exercise would be slowly bringing Consciousness closer to awareness of Itself, the Center. The situation, however, is a bit more complicated since we know today that consciousness may reside anywhere in the body. As Wilder Penfield, a neurophysiologic researcher, writes:

> To suppose that consciousness or the mind has
> localization is a failure to understand neurophysiology.
> ([P10],109).

Fred Wolf, a physicist, deduces from this quote that

> . . . mind appears to be everywhere. It is observing on
> the scale of atoms and molecules, neurons, cells,
> tissues, muscles, bones, organs - in other words, it is
> observing on all scales of physical existence.
> ([W10],244).

Wolf's statement easily follows if Consciousness and Center, the Ultimate Ground of Existence, are one and the same.

Even if Awareness doesn't actually reside near the pineal gland, something crucial does seem to be going on in that part of the body. Penfield continues:

> The great mathematician and philosopher, Rene Descartes (1596-1650), made a mistake when he placed it in the pineal gland. The amusing aspect is that he came so close to that part of the brain in which the essential circuits of the highest brain-mechanism must be active to make consciousness possible. ([P10],109).

It seems consciousness somehow connects to body there.

> . . . [I]t became quite clear in neurosurgical experience, that even large removals of the cerebral cortex could be carried out without abolishing consciousness. On the other hand, injury or interference with function in the higher brain-stem, even in small areas, would abolish consciousness completely. ([P10],18).

So, even if our exercise doesn't bring us closer to the actual seat of consciousness, and even if there is no fixed seat of consciousness in the body, our exercise does bring us closer to a point to which consciousness enjoys some sort of special relation.

Right Concentration

The affirmative way, the negative way, the moral virtues, the other steps along the path, and right mindfulness all aim towards right concentration, towards gnosis. If we think of the Eternal immanently, then they all aim at freeing an Awareness from involvement with the relative, the transitory, and aim instead to have It become aware of Itself. For when a mystic's

> . . . consciousness of I-hood and consciousness of the world disappear, the mystic is conscious of being in immediate relation with God Himself; of participating in Divinity. (Delacroix, *Etudes sur le Mysticisme*, p.370 in [U01],330).

There are degrees of right concentration. At first, the mystic may draw closer to God in illuminative experiences. These experiences change the mystic even as wood drawing closer to a fire begins to take on some of fire's qualities: it glows, or even burns with the fire's flame. If the wood draws close enough, it eventually enters the fire.

> When the soul is plunged in the fire of divine love, like
> iron, it first loses its blackness, and then growing to
> white heat, it becomes like unto the fire itself. And
> lastly, it grows liquid, and losing its nature is
> transmuted into an utterly different quality of being.
> ([U01],421).

The wood, glowing and burning even as the fire glows and burns, may begin to see its real self as indistinguishable from the fire's. Its flame merges with the fire's flame. Similarly, as the mystic lives more and more in awareness of the Eternal, a like merging of consciousness may occur. In the words of Symeon:

> 'It is a truly divine fire, uncreated and invisible, eternal
> and immaterial, perfectly steadfast and infinite,
> inextinguishable and immortal, incomprehensible,
> beyond all created being.' This light 'has separated me
> from all being visible and invisible, granting me a
> vision of the uncreated One. . . . I am united with the
> One who is uncreated, incorruptible, infinitely invisible
> to all.' ([L09],118).

Eventually, the wood is transformed into fire. Similarly, if mystical experiences reoccur, the mystic may have

> . . . more and more the impression of being that which
> he knows, and of knowing that which he is. (Delacroix,
> *Etudes sur le Mysticisme*, p.370 in [U01],330).

The mystic may begin to see their real self as no different from the Self, the Eternal. Such a mystic may say:

> I am, verily, that supreme *Brahman* which is eternal,
> pure, free and one, impartite bliss, non-dual, and
> existence, consciousness and infinite. ([S09],54)

Or, as the writer of the taittiriya upanishad, expressed it:

> I am established in the purity of Brahman. I have
> attained the freedom of the Self. I am Brahman, self-
> luminous . . . I am immortal, imperishable. ([U04],54).

Complete union, however, isn't fully achieved as long as the mystic experiences any sort of duality. Feeling identical to the One is not the same as actual perfect conscious union with the One, since the duality of knower, knowing, known still exists. Thinking or even knowing the experience is of the Self is not sufficient for union. For as along as the mystic doesn't *experience* the Eternal as their own Self, their gnosis remains first-hand, not unitive.

> The lower states of samadhi . . . lack . . .
> completeness of union. Ramakrishna *knew* that
> Mother Kali was not other than Brahman; yet, because
> of his great love for her, he was at first unable to
> accept this fact completely. . . . Ramakrishna's love for
> Kali was the last-remaining trace of dualism in his
> mind. When he could go beyond that, he could attain
> union with Brahman. ([I04],118-9).

For a while, Ramakrishna's love of Kali as something other than himself, as "the effulgence of pure consciousness", kept him in duality. He couldn't free himself from seeing the Eternal as something other than himself. He worshiped the One as Kali, the Mother of the universe, the Uncreated Light, and couldn't rise above this duality.

> The mind withdrew itself easily from all other things
> but, as soon as it did so, the intimately familiar form of
> the universal Mother, consisting of the effulgence of
> pure consciousness, appeared before it as living and
> moving . . . ([S01],225).

Knower, knowing, and known still remained.

Union

Ramakrishna was eventually able to rise above all duality.

> With a firm determination I sat for meditation again
> and, as soon as the holy form of the divine Mother
> appeared now before the mind as previously, I looked
> upon knowledge as a sword and cut it mentally in two
> with that sword of knowledge. There remained then no
> function in the mind, which transcended quickly the
> realm of name and forms . . . ([S01],225).

Ceasing to experience the Uncreated Light as something other than himself, he now saw It as his true Self. Losing all distinction of knower, knowing, known he united with It. A monk of the Ramakrishna order describes what happened next:

> In that rapturous ecstasy the senses and mind
> stopped their functions. The body became motionless
> as a corpse. The universe rolled away from his vision -
> even space itself melted away. Everything was
> reduced to ideas, which floated like shadows in the
> dim background of the mind. Only the faint
> consciousness of "I" repeated itself in dull monotony.

> Presently that too stopped, and what remained was
> Existence alone. The soul lost itself in the Self, and all
> idea of duality, of subject and object, was effaced.
> ([L07],161).

Notice that with this kind of experience, unconsciousness of the external world is due to the nature of the experience itself, rather than any physical weakness. For self-referential awareness is, by definition, Consciousness aware of Itself, rather than the exterior world. It's also been called pure consciousness since for an Awareness aware only of Itself there's no mixture or combination of It and something else.

Another description of the unitive experience Ramakrishna enjoyed is as follows:

> Beyond the realm of thought, transcending the domain
> of duality, leaving Maya with all her changes and
> modifications far behind, towering above the delusions
> of creation, preservation, and destruction, and
> sweeping away with an avalanche of ineffable Bliss all
> relative ideas of pain and pleasure, weal and woe,
> good and evil, shines the glory of the Eternal
> Brahman, the Existence-Knowledge-Bliss Absolute, in
> the Nirvikalpa Samadhi. Knowledge, knower, and
> known dissolve in the menstruum of One Eternal
> Consciousness; birth, growth, and death vanish in that
> infinite Existence; and love, lover, and beloved merge
> in that unbounded ocean of Supreme Felicity. . . .
> Space disappears into nothingness, time is swallowed
> up in Eternity, causation becomes a dream of the past,
> and a tremendous effulgence annihilates the
> oppressive darkness of sense and thought. . . . [O]nly
> Existence is. . . . His illumination is steady, his bliss
> constant, and the oblivion of the phenomenal universe
> is complete. ([L07],153).

It's claimed Ramakrishna's body, became ([L07],161) "motionless as a corpse." It's also claimed his breathing and heart stopped!

> . . . [T]here was not the slightest function of the vital
> force in his body . . . [H]is face was calm and sedate
> and full of effulgence. . . . completely dead to the
> external world . . . [H]is mind, merged in Brahman,
> was calm and motionless like an unflickering lamp in a
> windless place. ([S01],256)

and
> . . . when he ascended the highest plane of non-dual
> consciousness his heart-beats and functions of the
> senses were stopped . . . and the body lay like a
> corpse; all the modifications of his mind, such as
> thought, imagination etc., came to a stand-still and he
> dwelt in absolute Existence-Knowledge-Bliss.
> ([S01],589).

Are such claims accurate? We'll return to this question later.

The Result of Union: Annihilation or Return

Many mystics ([S01],151) never return from the unitive state; they merge with the Eternal and leave the external world behind, forever. The wood is consumed; the finite, separate identity vanishes without a trace. As a drop of water enters the ocean and loses its individual identity, Consciousness merges back with the One.

Deprived of soul, the body ([L07],154) eventually dies. Although Ramakrishna believed suicide ([G03],164) a "heinous sin," he didn't consider a gnosis experience which resulted in death of the physical body as suicide. Indeed, if gnosis is the aim of human life, then the physical body has fulfilled its purpose once permanent union is established. Once someone reaches the other shore, they may discard the boat.

It's said, however, that some people do return; in fact, this was said of Ramakrishna. Such a person, however, is forever transformed. Like wood almost totally transformed by fire, the ego remains as a kind of ash, an insubstantial residue. For such a mystic,

> . . . that 'I'-ness of theirs lives in constant unbroken
> consciousness of an intimate relation with God, such
> as 'I am a servant, a child, or a part of Him'! . . .
> Knowing that God is the quintessence of everything,
> that "I" does not any more hanker after the enjoyment
> of worldly objects such as sight, taste, etc. . . . Those
> who were till then in worldly bondage, but have now
> attained perfection . . . and are living the rest of their
> lives in some loving relation with the divine Lord are
> known to be 'liberated-in-life.' ([S01],356).

Now, living beyond the show world and seeing all things as one, that person sees

. . . all things equal . . .
Absorbed in Brahman
He overcomes the world
Even here, alive in the world. ([S18],60).

For them, illumination is permanent; they *see* that the Eternal One has became all. For them, the world's people and objects, their own selves and Self, and God, are one and the same; for them, there is only One.

We know a man, a cow and a mountain merely as such. He saw that the man, the cow and the mountain were indeed a man, a cow and a mountain; but at the same time, he saw that the indivisible Existence-Knowledge-Bliss, the cause of the universe, was gleaming through them. ([S01],589).

For such people, all is pure, all is God.

15

- Towards An Exact Science -

Chapter Summary: This chapter discusses if a scientific religion could ever be an exact science. Various ways of judging a person's mystical state are explored.

Teachers and Guides

Those who have united with the Eternal Substance have transcended the play of Maya. Since they're indifferent to pleasure and pain and aloof from the concerns of practical people, such people may seem foolish. Yet since they're obviously God-centered they're called "God's fools" or even "God-men" or "God-women." Eckhart writes:

> That person who has renounced all visible creatures and in whom God performs His will completely - that person is both God and man. His body is so completely penetrated with Divine light . . . that he can properly be called a Divine man. For this reason, my children, be kind to *these men*, for they are strangers and aliens in the world. ([J03],223).

Such people are consciously united with the Ultimate. Could they help others to union? Eckhart continues:

> Those who wish to come to God have only to model their lives after these men. . . . Those who are on the way to the same God and have not yet arrived will do well to become acquainted with these people who have attained. ([J03],223).

Of course, every seeker doesn't need so advanced a guide. Those just beginning can obtain help and advice from more advanced seekers, or from books. But with all the traps and pitfalls along the path, it's not surprising an evolved teacher, a spiritual director and advisor, is often recommended.

> If a man is unlikely to take an unexplored path without a true guide; if no one will risk going to sea without a skilful navigator; if no man will undertake to learn a

369

> science or an art without an experienced teacher, who
> will dare attempt . . . to enter the mysterious path
> leading to God, and venture to sail the boundless
> mental sea . . . without a guide, a navigator and a true
> and experienced teacher? ([S12],159).

The Hesychast tradition advises a seeker to find a teacher and guide
who is

> . . . a man bearing the Spirit within him, leading a life
> corresponding to his words, lofty in vision of mind,
> humble in thought of himself, of good disposition . . .
> ([W11],174).

Having found such a guide, the seeker is advised to offer ([W11],174)
"total and unquestioning obedience."

The dangers of such obedience are probably obvious. What if the
teacher is misled or corrupt? Wouldn't such obedience and adulation
be liable to corrupt even a worthy teacher? Would disciples find
themselves working to advance the fortunes and fame of the teacher
rather than their own journey toward gnosis?

Someone whose object is gnosis may avoid dangers and progress
faster if they obtain the guidance of a teacher who is genuine, sane,
and more advanced - but how can genuineness, sanity, and progress
be accurately judged? How can a teacher and guide be identified?
What qualities should a teacher have? How can charlatans and fakes
be avoided? This chapter discusses these questions.

Physical Effects of Gnosis

Just as closeness to a fire may cause a physical object to change, to
melt, for example, first-hand experience of God can have powerful
effects upon the body.

> The spiritual moods arising out of intense love of God .
> . . cause extraordinary physical changes. ([S01],366).

The phenomena of heat, for example, is mentioned in the
Christian medieval *The Scale Of Perfection.*

> Not all those who speak of the fire of love understand
> properly what it is. . . . [T]he presence of this fire in the
> soul may produce bodily heat . . . ([H06],38).

A footnote reads "That a feeling of bodily heat accompanies certain
mystical experiences is well known." For example, Schimmel
([S04],170) recounts Islamic and Hindu analogues. It seems it's

possible to control this heat which Tibetan Buddhist call *tumo*. Tibetan monks sometimes compete to see whose *tumo* can dry the most freezing wet blankets.

Sometimes drops of blood, like perspiration, come from the mystic as they did ([S01],151) from Ramakrishna, and from Jesus before the passion. Also, Ramakrishna's

> . . . chest appeared constantly reddish and the eyes became sometimes suddenly full of tears. ([S01],142),

a phenomena, perhaps, akin to the "gift of tears," a spiritual gift considered important by Symeon (213],30-1) and others Eastern Christian saints.

Yet another physical sign is the stigmata, those wounds of Christ evidenced by Saint Francis and, more recently, Therese Neumann and Padre Pio. Therese had the wounds for 11 years, yet they never ([S22],33,233) became inflamed or infected.

Therese Neumann was born in 1898 in Germany. Like Terese of Avila, she is also said ([S22],59) to have levitated, and to have been simultaneously present to different people in different, widely separated, places. The most substantiated claim, however, is that for decades she lived with no drink or food except a daily Communion wafer, a very thin piece of bread, 2 to 3 cm in diameter. Theresa first began eating and drinking less and less. Then

> [f]rom Christmas of 1926 she finally refused to take any further nourishment at all. She took only Holy Communion, every day . . . ([S22],27)

and a little water. She continued until September 1927 when she gave up water, too.

> From this time until the end of her life, a period of 35 years, Therese Neumann lived without taking any food and any drink: daily Communion was her only nourishment. ([S22],27).

In 1927, Therese underwent medical observation. Experts decided no one could live for more than 11 days without food and water. She was closely observed ([S22],29,228-230) for 15 days by four nurses who took turns watching her, two at a time. The doctor and nurses reported:

> In the period from July 14 to 28, 1927, Therese Neumann took no natural nourishment at all, either liquid or solid. . . . Her weight at the end of the

> examination . . . was the same as at the beginning . . .
> Neither during the period of examination, nor at its
> end, did any special states of exhaustion make their
> appearance. ([S22],228-9).

Therese's case at the time was well known, so much so that when Germany entered the Second World War and started food rationing, the government decided not to issue ([S22],30) her any ration stamps.

On what then did she survive? One author writes ([S22],29) she replied "On our Savior," referring to her daily Communion wafer which she believed the actually body of Christ. Another author, Paramahansa Yogananda, writes ([Y02],422) she told him she lived by "God's light."

Paramahansa Yogananda was himself a mystic. He was born in India but lived much of the later part of his life in California. After his death in 1952, a Los Angeles funeral director claimed his body remained unspoiled even after 21 days. No drying or decay of tissues was noticed; no mold grew on the skin; no odor was detected. He wrote ([Y02],575) "the case of Paramahansa Yogananda is unique in our experience."

But it's hardly unique in the literature of mysticism. Physical effects are common. In fact, Underhill writes that great Western contemplatives,

> . . . though almost always persons of robust
> intelligence and marked practical or intellectual ability .
> . . have often suffered from bad physical health. More,
> their mystical activities have generally reacted upon
> their bodies in a definite and special way; producing in
> several cases a particular kind of illness and of
> physical disability . . . ([U01],59).

She attributes such signs to

> . . . the immense strain which exalted spirit puts upon
> a body which is adapted to a very different form of life.
> ([U01],59).

Indeed, some indication of mystical experience's strain on the body is that gnosis often renders the mystic temporarily unconscious of the exterior world. Symeon, for example, describes a mystic to whom

> . . . a flood of divine radiance appeared from above
> and filled all the room. As this happened the young
> man lost all awareness . . . and . . . saw nothing but
> light all around him . . . [H]e was wholly in the

presence of immaterial light and seemed to himself to
have turned into light. Oblivious of all the world he was
filled with tears and with ineffable joy and gladness.
([S26],245-6).

If the experience is strong enough, the mystic may lose contact
with their body as well as the exterior world.

. . . [H]is life force is withdrawn from the body, which
appears "dead," or motionless and rigid. ([Y02],278).

Such "unconsciousness" occurred, for example, in the case of
Ramakrishna, and of Teresa whose

. . . ecstatic seizures . . . left the nun seemingly dead
to the world . . . ([C10],61).

Teresa's

. . . pulse would almost stop beating, her eyes remain
closed or open yet unseeing. . . . Whilst the body
remains in this trancelike state, the limbs rigid and
impervious to sensation and the pulse scarcely
perceptible, the soul seems 'raised above itself and all
earthly things'. ([C10],65).

Sometimes, the mystic's body eventually adapts to mystical
experience. Then the mystic

. . . communes with God without bodily fixation; and in
his ordinary waking consciousness, even in the midst
of exacting worldly duties. ([Y02],278).

During Teresa's last seven years, for instance, ([C10],66) trances were
rare.

Sometimes, however, the body doesn't adapt; some mystics can't
bear the strain.

When the powerful flood of divine moods comes on
human life unexpectedly, it cannot be suppressed or
concealed by thousands of efforts. . . . [T]he gross,
inert body very often fails to contain that powerful
onrush of divine emotion into the mind and is
completely shattered. Many sadhakas met with death
that way. A fit body is necessary to contain the
abounding surge of emotions born of perfect
knowledge or perfect devotion. ([S01],151).

(A "sadhaka" is a spiritual seeker.)

Since it can be so stressful to the body, it's perhaps fortunate that
mystical experience is often

> . . . a brief act. The greatest of the contemplatives
> have been unable to sustain the brilliance of this awful
> vision for more that a little while. ([U01],331).

However, a sudden achievement of *permanent* consciousness of union
with the Infinite (refer, for example, [T03],10-11) is not unknown.

"Awful," by the way, is doubly applicable: as we've seen, to those
who are not sufficiently prepared, vision of the Eternal may be
terrible and painful; to those sufficiently prepared, on the other hand,
such vision is full of awe. Indeed, some mystics regard such vision as

> . . . a foretaste of the Beatific Vision: an entrance here
> and now into that absolute life within the Divine Being,
> which shall be lived by all perfect spirits when they
> have cast off the limitations of the flesh and re-entered
> the eternal order for which they were made.
> ([U01],424).

The Significance of Signs

Ramakrishna's heart and breath, it's said, stopped during his unitive
experience. Did it? Teresa of Avila and others claimed ([C10],63) she
floated above the ground? Did she? Did Therese actually live without
food? Did Yogananda's body actually resist decay?

Are the physical phenomena - cessation of heart and breath, living
without food, levitation, precognition, healing powers, etc. - which
sometimes accompany mystical experience really possible? Some
people answer "yes, they are possible" with great conviction, or "they
are certainly not possible" with firm certitude. Yet they answer out of
faith, not knowledge. Science has a better way: it impartially
investigates such claims and rejects the phony. And, if any prove
genuine, then science makes room for them and adjusts its theories
until they better describe reality. Otherwise it's not science but just
another variety of ungrounded belief. My belief happens to be that
some of the phenomena we've discussed are possible, even if they're
unexplainable to science today.

But another, more important question is this: are such phenomena
significant? Are they important? Religious literature often describes
the unusual phenomena which may accompany mysticism as
valueless in themselves, and warns against seeking the "miraculous"
for its own sake. A story I once heard illustrates.

Two monks were traveling and came to a river. One paid a dollar and was ferried across. The other proudly walked across the river. When they rejoined on the other side, one asked the other how long he'd practiced to walk on water. Ten years, was the answer. So, said the first monk, your ten years of practice have just returned a dollar. A poor return!

In Christianity, St. Paul also points to the worthlessness of abnormal phenomena as a substitute for true love of God.

> Though I speak with the tongues of men and of angels, and have not charity, I am become *as* sounding brass, or a tinkling cymbal. And though I have *the gift of* prophecy, and understand all mysteries, and all knowledge; and though I have all faith, so that I could remove mountains, and have not charity, I am nothing. ([H08], 1Cor, 13:1-2).

A chess master once advised students to always remember the object is to capture the opponent's king. The advice is both trite and profound. It's trite since it's a basic rule, one of the first rules you learn: capture your opponent's king and you've won. It's profound since it's easy to become so concerned with subordinate objectives - controlling the centerboard, building a strong defense, etc. - that the paramount objective is forgotten. If you capture your opponent's king then you've won; at that point, the position of your defense and degree of control of the centerboard are irrelevant.

Similarly, the essential thing in mysticism is experience of and union with God; subordinate objectives are ultimately irrelevant. Suppose someone wishes to fast as an aid to prayer and union with God. Fine. But suppose they become so concerned with fasting, or even praying deeply, that they lose sight of their fundamental objective: union with God. Then they may "win the battle but lose the war." They may even fast until they're close to death but find themselves no closer to God. Buddha seems to have experienced this.

A human being cannot, as a rule, walk on water, live without food, generate great quantities of bodily heat, etc. When we die our bodies usually decay. But it's easy to imagine another species who can do all these tricks and more. That doesn't necessarily mean they're closer to direct experience of the Ultimate than we are.

So the seeker should always keep in mind That which they are seeking and not let mastering some trick become their goal. To the mystic, extraordinary phenomena are worthless in themselves.

But suppose such phenomena invariably happened to accompany mystical awareness, and suppose they never appear in any other circumstance. Then, although worthless to the mystic, they would have a very definite value for others. If certain physical signs always accompanied mystical evolution, then there would be an objective way of discriminating the genuine mystic from the bogus, the healthy from the unhealthy. A scientific religion could judge not only their statements, but also the mystics and alleged mystics themselves. Supposed mystics could be subjected to scientific verification.

So although the true mystic might ignore any physical signs, see them as worthless, and refuse to become enamored with them, the scientist can still examine them, attempting to determine how well they correlate with mystical enlightenment.

I don't expect a simple answer. I don't suppose the Tibetan monk who can generate the most *tumo* is always the most evolved mystic. But I do expect that any scientific religion would thoroughly investigate this question. In fact, science has already begun to do this. We'll discuss what it's found after we examine a more basic question.

Holism and Measuring Inner States
Many things once thought unmeasurable are indeed measurable, with the proper instruments. Centuries ago, changes in the position of the sun and moon objectively indicated longer time spans. A day could objectively be defined as sunrise to sunrise, or sunset to sunset. But short time spans - a moment, a second, or a while - probably seemed permanently unmeasurable, subjective quantities. We are all familiar with how slowly time seems to pass when we're bored, how quickly when we're entertained. Eventually, timepieces were devised which accurately measured shorter time intervals, and today they are universally available.

The feeling of temperature, too, was once subjective. Many a husband and wife could argue over whether a room *was* too hot or too cold. Today, thermometers have settled this question. Of course, whether a room feels too hot or too cold is another matter.

So clocks and thermometers have attached numbers to temperature and time, transforming them into purely objective entities. Time and temperature have become operationally defined quantities.

But time and temperature are external to ourselves. Can a person's inner characteristics be measured too? Family traits are intimate to the person. Once they were entirely subjective; again, many a husband and wife argue over who a child takes after. (Generally, desirable traits descend from one's self, while undesirable characteristics clearly come from one's spouse.) Then science discovered genes. Today, many physical traits can be scientifically measured.

Inherited traits, however, are static; they remain throughout one's life. Is it possible to measure changing emotional and mental states? It seems that it is.

When we previously discussed personal identity, we mentally divided a human being into body, emotion, intellect, and soul. If a person were actually composed of entirely separate parts - body, emotion, intellect, and soul, for example - then emotion and body, or intellect and body, and certainly Consciousness and body, might be entirely unrelated. There would be no reason to expect mystical states, which are states of consciousness, to manifest in the body, emotions, or intellect.

Yet, a human being is a holistic entity, even as we acknowledge in a previous chapter. The division into body, emotion, intellect, and soul is mental, not absolute. The intimate, holistic connection between soul, intellect, emotion, and body offers some hope that there are physical, emotional, and intellectual phenomena which correlate with mystical states.

Let's begin by discussing some correlates between the body and emotions. There's a well-known link between emotions and physical signs. Everyone is acquainted with the smile of the happy person, the redden face of the angry one. Indeed,

> . . . facial representations of sadness, fear, anger, disgust and other emotions are remarkably constant and recognizable around the globe. ([U05],58).

Recently, scientists have used the EEG to uncover more subtle manifestations of emotions.

> . . . [Infants more prone to distress when separated from their mothers show increased activity in the right

frontal lobe, as do people with a more pessimistic outlook. People who have at some point in their lives been clinically depressed show decreased left frontal lobe activity compared with subjects who have never been depressed. ([U05],58).

And

[w]hen people are anxious, cerebral blood flow - a measure of brain cell activity - increases in the area at the tips of the brain's temporal lobes just behind the eyes . . . When subjects report feeling emotions such as fear and disgust, their right frontal lobes show increased electrical activity . . . Sadness seems to diminish activity in the left frontal lobe as measured by an electroencephalogram (EEG), while certain positive emotions like happiness and amusement increase it. ([U05],57-8).

The electroencephalograph (EEG) measures electrical activity in particular portions of the brain - the right frontal lobe, for example, or the temporal lobes. The brain's electrical activity - its "brain waves" - is roughly classified (refer, for example, [C01],24-5) by frequency. Brain waves seem to correlate with certain mental states. Frequencies higher than about 13 cycles per second (or 13 Hertz, abbreviated 13 Hz) are called beta waves, and indicate attention. Alpha waves are 8 to 13 Hz and correlate with a calm, relaxed state. Theta waves range from 4 to 7 Hz and are associated with the dream state. Delta waves are produced in deep sleep and are below 4 Hz.

There are, of course, other devices which scientists use to measure inner states. The polygraph ("lie-detector") is a familiar one. It measures certain bodily characteristics believed to vary with truth or falsehood. Other devices are the electrical skin resistance (ESR) meter which, as the name suggests, measures the electrical resistance of the skin, and the electrocardiograph (EKG) which measures heart function. Lastly, the electromyograph (EMG) measures voltages which correspond with muscle tension. One researcher has found the EMG to be

. . . of considerable importance in the development of self-awareness. Early emotional conflicts are often reflected in the body armor a person has built - permanently tense muscles intended as body defense. The electromyograph facilitates specific therapy for

> these states, and the ability to exercise fine control
> over muscle tension may be one of the best indicators
> of the subject's ability to relax at will, which is the
> gateway to meditation as well as to improved general
> health. ([C01],14-5).

With such instruments, science has been able to measure not only physical and emotional states, but intellectual and perhaps spiritual states as well. A few examples follow.

Measures of Meditative Mental States

A goal of Zen Buddhism is an increased yet disinterested awareness of the external world. Zen adapts, therefore, should possess such awareness in some degree. It seems that they do.

Imagine someone sitting relaxed in a quiet room. An EEG machine indicates their brain is emitting alpha, indicating their mind is in a quiet, relaxed state. There's a click, which draws their attention to the room. The EEG machine detects their increased attention and indicates their brain is now emitting beta waves. Soon, however, they're relaxed again, and the EEG machine indicates the alpha state. A minute later, there's another click. Again, their mind goes from alpha to beta, but the beta frequency is not as high. The regular clicking sound is moving to the edge of their awareness. As the clicks repeat, they notice them less and less. Eventually they habituate: they no longer hear the click; their minds remain in alpha.

Habituation is a familiar phenomena to anyone who has ever lived near a railroad track. At first, you hear every train; after a month, you rarely notice any.

When a Zen master in a state of Zen meditation, called *zazen*, was the subject ([K02]) in the experiment no habituation was detected. The Zen master heard each click.

> This non-habituation . . . in response to click stimuli
> during *zazen* is consistent with the description by one
> Zen master of the state of mind, cultivated in *zazen*, of
> "noticing every person one sees on the street but not
> looking back with emotional curiosity." ([F07],118).

In contrast, control subjects - people who weren't doing zazen meditation - habituated.

Of course, we might wonder if lack of habituation is desirable. Many people living near a railroad track wouldn't want to hear each and every train, day after day. The point is, however, science appears to have objectively verified a certain type of mental awareness which Zen meditation masters claim to possess. Rather than taking on faith the claim that someone has achieved a certain level of success in Zen meditation, science can test for itself. And anyone who studies Zen might wonder how their teacher would perform in the above experiment.

In contrast to Zen, many types of Yoga advocate quite a different goal, replacement of attachment to the external world with attachment to its eternal Basis, an attachment implying a measure of indifference to the external world. In one study ([A05]), yogis were subjected to various external stimuli such as

> . . . turning on a strong light, banging on an object,
> vibrating a tuning fork, and touching the yogis with a
> hot glass tube. ([W06],234-5).

They reacted normally, except during meditation when they evidenced insensitivity, i.e., their alpha patterns remain undisturbed. Therese Neumann underwent a similar test (minus the alpha pattern monitoring) with 5,000 watt carbon arc lamps while she was in a mystical state.

> The lamps were focused directly on her open eyes
> during the ecstasies. . . . Therese did not even blink.
> This was proof that she was completely insensitive to
> external influences when she was in a state of
> visionary contemplation . . . ([S22],30).

In another yoga study

> . . . two other yogis, who claimed to have developed
> high pain thresholds, were able to keep one hand in 4
> degree C. water for 45-55 minutes with no EEG
> disturbance or apparent discomfort. ([W06],235).

On the other hand (no pun intended) there's a type of yoga, kriya yoga, which claims to activate and channel the meditator's kundalini energy, supposedly a potent, highly active spiritual force. Kundalini energy is worshiped as the divine by its devotees. Although still aiming at detachment from the external world, kriya yogis seek attachment to their own internal, dynamic kundalini energy. They were found ([D02]) to have

> . . . extremely fast beta activity (indicative of high
> arousal) with high amplitude waves (frequency up to
> 40 hz, amplitude 30-50 microvolts) . . . ([W06],236-7).

Moreover,

> . . . various stimuli applied during meditation had
> absolutely no effect on the EEG. . . The latter finding is
> again strong evidence of withdrawal from the
> environment . . . ([W06],237).

Here again, science tested and verified a religious inner state: withdrawn from the external and attachment to some highly active internal energy.

A neurochemical basis involving the pineal gland and conversion of the body's melatonin into an hallucinogen ([K10],297) has been suggested as the basis of kundalini consciousness. If true, then measurement of brain chemistry might yield another objective measure of this kind of religious consciousness.

An Exact Science?

In India, "Nirvikalpa" refers to union with God. A swami writes:

> . . . a man cannot be fit to realize the eternal peace, till
> he reaches the Nirvikalpa state through the cessation
> of all mental modifications and the non-dual state of
> consciousness becomes natural to him. ([S01],355).

Can science now verify if a person has reached the state of "cessation of all mental modifications"? Perhaps. Could it ever verify if someone has truly reached union with God? I don't know.

But science could certainly investigate any physical, emotional, or mental phenomena that seem associated with mystical awareness. It might find, of course, that such by-products fail to correlate with mystical evolution: that they don't always occur in genuine mystics, and sometimes occur in non-mystics. If this was the case, then scientific religions might never be able to become exact sciences.

On the other hand, some reliable physical, emotional, or mental by-products might be found. Some signs might be discovered which always indicate a certain level of mystical insight. If so, objective indicators of mystical states would exist.

Someday, one's state of mind, be it holy, worldly, or profane, may be objectively measurable. It may be an operationally definable quantity. This would greatly help the identification of true mystics from others, as well as truly mystical declarations from other kinds of statements. Moreover, if these indicators were quantifiable as well, then, perhaps, some sort of exact relation might be found between them. Scientific religions might be exact sciences.

The possibility that, someday, a scientific religion may exist which is not only descriptive and experimental, but exact as well is (to me, at least) exciting. If, however, the nature of things does not allow an exact scientific religion, it should nonetheless be as fully science and as fully religion as possible.

Conclusion

- Conclusion -

Chapter Summary: This chapter concludes the book with a brief summary. Lastly, various futures are described.

Looking Back

Years ago I began recording the thoughts and ideas that make up the world view presented in this book. These thoughts and ideas represent my effort to make sense of my life, the world, and my place in it. They aren't dogma - not even for me - and are subject to change and modification. In fact, I don't consider this book entirely finished. I plan to continue working on it - clarifying, refining, and perhaps adding some new material - until I feel it's the best I can do.

Yet as I review what's written, I'm satisfied. I've more or less said what I wanted to say. And more. For at times, in writing down A, B, and D, I've realized C was missing or badly thought out. This prompted me to think and read more about C. As a result, I've explored a few topics I hadn't previously thought about, or had thought about only vaguely.

Even though I've said most of what I wanted to say, I haven't necessarily said it as well as it can be said, or cast it in the most direct, lucid prose possible. No doubt by many standards I've failed miserably.

Strange how when we seek mere entertainment and diversion, we demand perfection. We have no place for the slightly off tune singer, the ham actor, the bush league baseball player. How much less is there for the amateur philosopher and writer? On the other hand, notice how when the thing is of real importance we tolerate the less than perfect. A young boy runs up and says "Mister, your car . . . down hill . . . car . . . rolling!" Do we correct his sentence structure, giving him a lesson in subjects, verbs, and objects? Or do we understand what he is trying to say and act appropriately?

Not that I have contempt for the art of writing. I've tried to write the best book I could. But I'm not an accomplished writer, philosopher, scientist, mystic, metaphysician, etc., and I'm sure it's shown at times.

So the only important criterion for me is this: is what I've said true? Obviously you must answer that question for yourself. Whatever your answer, I hope you've found what's been said worth the reading.

Let's close by briefly reviewing what's been said, and entertaining a few speculations.

In the first part, we examined two different ways of knowing, two different epistemological methods: the scientific way of knowing and the revelational way of knowing so often used by religion. We found the scientific way of knowing superior.

Then we investigated science's domain of knowing, focusing on an entity which science believes is eternal. We saw how this entity deserved to be called "the eternal substance" and many other names. Then we saw how religions apply many of the same terms to God. Yet religions often speak of God in an entirely different way, as a Person. So we decided to use two distinct terms - "God who is a Person" and "God which is not a Person" - to label these two very different ideas of God. We then introduced a class of persons, the mystics, who seem to have direct experience of God, and demonstrated how many of them have experienced the God which is not a Person.

Next, we discussed how applying the scientific way of knowing to the religious domain would create a scientific religion, and saw that such a philosophy and way of life would be fully science and fully religion, even though it wouldn't contain any dogma or Gods who are actually Persons.

In the second part, we began constructing a world view, based on the declarations of the mystics, compatible with science, yet religious in the deepest sense. We saw how the universe can be seen as a "mode of light" or in various dualistic ways. We adopted the yang/yin model, an inseparable dualistic model where entities exist "in themselves" as modes of light, but may be seen by the observer in various dualistic ways. We then discussed our own personal identity, and examined the surprising mystical idea that our deepest self is either the Ultimate Ground of Existence, or doesn't really exist. We came to a similar conclusion about Gods who are Persons. We also saw that relating to the God which is not a Person as if It were a Person is very similar to regarding God as an actual Person. We saw that the idea of Gods who are Persons may have originally derived from the God who is not a Person. Finally, we discussed various concepts - component entity, relative and absolute existence, actions, voidness and emptiness - which apply in general to the universe, ourselves, and Gods who are Persons.

Part III concerned the practical application of the ideas we'd developed. We saw how a person could decide to make gnosis one of their life's goals. We described some of the values, ethics, and morals that might follow from such a decision. Then, we discussed some of the practical acts and attitudes such a person might adopt, the actual way of life they might follow. Next we discussed the ultimate goal of such a path, union with God. Finally, we wondered if science could ever objectively measure the mystical evolution of an individual, and if a scientific religion could ever be an exact science.

Looking Further Back
There is a certain satisfying symmetry in coming full circle, arriving at journey's end to the point from which we began. The feeling is one of returning, of coming home, and yet may also be one of seeing for the first time what was previously present, but unnoticed. To conclude in this manner, to reach the point from which we began, I'll offer an historical parallel, a modest outline not of "the shape of things to come" but merely of "the shape of things which may come."

About two thousand years ago, people in the lands around the Mediterranean sea were under Roman rule. These peoples spoke different languages, had different cultures and customs, and were mostly free to follow their own religious beliefs and practices. These practices included: the worship of Mithras, who was born on December the twenty-fifth, whose holy day was Sunday, the day of the Conquering Sun, and whose followers celebrated a Eucharist feast; the worship of Osiris who was born in a cave and died for the salvation of his people; and the worship of the virgin Earth Mother and her son, the vegetation that dies each winter only to rise from the dead the following spring. These peoples shared the common Roman political and economic system, and a common Greco-Roman culture. In the Roman State, Egyptian and Greek, Syrian and German could travel, communicate and trade.

A few hundred years later, they shared a common religious system as well. There had formed a new sacred revelation, which included many older writings and many contemporary ideas and insights. Religions based on this scripture still exist, and are called Christian.

The state of the world today is in many ways similar.

True, a single world government doesn't exist as it did then. But common problems, such as the containment of nuclear weapons and the protection of the world's single ecosystem, often compel cooperation between governments. Our economic systems are closely linked too; most nations, their differing political systems notwithstanding, trade and communicate extensively. More and more, we live in a "Global Village."

And just as the Greco-Roman culture existed throughout the Mediterranean world, the culture of science and technology can be found today the world over. Science's values, methodology, and world view are found everywhere and possess a unity which transcends national boundaries. A discovery by a French, Russian, Chinese, English, Hungarian, or American scientist, once verified, is accepted by scientists everywhere. The scientific world view and some of its practical consequences - electricity, radios, televisions, automobiles, trains, telephones, computers - are ubiquitous.

We are currently in an era of increasing scientific knowledge and technological progress, an era which began about the time of Newton, a few hundred years ago. Historically, we know that such eras don't last forever. What happens when they end? Examining a similar time about 1,000 years ago may offer some clues.

From about 700 to 1100 C.E., Islamic civilization experienced a ([D10],240) "vibrant awakening," a ([D10],240) "flowering of science, literature, and art." In fact,

> [o]nly at the peaks of history has a society produced, in an equal period, so many illustrious men - in government, education, literature, philology, geography, history, mathematics, astronomy, chemistry, philosophy, and medicine - as Islam . . . ([D10],343)

in that period. Islamic scholars knew ([D10],329) the earth was spherical; one proposed ([D10],328) in 1122 C.E. a theory of universal gravitation, anticipating Newton by many centuries; another ([D10],329) wrote a botanical reference which remained the standard for 300 years; yet another wrote ([D10],330) "the most complete treatment of agricultural science in the whole medieval period." And

Omar Khayyam wrote the *Rubaiyat*, a book of poetry still in print 900 years later.

Art, literature, and science thrived. But traditional religious faith did not.

Science and Religion

Often, a scientific world view and an orthodox religious world view don't mix. As a consequence, science may suffer. For example, in Islamic Spain

> [s]cience and philosophy . . . were largely frustrated by the fear that they would damage the people's faith. ([D10],305).

Certainly, religions have sometimes made room for science; but usually at their own expense. The Roman Catholic Church, for example, now teaches its scriptures are not necessarily free from errors in natural science. How much greater was the belief of the common man and woman in God and religion before this was so!

Conversely, when science does advance it's often at the expense of traditional religious faith. The increasing acceptance of a scientific world view is often accompanied by a growing disbelief in the orthodox God or Gods. Science either promotes no religion at all, or in effect promotes a practical atheism which

> . . . consists of ignoring or neglecting any relationship to God in one's actions, or in living as if God did not exist . . . Practical atheism, therefore involves the orientation of one's life exclusively toward the attainment of earthly goals. ([N05],v2,259).

The spread of science is often accompanied by a lessening of religious belief or, if you will, by conversion to the only religion which science, however indirectly, promotes - *de facto* atheism, an atheism in practice if not in theory.

Eventually, religion reasserts itself. The need for religion has been, and probably still is, inherent and primal. The need for an all-embracing world view that makes sense of it all, for comfort in time of sorrow, for self-transcendence, can be pushed aside for a while, but sooner or later becomes irresistible. An example, again from medieval Islam, will be useful.

Ghazzali, who we met earlier, studied ([D10],331) philosophy, law, and theology, eventually achieving great renown in law. Then he

([D10],331) "lost belief in the capacity of reason to sanction the Mohammedan faith" and had a mental break down. A mystical experience, however, restored his faith in traditional Islam and led him to teach that

> . . . reason leads to universal doubt, intellectual bankruptcy, moral deterioration, and social collapse. ([D10],332).

His teaching pleased many people who were troubled by their age's increasing skepticism and agnosticism toward religion. So successful was Ghazzali that

> [w]hen he died (1111), the tide of unbelief had been effectually turned. ([D10],332).

The result?

> After him . . . philosophy hid itself in the remote corners of the Moslem world; the pursuit of science waned; and the mind of Islam more and more buried itself in the . . . Koran. ([D10],332).

Science and Mysticism

Evelyn Underhill writes:

> . . . the great periods of mystical activity tend to correspond with the great periods of artistic, material, and intellectual civilization. As a rule, they come immediately after, and seem to complete such periods . . . When science, politics, literature, and the arts . . . have risen to their height and produced their greatest works, the mystic comes to the front . . . It is almost as if he were humanity's finest flower; the product at which each great creative period of the race had aimed. ([U01],453).

Indeed, in Islam after Ghazzali

> . . . the field of religious thought was yielded to Sufi monks and saints. . . . Sufi devotees now abandoned family life, lived in religious fraternities . . . Some by prayer and meditation, some by ascetic self-denial . . . sought to transcend the self and rise to a . . . unity with God. ([D10],332).

But if it's true that mystical periods often conclude and complete golden periods of material, artistic, and intellectual progress, then it's just as true that they often precede dark periods of material, artistic, and intellectual regression; that they're often watershed events, separating periods of progress from periods of decline. So a question naturally follows: is it possible that an outbreak of mystical activity can somehow *cause* a decline in material, artistic, and intellectual activity?

I think not. I believe rather that the incompatibility of a scientific and religious world view forces the choice of one at the expense of the other. Like Ghazzali, an entire civilization can chose science and reason for a while. Eventually, however, the need for religion becomes desperate. The scientific civilization "breaks-down", decides that "reason leads to universal doubt, intellectual bankruptcy, moral deterioration, and social collapse" and turns to God. For a while, perhaps, that civilization enjoys the best of both worlds; its new-found interest in God, and its old, established, not yet fully renounced concern with science give it a wonderful balance. Eventually, however, it falls. Unable to balance science with a concern for God, it loses its scientific ability.

Eventually, after hundreds of years, an exclusively religious orientation proves as unsatisfying as an exclusively scientific orientation. Art, literature, and science struggle to reassert themselves.

And so the cycle begins again. We limp with one leg, then the other, unable to walk on both, except during brief transition periods. Science and religion, it seems, form an unstable balance: we continually fall into an acceptance of one at the expense of the other. A balanced, whole-hearted belief in both seems difficult - as it should be if they're basically incompatible ways of looking at the world.

> For a long time we have been accustomed to the compartmentalization of religion and science as if they were two quite different and basically unrelated ways of seeing the world. I do not believe that this state of doublethink can last. . . . ([W03],xviii).

We stand with many centuries of scientific progress behind us. Is it time now for older religions, with their frozen scriptures and time-bound belief systems, to reassert their influence in everyday life? Or is the world approaching the creation of a new religion, a religion with a still growing and pliable set of "revelations," a religion better able to incorporate contemporary wisdom?

> . . . It must eventually be replaced by a view of the
> world which is neither religious nor scientific but simply
> our view of the world. . . . ([W03],xviii).

Suppose the future holds a new religion, one which will achieve dominance in our world, just as Christianity achieved dominance in the ancient Roman world. If this new religion isn't compatible with science, then one or the other will suffer. If the new religion triumphs at science's expense then another "dark age" may follow.

If, on the other hand, the new religion is a scientific religion, then we would finally have an integrated, harmonious world view. We would finally walk on both legs. Science and religion would be integrated, not merely in a state of "peaceful co-existence," but truly united. There'd be a deeply scientific world view which was at the same time deeply religious.

> . . . More exactly, it must become a view of the world
> in which the reports of science and religion are as
> concordant as those of the eyes and the ears.
> ([W03],xviii).

Perhaps then the cycle of scientific progress at the expense of religion and religious faith at the expense of science would end.

And so we conclude by coming full circle, by arriving at the last quote of this book which was also its first. Yet now the quote serves not as introduction and foreshadowment, but as warning, prophesy - and challenge.

- Bibliography -

[A01] - Abramson, Paul R. *Personality* (Holt, Rinehart and Winston, New York, 1980) {A college psychology textbook}

[A02] - *African Ideas of God*, Ed. by Edwin W. Smith (Edinburgh House Press, London, 1966)

[A03] - *Al-Ghazzali's Mishkat Al-Anwar ("The Niche for Lights")*, Trans. by W.H.T. Gairdner (Ashraf Press, Lahore, 1952) {First published as Monograph Vol. XIX by the Royal Asiatic Society, London, 1924}

[A04] - Alpert, Richard, *Be Here Now* (Crown Publishing, New York, 1971) {Based on a lecture by Alpert recorded as *The Evolution of Consciousness*}

[A05] - Anand, B. K., G. S. Chhina, and B. Singh. *Some Aspects of Electroencephalographic Studies in Yogis*. (Electroencephalography and Clinical Neurophysiology, 13, 1961, 452-456)

[A06] - Anonymous. *The Prayer of Love and Silence* (Dimension Books, Wilkes-Barre, Pa., 1962) {Sermons of a Roman Catholic Carthusian monk}

[A07] - Aquinas, Thomas. *Summa Theologia* (Blackfriars, Great Britain, 1964)

[A08] - Arberry, A.J., *Sufism. An Account of the Mystics of Islam* (George Allen & Unwin Ltd., London, 1950)

[A09] - *Arsenal For Skeptics*, Ed. by Richard W. Hinton (A. S. Barnes and Company, New York, 1961; Alfred A. Knopf, 1934) {Includes excerpts from Paine, Gibbon, and Nietzsche}

[A10] - *Ashtavakra Gita*, trans. by Hari Prasad Shastri (Shanti Sadan, London, 1961) {"a favourite text among the Mahatmas of the Himalayan regions."}

[A11] - Asimov, Isaac. *The Universe* (Avon Books, New York, 1966) {Mr. Asimov was a well-known science and science fiction writer, and a professor at Boston University}

[A12] - Attar, Farid ud-Din. *The Conference of the Birds (Mantiq Ut-Tair)*, Trans by S. C. Nott (The Janus Press, London, 1954)

[A13] - Augustine, St. *The City of God*, Trans. by Marcus Dods, D.D. (The Modern Library, New York, 1950)

[A14] - Augustine, St. *The Confessions of Saint Augustine*, Trans. by E.B. Pusey (Peter Pauper Press, Mount Vernon) {First published in 1838, based on an earlier translation}

[A15] - *The Awakening of Faith*, trans. by Yoshito S. Hakeda (Columbia University Press, New York, 1967)

[B01] - Bach, Richard. *Jonathan Livingston Seagull* (The Macmillan Company, New York, 1970)

[B02] - Benson, Herbert and Miriam Z. Klipper, *The Relaxation Response* (Avon Books, New York, 1975) {Bestseller by Harvard M.D. about medical benefits of meditation}

[B03] - Berryman, Phillip. *Liberation Theology*, (Pantheon Books, New York, 1987)

[B04] - Bilinkoff, Jodi. *The Avila of Saint Teresa*, Cornell University Press, Ithaca, N.Y., 1989)

[B05] - *The Book of Angelus Silesius*, Trans. by Frederick Franck (Vintage Books, New York, 1976) {Writings of the Christian mystic Angelus Silesius presented and compared with Zen Buddhist writings}

[B06] - *The Book of the Gradual Sayings (Anguttara-Nikaya) Vol II*, Trans. by F. L. Woodward, (Oxford University Press, London, 1933) {Pali Text Society, No. 24}

[B07] - *The Book of the Kindred Sayings (Sanyutta-Nikaya) Part II*, Trans. by Rhys Davis, (Luzac & Company Ltd., London, 1952) {Pali Text Society, No. 10}

[B08] - *The Book of the Kindred Sayings (Sanyutta-Nikaya) Part III*, Trans. by F. L. Woodward, (Luzac & Company Ltd., London, 1954) {Pali Text Society, No. 18}

[B09] - *The Book of the Kindred Sayings (Sanyutta-Nikaya) Part IV*, Trans. by F. L. Woodward, (Luzac & Company Ltd., London, 1956) {Pali Text Society, No. 14}

[B10] - *The Book of Middle Length Sayings (Majjhima-Nikaya) Vol III*, Trans. by I. B. Horner, (Luzac & Company Ltd., London, 1967) {Pali Text Society, No. 31}

[B11] - Borges, Jorge Luis. *Labyrinths, Selected Stories & Other Writings*, Ed. by Donald A. Yates and James E. Irby (New Directions Publishing Corporation, New York, 1964) {"recognized all over the world as one of the most original and significant figures in modern literature."}

[B12] - Broad, William and Nicholas Waye. *Betrayers of the Truth - Fraud and Deceit in the Halls of Science* (Simon & Schuster, New York, 1982) {"Utterly fascinating reading." - Science 1983}

[B13] - Buddhadasa. *Toward the Truth*, Ed. by Donald K. Swearer (The Westminster Press, Philadelphia, 1971) {Mr. Swearer is a professor in the Department of Religion at Swarthmore College. Clearest book on Buddhism I've found}

[B14] - *A Buddhist Bible*, Ed. by Dwight Goddard (Beacon Press, Boston, 1966) {Huston Smith: "for the reader who is looking for scholarship *and* meaning, coverage *and* control - all four - I know no alternative that is its equal"}

[B15] - Burroughs, William and Allen Ginsberg. *The Yage Letters*, (City Light Books, San Francisco, 1963 & 1975)

[B16] - Bush, Richard C. et al, *The Religious World: Communities of Faith* (Macmillan Publishing Co., New York, 1982) {College textbook}

[B17] - Butler, Edward Cuthbert. *Western Mysticism*, (Constable, London, 1922) {"The Teaching of Augustine, Gregory, and Bernard on Contemplation and the Contemplative Life}

[C01] - Cade, C. Maxwell and Nona Coxhead. *The Awakened Mind: Biofeedback and the Development of Higher States of Awareness*, (Delacorte Press, New York, 1979)

[C02] - Calder, Nigel. *Einstein's Universe* (Penguin Book, New York, 1979) {Christian Science Monitor calls Mr. Calder "One of the world's best science writers"}

[C03] - Capra, Fritjof. *The Tao of Physics*, (Shambhala, Boulder Colorado, 1975)

[C04] - Carus, Paul. *The Gospel of Buddha* (The Open Court Publishing Company, LaSalle, IL 1973) {"Complied from ancient records . . . first edition in 1894 . . . officially introduced in Buddhist schools and temples in Japan and Ceylon"}

[C05] - Cerf, Christopher and Victor S. Navasky. *The Experts Speak, The Definitive Compendium of Authoritative Misinformation* (Pantheon Books, New York, 1984) {Entertaining; scary too - "Nuclear weapons are so terrible and destructive that war has been rendered impossible" was essentially said of dynamite, machine guns, submarines, and airplanes. refer 254-5}

[C06] - Christian, James L. Philosophy: *An Introduction to the Art of Wondering*, (2nd ed., Holt, Rinehart and Winston, 1977) {College level text}

[C07] - Christoforidis, A. John, *Atlas of Axial, Sagittal, and Coronal Anatomy with CT and MRI* (W.B. Saunders Company, Philadelphia, Pa., 1988) {Medical reference book}

[C08] - *Classic Philosophical Questions*, Ed. by James A. Gould (Charles E. Merrill Publishing, Columbus, Ohio, 1971)

[C09] - Climacus, St. John. *The Ladder of Divine Ascent* (Holy Transfiguration Monastery, Boston, Ma., 1978) {A classic Orthodox Christian ascetic treatise}

[C10] - Clissold, Stephen. *St Teresa of Avila*, (Seldon Press, London, 1979)

[C11] - Coffey, P., *Ontology, or the Theory of Being*, (Peter Smith Press, Gloucester, Mass., 1970) {Irish professor of Logic and Metaphysics - first published in 1912}

[C12] - Cohen, J.M. and J-F Phipps. *The Common Experience* (J.P. Tarcher, Los Angeles, 1979) {Similar to Aldous Huxley's *The Perennial Philosophy*}

[C13] - *The Collection of the Middle Length Sayings (Majjhima-Nikaya) Vol. I*, Trans. by I. B. Horner, (Luzac & Company Ltd., London, 1954) {Pali Text Society, No. 29}

[C14] - *Collier's Encyclopedia* (Crowell-Collier Educational Corporation, 1968)

[C15] - Compton, Arthur Holly. *The Cosmos of Arthur Holly Compton* (Alfred A Knopf, New York, 1967) {Won the Nobel Prize for discovering the Compton X-ray effect}

[C16] - *Contemplative Review*, Spring 1984

[C17] - Cott, Allan, Jerome Agel, and Eugene Boe. *Fasting: The Ultimate Diet*, (Bantam Books, New York, 1975) {Mr. Cott is an M.D.}

[D01] - Dampier, Sir William Cecil. *A History of Science and Its Relations with Philosophy and Religion*, 4th ed., (Macmillan Co., New York, 1949) {"considered one of the outstanding histories of science")

[D02] - Das, N. N. and H. Gastuat. *Variations de l'activite electrique du cerveau, du coeur, et des muscles squelettiques au cours de la meditation et de l'extase yogique.* (Electroencephalogrphy and Clinical Neurophysiology, 1957, supp. 6, 211-19)

[D03] - Davies, A. Powell. *The Meaning of the Dead Sea Scrolls* (Mentor Books, New American Library, 1956) {Mr. Powell was the pastor of the All Souls Church in Washington, D.C.}

[D04] - Davis, Philip J. and Reuben Hersh, *The Mathematical Experience* (Birkhauser, Boston, 1981)

[D05] - Dawson, Miles Menander. *The Ethical Religion of Zoroaster* (AMS Press, New York, 1969)

[D06] - Dhalla, Maneckji Nusservanji. *Zoroastrian Theology, From the Earliest Times to the Present Day* (AMS Press, New York, 1972)

[D07] - *The Dialogues of Plato*, Trans. by B. Jowett (Oxford at the Clarendon Press, Oxford, 1953) {Author was a Doctor of Theology and a professor of Greek}

[D08] - Dionysius, The Areopagite. *The Divine Names and The Mystical Theology*, Trans. by C.E. Rolt (S.P.C.K., London, 1940) {"His Christianized Neoplatonism had an immense influence on subsequent Christian thought. The medieval mystics are deeply indebted to him, and St Thomas Aquinas used him as authority."}

[D09] - *Divine Revelation (Pamphlet #81)* (Catholic Information Service, Knights of Columbus, New Haven, Conn.)

[D10] - Durant, Will. *The Age of Faith*, (Simon and Schuster, New York, 1950) {A history of medieval civilization - Christian, Islamic, and Judaic - from Constantine to Dante}

[E01] - Eddington, Sir Arthur. *The Nature of the Physical World* (Ann Arbor Paperbacks, The University of Michigan Press, 1958)

[E02] - Eddy, Mary Baker. *Science and Health with Key to the Scriptures,* Auth. Ed. (Boston, 1934)

[E03] - Einstein, Albert. *Ideas and Opinions* (Crown Publishers, New York, 1954) {"most definitive collection of Albert Einstein's popular writings, gathered under the supervision of Professor Einstein, himself."}

[E04] - Epstein, Perle. *Kabbalah, The Way of the Jewish Mystic* (Doubleday, New York, 1978)

[E05] - Erigena, Johannes Scotus. *Periphyseon (The Division of Nature),* Trans. by I.P. Sheldon-Williams, Rev. by John J. O'Meara (Dumbarton Oaks, Washington D.C., 1987)

[E06] - Ernst, Carl W. *Words of Ecstasy in Sufism* (State University of New York Press, Albany, 1985)

[F01] - *Faith and Practice,* (Philadelphia Yearly Meeting, 15 & Cherry Sts., Phila, Pa. 19102) {Statement of various Quaker beliefs}

[F02] - Fernando, Anthony with Leonard Swidler, *Buddhism Made Plain, An Introduction for Christians and Jews* (Revised Edition) (Orbis Books, Maryknoll, New York, 1985)

[F03] - Ferris, Timothy. *Coming of Age in the Mikly Way* (Doubleday, New York, 1988) {"Editors' Choice: The Best Books of 1988" New York Times}

[F04] - *Flesh of the Gods; The Ritual Use of Hallucinogens,* Ed. by Peter T. Furst (Praeger Publishers, New York, 1972)

[F05] - Foster, Genevieve W. *The World Was Flooded with Light; A Mystical Experience Remembered* (Univ. of Pittsburgh Press, Pittsburgh, 1985)

[F06] - Frost, S.E. Jr. *Basic Teachings of the Great Philosophers* (Doubleday, New York, 1962) {Assistant Professor of Education, Brooklyn College; First published in 1942}

[F07] - Funderburk, James. *Science Studies Yoga: A Review of Physiological Data.* (Himalayan International Institute of Yoga Science & Philosophy of USA, Honsdale, PA, 1977) {Professor of mathematics}

[F08] - *Funk & Wagnalls Standard College Dictionary* (Funk & Wagnalls, New York, 1968)

[G01] - Goodenough, Erwin R. *By Light, Light; The Mystic Gospel of Hellenistic Judaism* (Yale University Press, New Haven, 1935) {Professor of the History of Religion, Yale University}

[G02] - *Good News Bible, Today's English Version* (American Bible Society, New York, 1976) {"attempts . . . to set forth the Biblical content and message in a standard, everyday, natural form of English}

[G03] - *The Gospel of Sri Ramakrishna,* Trans. by Swami Nikhilananda (Ramakrishna-Vivekananda Center, New York, 1977) {Sri Ramakrishna is considered in India an Incarnation of God}

[G04] - Gribbin, John. *In Search of Schrodinger's Cat* (Bantam Books, New York, 1984)

[G05] - Gunther, R. T. *Early Science in Oxford; Vol. VI, The Life and Work of Robert Hooke (Part I)*, (Oxford University Press, 1930)

[H01] - Hartmann, William K. *Astronomy: The Cosmic Journey* (Wadsworth Publishing company, Belmont, California, 1982, 2ed.) {An astronomy textbook}

[H02] - Hawking, Stephen W. *A Brief History of Time: From the Big Bang to Black Holes* (Bantam Books, New York, 1988) {"widely regarded as the most brilliant theoretical physicist since Einstein"}

[H03] - Heisenberg, Werner. Across the Frontiers, Trans. by Peter Heath (Harper & Row, New York, 1974) {Heisenberg won the Nobel prize in physics in 1932}

[H04] - Hesse, Hermann. *Magister Ludi*, Trans. by Richard and Clara Winston (Bantam Book, New York, 1969) {Won Hesse the Nobel prize for literature}

[H05] - Hesse, Hermann. *Siddhartha*, Trans. by Hilda Rosner (Bantam Book, New York, 1971) {"His most famous novel"}

[H06] - Hilton, Walter. *The Scale of Perfection*, {Burns Oates, London, 1953) {Trans. by Gerard Sitwell}

[H07] - *A History of the Private Life*, Ed. by Paul Veyne, (The Belknap Press of Harvard University Press, Cambridge, Mass., 1967)

[H08] - *Holy Bible, King James Version*

[H09] - *How to Know God, the Yoga Aphorisms of Patanjali*, trans. by Swami Prabhavananda and Christopher Isherwood (New American Library, New York, 1953)

[H10] - Huxley, Aldous. *The Doors of Perception and Heaven and Hell* (Harper Colophon Books, New York, 1963)

[H11] - Huxley, Aldous. *The Perennial Philosophy* (Harper & Brothers, New York, 1945)

[H12] - *The Hymnbook* (Presbyterian Church in the United States, Copyright by John Ribble, 1955)

[H13] - *Hymns of Guru Nanak*, Trans. by Khushwant Singh (Orient Longmans, New Delhi, 1969)

[I01] - *Ibn al-'Arabi, The Bezels of Wisdom*, Trans. by R.W.J. Austin (Paulist Press, New York, 1980)

[I02] - *I Ching, The Chinese Book of Changes*, Arr. by Clae Waltham (Ace Books, New York, 1969) {"perhaps the oldest extant writing in the world"}

[I03] - *Inerrancy And The Church*, Ed. John D. Hannah (Moody Press, Chicago, 1984) {Part of a series published by the International Council on Biblical Inerrancy}

[I04] - Isherwood, Christopher. *Ramakrishna and His Disciples* (Vedanta Press, Hollywood, California, 1965) {"fresh and important contribution to the history of a religious mysticism" - New York Times}

[J01] - Jeans, James. *The Mysterious Universe* (The Macmillan Company, New York, 1944) {Sir James Jeans was a physicist and a popular philosopher of science}

[J02] - Jones, Rufus M. *The Luminous Trail* (The Macmillan Company, New York, 1947) {Quaker professor at Haverford College}

[J03] - Jones, Rufus M. *Studies in Mystical Religion* (Russell & Russell, New York, 1970) {First published in 1909}

[J04] - Jung, Carl G., et al. *Man and his Symbols* (Anchor Press, New York, 1964) {"only work in which . . . the world-famous Swiss psychologist, explains to the general reader his greatest contribution to our knowledge of the human mind"}

[J05] - *The Journal of George Fox*, Revised Edition by John L. Nickalls (Cambridge at the University Press, 1952)

[J06] - Julian of Norwich, *Revelations of Divine Love*, Trans. by Clifton Wolters (Penguin Books, Baltimore, 1966)

[J07] - Julian of Norwich, *A Shewing of God's Love*, Ed. by Anna Maria Reynolds (Longmans, Green and Co., London, 1958)

[K01] - Kapitza, Sergei. *Antiscience Trends in the U.S.S.R.*, (Scientific American, Scientific American, Inc., New York, Vol 265, No. 2, 1991)

[K02] - Kasamatsu, A. and T. Hirai. *An Electroencephalographic Study on the Zen Meditation (Zazen)*. (Folio Psychiatrica et Neurologica Japonica 20, 1966) {Reprinted in *Altered States of Consciousness*, ed. by C. T. Tart. New York, John Wiley & Sons, 1969}

[K03] - Kempis, Thomas. *The Imitation of Christ*, Trans. by Leo Sherley-Price, (Penguin Books, Baltimore, Maryland, 1952) {"Aside from the Bible, perhaps the best known and loved Christian book"}

[K04] - Kleps, Art. *Millbrook* (The Bench Press, Oakland, 1975)

[K05] - Kook, Abraham Isaac. *Abraham Isaac Kook - The Lights of Penitence, the Moral Principles, Lights of Holiness, Essays, Letters, and Poems*, Trans. by Ben Zion Bokser (Paulist Press, New York, 1978)

[K06] - Kook, Abraham Isaac. *The Essential Writings of Abraham Isaac Kook*, Ed. and Trans. by Ben Zion Bokser (Amity House, Warwick, New York, 1988)

[K07] - *The Koran*, Trans. by N.J. Dawood (Penguin Books, London, 1974) {The Koran is the holy book of Islam}

[K08] - Kotwal, Firoze M. and James W. Boyd. *A Guide to the Zoroastrian Religion, A Nineteenth Century Catechism with Modern Commentary* (Scholars Press, Chico, California, 1982)

[K09] - Kuhn, Thomas S. *The Structure of Scientific Revolutions* (New American Library, New York, 1986)

[K10] - *Kundalini, Evolution and Enlightenment*. Ed. by John White (Anchor Books, Garden City, New York, 1979) {Editor is director of education at The Institute of Noetic Sciences in Palo Alto, California}

[K11] - Kushner, Harold S. *When Bad Things Happen to Good People* (Avon books, New York, 1981) {"Acclaimed National Bestseller"; exploration of the problem of evil by a Massachusetts rabbi}

[L01] - Lao Tsu, *Tao Te Ching*, Trans. by Gia-Fu Feng and Jane English (Vintage Books, New York, 1972){China's ancient Taoist scripture}

[L02] - Lapp, Ralph E. and the Editors of LIFE, *Matter* (Time, New York, 1965) {A *Time-Life Life Science Library* book}

[L03] - Lappe, Frances Moore. *Diet For A Small Planet* (Ballantine Books, New York, 1975) {Best selling cookbook with a social conscience}

[L04] - Leary, Timothy. *The Politics of Ecstasy* (College Notes & Texts, New York, 1965)

[L05] - Leary, Timothy, Ralph Metzner and Richard Alpert, *The Psychedelic Experience* (University Books, New York, 1964) {An LSD tripping manual adapted from *The Tibetan Book of the Dead*}

[L06] - Lee, Martin A. and Bruce Shlain, *Acid Dreams - The CIA, LSD and the Sixties Rebellion* (Grove Weidenfeld, New York, 1985) {"A fascinating social history." - Publishers Weekly}

[L07] - *The Life of Sri Ramakrishna* (Pub. by Swami Budhananda, Advaita Ashrama, Calcutta, 1971) {"interesting and well-written" - Manchester Guardian}

[L08] - Lossky, Vladimir. *The Mystical Theology of the Eastern Church* (St. Vladimir's Seminary Press, Crestwood, NY, 1976)

[L09] - Lossky, Vladimir, *The Vision of God*, Trans by Asheleigh Moorhouse (American Orthodox Press, Clayton, Wis., 1963)

[M01] - Machiavelli, Niccolo. *The Prince and Other Works*, Trans. by Allan H. Gilbert (Hendricks House, New York, 1964)

[M02] - Maloney, George A. *Russian Hesychasm, The Spirituality of Nil Sorskij* (Mouton, The Hague, 1973)

[M03] - Maloney, George A. *A Theology of Uncreated Energies* (Marquette University Press, Milwaukee, Wis, 1978)

[M04] - Margenau, Henry, David Bergamini and the Editors of LIFE, *The Scientist* (Time, New York, 1964) {A Time-Life Life Science Library book}

[M05] - Masters R.E.L. and J. Houston, *The Varieties of Psychedelic Experience* (Holt, Rinehart and Winston, New York, 1966)

[M06] - Mbiti, John S. *African Religions and Philosophy* (Doubleday & Company, New York, 1970) {Author has a doctorate in theology from Cambridge University}

[M07] - McBrien, Richard P. *Catholicism* (Winston Press, Minneapolis, Mn., 1980) {Chariman of the Department of Theology, University of Notre Dame}

[M08] - McCoy, Brad. *136 Biblical "Contradictions" . . . Answered* (Box 78512-F, Shreveport, LA 71137, 1985) {Author has a Th.M. degree with honor from Dallas Theological Seminary}

[M09] - McLeod, W.H. *Guru Nanak and the Sikh Religion* (Oxford University Press, London, 1968)

[M10] - *The Meaning of the Glorious Koran*, Trans. by Mohammed Marmaduke Pickthall (New American Library, New York, 13th printing, no date) {A translation of the Koran}

[M11] - *Meister Eckhart*, trans. by Raymond B. Blakney (Harper & Row, New York, 1941) {Eckhart is a "A central source and inspiration of dominant currents in philosophy and theology since Aquinas."}

[M12] - *Meister Eckhart; Mystic and Philosopher*, Trans. by Reiner Schurmann (Indiana University Press, Bloomington, 1978) {Translation of *Maitre Eckhart ou la joie errante*}

[M13] - *Merit Students Encyclopedia* (Crowell-Collier Educational Corporation, 1967)

[M14] - Meyendorff, John. *St. Gregory Palamas and Orthodox Spirituality* (St. Vladimir's Seminary Press, 1974)

[M15] - Milne, Hugh. *Bhagwan: The God That Failed*, (St. Martin's Press, New York, 1986) {Written by former disciple of Bhagwan Shree Rajneesh}

[M16] - Moody, Raymond A. Jr. *Life After Life* (Stackpole Books, Harrisburg, Pa., 1976) {An investigation of near-death experiences}

[M17] - Morrison, Philip and Phylis, and The Office of Charles and Ray Eames. *Powers of Ten* (Scientific American Books, New York, 1982)

[M18] - *Muslim Saints and Mystics*, Trans. by A.J. Arberry (Routledge & Kegan Paul, London, 1966) {Episodes from *Tadhkirat al-Auliya*, "Memorial of the Saints," by Farid al-Din Attar}

[M19] - *Mystic Places*, The Editors of Time-Life Books, (Time-Life Books, Alexandria, Virginia)

[N01] - *National Geographic, Vol. 164, No. 6, December 1983* (National Geographic Society, Wash. D.C.)

[N02] - *New American Bible* (Catholic Book Publishing Co., New York, 1970) {"new Catholic version of the Bible in English"}

[N03] - *New Catholic Encyclopedia*, (The Catholic University of America, Washington D.C., 1967)

[N04] - *The New Encyclopaedia Britannica*, 14th Edition (Encyclopaedia Britannica, Chicago, 1969)

[N05] - *The New Encyclopaedia Britannica*, 15th Edition (Encyclopaedia Britannica, Chicago, 1983)

[N06] - *The New Encyclopaedia Britannica*, 15th Edition (Encyclopaedia Britannica, Chicago, 1984)

[N07] - *The New Encyclopaedia Britannica - Micropaedia*, 15th Edition (Encyclopaedia Britannica, Chicago, 1983)

[N08] - *The New Saint Joseph Baltimore Catechism, No. 1*, Official Revised Edition (Catholic Book Publishing Co., New York, 1964) {Roman Catholic elementary school catechism}

[N09] - *The New Saint Joseph Baltimore Catechism, No. 2*, Official Revised Edition (Catholic Book Publishing Co., New York, 1969-1962)

[N10] - *The New Testament Study Bible, Mark.* (The complete Biblical Library, Springfield, Missouri, 1986)

[N11] - Nicholson, Reynold A. *The Mystics of Islam* (Arkank, Viking Penguin, New York, First Published 1914) {"In this classic work Dr Nicholson provides . . . an easy approach to . . . Islamic mysticism."}

[N12] - Nietzsche, Friedrich W. *Thus Spoke Zarathustra* (Penguin Books, Baltimore, Md., 1969)

[N13] - Nigosian, Solomon. *Judaism, The Way of Holiness* (The Aquarian Press, Great Britian, 1986) {Mr. Nigosian is a historian of religion}

[O01] - *136 Biblical Contradictions* (Crusade Publications, P.O. Box 200, Redmond, WA 98052, 1981)

[P01] - Pagels, Elaine. *The Gnostic Gospels* (Vintage Books, New York, 1981) {Winner of the National Book Critics Circle award}

[P02] - Panati, Charles. *Panati's Extraordinary Endings of Practically Everything and Everybody* (Harper & Row, New York, 1989)

[P03] - Patterson, Richard S. and Richardson Dougall, *The Eagle and the Shield - A History of the Great Seal of the United States* (Office of the Historian, Department of State, U.S. Government, 1978)

[P04] - Paramananda, Swami. *Book of Daily Thoughts and Prayers* (Vedanta Centre Publishers, Cohasset, Mass., 1977)

[P05] - Parrinder, Geoffrey. *Mysticism in the World's Religions* (Sheldon Press, London, 1976) {University of London Professor of the Comparative Study of Religions}

[P06] - Partee, Phillip. *The Layman's Guide to Fasting and Losing Weight*, (Sprout Publications, 1979)

[P07] - Payne, Franklin E. Jr. M.D. *Biblical Medical Ethics*, (Mott Media, Milford, Mich., 1985)

[P08] - Peck, George T. *What is Quakerism? A Primer* (Pendell Hill Pamphlet 227, Pendel Hill, Wallingford, Pennsylvania, 1988)

[P09] - Peifer, Claude J. *Monastic Spirituality*, (Sheed and Ward, New York, 1966) {Written by a Roman Catholic monk}

[P10] - Penfield, Wilder. *The Mystery of the Mind, A Critical Study of Consciousness and the Human Brain* (Princeton University Press, Princeton, 1975) {Written by a neurophysiologic researcher}

[P11] - Pennington, Basil M. *Centering Prayer* (Image Books, Garden City, N.Y., 1980) {Trappist monk at St. Joseph's Abbey in Spencer, Massachusetts}

[P12] - Perry, Whitall N., *A Treasury of Traditional Wisdom* (Harper & Row, San Francisco, 1971) {"attempt to collect from the writings of world religions a world bible"}

[P13] - *The Philokalia - The Complete Text*, Trans by Palmer, G.E.H.; Sherrard, Philip; and Ware, Kallistos (Faber and Faber, London, 1979) {Collection of writings of Christian saints}

[P14] - *The Practical Vedanta of Swami Rama Tirtha* (The Himalayan International Institute, Honesdale, Pa., no date) {Rama was a professor of mathematics. Clear and direct talks}

[P15] - *Principles of Catholic Theology*, Ed. by Edward J. Gratsch, S.T.D. (Alba House, New York, 1981) {"A synthesis of dogma and morals"}

[Q01] - *Quantum Questions, Mystical Writings of the World's Great Physicists*, Ed. by Ken Wilber (Shambhala Publications, Boulder, Co., 1984) {"Mystical writings of the world's great physicists"}

[R01] - *The Random House College Dictionary*, Revised Edition (Random House, New York, 1984)

[R02] - *Remington's Pharmaceutical Sciences* (Mack Publishing Co., Easton. Pa, 1970) {written by 300 editors, associated editors, and contributors}

[R03] - Rist, J. M. *Plotinus, The Road to Reality*, (Cambridge University Press, Cambridge, 1967)

[R04] - *A Rosary of Islamic Readings, 7th to 20th Century*, Compiled and edited by G. Allana (National Publishing House, Ltd., Karachi-Rawalpindi, 1973)

[R05] - *The Rubaiyat of Omar Khayyam*, Trans. by Edward J. Fitzgerald (Airmont Publishing, New York, 1970) {Famous 800 year old poem}

[R06] - Rucker, Rudy. *Infinity and the Mind: The Science and Philosophy of the Infinite* (Birkhauser, Boston, 1982) {"Informal, amusing, witty, profound . . . extraordinary burst of creative energy" - Martin Gardner}

[R07] - *John Ruusbroec: The Spiritual Espousals and Other Works*, Trans. by James A. Wiseman, O.S.B., (Paulist Press, New York, 1985) {Works of a medieval Christian mystic; often spelt "Ruysbroeck"}

[R08] - Ryan, Thomas. *Fasting Rediscovered*. (Paulist Press, New York, 1981) {Religious treatment of fasting by a Catholic priest}

[R09] - Rydnik, V. *ABC's of Quantum Mechanics* (English language version) (Mir Publishers, Moscow, 1978) {Straightforward, layman account of quantum mechanics}

[S01] - Saradananda, Swami. *Sri Ramakrishna, The Great Master* (Sri Ramakrishna Math, Mylapore, Madras, India) {The Swami was a direct disciple of Ramakrishna}

[S02] - Satris, Stephen. *Taking Sides, Clashing Views on Controversial Moral Issues*, (Dushkin Publishing Group, Guilford, Conn., 1988)

[S03] - Scheffler, Israel. *Science and Subjectivity*, (Hackett Publishing company, Indianapolis, 2nd ed. 1982)

[S04] - Schimmel, Annamarie. *Mystical Dimensions of Islam* (The University of North Carolina Press, Chapel Hill, NC, 1975) {"will surely be the standard treatment of Sufism for a long time to come" - America}

[S05] - Schrodinger, Erwin. *Nature and the Greeks* (Cambridge at the University Press, 1954) {Schrodinger's famous equations are basic to Quantum Mechanics and won him the Nobel prize in Physics}

[S06] - Schrodinger, Erwin. *Science and Humanism, Physics in Our Time* (Cambridge at the University Press, 1961)

[S07] - Schrodinger, Erwin. *What is Life? & Mind and Matter* (Cambridge at the University Press, 1967) {Two seperate works in one volume}

[S08] - *Selected Treatises of St. Athanasius, vol II*, Trans. by John Henry Cardinal Newman (Longmans, Green, and Co., London, 1890) {St. Athanasius, 4th century C.E. Patriarch of Alexandria}

[S09] - *Self-Knowledge (Atma-Bodha) of Sri Sankaracarya*, Trans. by T.M.P. Mahadevan (Arnold-Heinemann, New Delhi, 1975) {Author is Director, Centre for Advanced Study in Philosophy, University of Madras)

[S10] - *Seventh-day Adventists Believe . . .* (Ministerial Association General Conference of Seventh-day Adventists, Review and Herald Publishing Association, Hagerstown MD, 1988) {"A Biblical Exposition of 27 Fundamental Doctrines"}

[S11] - *Shankara's Crest-Jewel of Discrimination*, Trans. by Swami Prabhavananda and Christopher Isherwood (New American Library, New York, 1947) {Translation of Shankara's *Vivek Chudamini*}

[S12] - Sherrard, Philip. *Athos, The Holy Mountain* (The Overlook Press, Woodstock, New York, 1982) {Photographs by Takis Zervoulakos}

[S13] - Silesius, Angelus. *The Cherubinic Wanderer*, Trans. by Willard R. Trask, (Pantheon Books, Inc., New York, 1953) {"Angelus Silesius" was the pen name of Johann Scheffler}

[S14] - Smith, Huston. *Forgotten Truth, The Primordial Tradition* (Harper & Row, New York, 1976).

[S15] - Smith, Huston. *The Religions of Man* (Harper & Row, New York, 1958) {Classic by former Professor of Philosophy, Massachusetts Institute of Technology}

[S16] - Smith, Wilfred Cantwell. *Islam in Modern History* (Princeton University Press, Princeton NJ, 1957)

[S17] - Sneader, Walter. *Drug discovery: The Evolution of Modern Medicines* (John Wiley & Sons, Chichester,)

[S18] - *The Song of God: Bhagavad-Gita*, Trans. by Swami Prabhavananda and Christopher Isherwood (New American Library, New York, 1972) {"Gospel of Hinduism, and one of the great religious classics of the world"}

[S19] - *The Spiritual Teaching of Ramana Maharshi*, (Shambhala Publications, Inc., Boulder, Co., 1972) {Introduction by Carl Jung}

[S20] - Stace, W. T. *Mysticism and Philosophy*, (Jeremy P. Tarcher, Inc., Los Angeles, 1960) { a "landmark book"- Huston Smith, Foreword. Stace was a Professor Emeritus in Princeton University's philosophy department.}

[S21] - Stein, George J. *Biological Science and the Roots of Nazism*, (American Scientist, New Haven, CT 06511, Vol 76, No. 1, Jan-Feb 1988)

[S22] - Steiner, Johannes. *Therese Neumann; A Portrait Based on Authenetic Accounts, Journals and Documents*, (Alba House, Staten Island, N.Y. 10314) (Originally published in German as *Theres Neumann von Konnersreuth*)

[S23] - Stevenson, Ian. *Children Who Remember Previous Lives* (University Press of Virginia, Charlottesville, 1987)

[S24] - Stevenson, Ian, *Twenty Cases Suggestive of Reincarnation* (2nd Ed., University Press of Virginia, Charlottesville, 1974) {1st Ed. pub. by American Society for Psychical Research, 1966}

[S25] - *Symeon, The New Theologian*, Trans. by Paul McGuckin (Cistercian Publications, Kalamazoo, Michigan, 1982)

[S26] - *Symeon, The New Theologian - The Discourses*, Trans by C. J. deCatanzaro, (Paulist Press, New York, 1980)

[T01] - *Tao Teh King*, Trans. by Archie J. Bahm (Continuum, New York, 1986)

[T02] - Taylor, F. Sherwood, *A Short History of Science and Scientific Thought* (W.W. Norton & Company, New York, 1949) {Published in England as *Science Past and Present*; Mr. Sherwood was curator of the Museum of history of science at Oxford}

[T03] - *The Teachings of Ramana Maharshi*, Ed. by Arthur Osborne (Samuel Weiser, New York, 1978) {"certainly regarded as the most important Indian Saint and Sage of the present century."}

[T04] - *Thus Spake Sri Ramakrishna*, Compiled by Swami Suddhasatwananda (Sri Ramakrishna Math, Madras, India, 1980) {collection Ramakrishna's sayings}

[T05] - *The Tibetan Book of the Dead*, Trans. by Lama Kazi Dawa-Samdup (Oxford University Press, London, 1927) {Compiled and edited by W. Y. Evans-Wentz)

[T06] - *The Tibetan Book of the Great Liberation*, Ed. by W.Y. Evans-Wentz (Oxford Press, London, 1954)

[T07] - Titus, Harold H., Marilyn S. Smith, Richard T. Nolan. *Living Issues in Philosophy*, 7th ed., (D. Van Nostrand Co., New York, 1979) {Popular introductory college philosophy text}

[U01] - Underhill, Evelyn. *Mysticism: A Study in the Nature and Development of Man's Spiritual Consciousness*, Revised Edition (E.P. Dutton and Company, New York - also New American Library, Penguin Books, New York) {"recognized as the classic study of . . . mysticism", New American Library edition}

[U02] - *Unseen Warfare; the Spiritual Combat and Path to Paradise of Lorenzo Scupoli*, ed. by Nicodemus of the Holy Mountain, rev. by Theophan the Recluse, trans. by E. Kadloubovsky and G. E. H. Palmer (Mowbrays, London, 1978) {First published by Faber and Faber, London, 1952}

[U03] - *The Upanishads*, Trans. by Swami Paramananda (Vedanta Centre Publishers, Cohasset, Ma., 1981) {"In the whole world there is no study so beneficial and so elevating as that of the Upanishads." - Schopenhauer}

[U04] - *The Upanishads, Breath of the Eternal*, Trans. by Swami Prabhavananda and Frederick Manchester (Mentor Press, New American Library, 1957) {"the core of India's most sacred scriptures"}

[U05] - *U.S. News & World Report* (U.S. News & World Report, Washington, D.C. June 24, 1991)

[V01] - *Vedanta for the Western World*, Ed. by C. Isherwood (The Viking Press, New York, 1945)

[W01] - Waley, Arthur. *The Way and Its Power* (The Macmillan Company, New York, 1934) {Translation with footnotes of the *Tao Te Ching*}

[W02] - Watts, Alan, W. *The Book: On the Taboo Against Knowing Who You Are* (Vintage Books, New York, 1972) {"Perhaps the most famous of all Watts' works"}

[W03] - Watts, Alan, W. *The Joyous Cosmology* (Vintage Books, New York, 1962) {"describes with startling clarity and poetic beauty his drug-induced experiences"}

[W04] - *The Way of a Pilgrim and the Pilgrim Continues His Way*, trans. by Helen Bacovcin. (Image Books, Garden City, New York, 1978) {Concerns the simple "Jesus prayer," a way of "praying without ceasing"}

[W05] - *Webster's Third New International Dictionary of the English Language, Unabridged* (Merriam-Webster, Springfield, MA, 1986)

[W06] - *What is Meditation?* Ed. by John White (Anchor Books, Garden City, New York, 1974)

[W07] - *The Wisdom of the Early Buddhists*, Compiled by Geoffrey Parrinder (New Directions Books, New York, 1977)

[W08] - *The Wisdom of the Jewish Mystics*, Trans. by Alan Unterman (New Directions, New York, 1976)

[W09] - White, Andrew Dickson, *A History of the Warfare of Science with Theology in Christendom* (George Braziller, New York, 1955) {Originally published by Appleton in 1896; also published by Prometheus Books in 1993}

[W10] - Wolf, Fred Alan. *Taking the Quantum Leap* (Harper & Row, New York, 1989) {American Book Award winner by former physics professor}

[W11] - *Writings from the Philokalia on Prayer of the Heart*, Trans by E. Kadloubovsky and G.E.H. Palmer (Faber and Faber, London, 1951) {Collection of writings on the Jesus prayer}

[Y01] - *Yoga, Union with the Ultimate; A New Version of the Yoga Sutras of Patanjali*, Trans. by Archie J. Bahm (Frederick Ungar Publishing Co., New York, 1967) {Professor of Philosophy at the University of New Mexico, Albuquerque}

[Y02] - Yogananda, Paramahansa. *Autobiography of a Yogi* (Self-Realization Fellowship, Los Angeles, California, 1977) {"A rare account" - New York Times}

[Z01] - Zaehner, R. C. *Drugs, Mysticism and Make-Believe* (Collins, St James's Place, London, 1972)

[Z02] - Zaehner, R. C. *Mysticism, Sacred and Profane* (Oxford at the Clarendon Press, 1957)

T

U

About the Author

Art D'Adamo has pursued a lifelong interest in science and religion through years of undergraduate and graduate education at three universities and in two religious communities. He meditates and has read much mystical literature, East and West. Art's book describes a synthesis of science and religion that combines the best of both fields, a scientific religion.

Printed in the United Kingdom
by Lightning Source UK Ltd.
122791UK00001B/73/A